THE AUDIO RECORDING HANDBOOK

THE COMPUTER MUSIC AND DIGITAL AUDIO SERIES

John Strawn, Founding Editor
James L. Zychowicz, Series Editor

DIGITAL AUDIO SIGNAL PROCESSING
Edited by John Strawn

COMPOSERS AND THE COMPUTER
Edited by Curtis Roads

DIGITAL AUDIO ENGINEERING
Edited by John Strawn

COMPUTER APPLICATIONS IN MUSIC:
A BIBLIOGRAPHY
Deta S. Davis

THE COMPACT DISC HANDBOOK
Ken C. Pohlman

COMPUTERS AND MUSICAL STYLE
David Cope

MIDI: A COMPREHENSIVE INTRODUCTION
Joseph Rothstein
William Eldridge, *Volume Editor*

SYNTHESIZER PERFORMANCE AND
REAL-TIME TECHNIQUES
Jeff Pressing
Chris Meyer, *Volume Editor*

MUSIC PROCESSING
Edited by Goffredo Haus

COMPUTER APPLICATIONS IN MUSIC:
A BIBLIOGRAPHY, SUPPLEMENT I
Deta S. Davis
Garrett Bowles, *Volume Editor*

GENERAL MIDI
Stanley Jungleib

EXPERIMENTS IN MUSICAL INTELLIGENCE
David Cope

KNOWLEDGE-BASED PROGRAMMING FOR
MUSIC RESEARCH
John W. Schaffer and Deron McGee

FUNDAMENTALS OF DIGITAL AUDIO
Alan P. Kefauver

THE AUDIO RECORDING HANDBOOK
Alan P. Kefauver

Volume 17 • The Computer Music and Digital Audio Series

THE AUDIO RECORDING HANDBOOK

Alan P. Kefauver

■

A-R Editions, Inc.
Middleton, Wisconsin

ISBN 0-89579-462-4

A-R Editions, Inc., Middleton, Wisconsin 53562
 © 2001 All rights reserved.
 Printed in the United States of America
 10 9 8 7 6 5 4 3 2 1

Contents

Preface		xi
Section I:	Measuring Sound	1
■ Chapter 1	**The Raw Materials of Sound**	3

    The Decibel 3
    The Logarithm 4
    Power 7
    Frequency 15

■ Chapter 2	**Sound Measuring Devices**	33

    The Volume Indicator (VU Meter) 33
    The Sound Level Meter 35
    The Peak Reading Meter 36
    Studio and Broadcast Reference Levels 38
    The Spectrum Analyzer 40
    The Oscilloscope 41
    Other Metering Standards 43

Section II:	Transducers: Microphones and Loudspeakers	45
■ Chapter 3	**Microphone Design**	47

    Dynamic Microphones 47
    Condenser Microphones 50

Acoustical Specifications 56
Distortion in Microphones 90
Microphone Noises: Sibilant, Plosive, and Vibration 91

■ Chapter 4 Microphone Technique — 97

Stereo Techniques (Binaural Systems) 101
Surround Sound Microphone Techniques 119
Multi-Microphone Techniques 123
Special-Purpose Microphones 133
Miking Musical Instruments 136

■ Chapter 5 Loudspeakers — 145

Loudspeaker Terminology 145
The Ideal Sound Source 147
The Practical Loudspeaker 148
Speaker Enclosure Systems 152
Multi-Speaker Systems 168
The Room/Speaker Interface 174
Summary 182

Section III: Sound Processing Devices — 185

■ Chapter 6 Time Domain Processing: Echoes and Reverberation — 187

Room Acoustics 187
Simulating Reverberation in the Studio 192
Special Effects 200

■ Chapter 7 Frequency Domain Processing: Equalizers — 207

Bandwidth and "Q" 209
Types of Equalizers 210
Applications 214

Effect of Equalization on Dynamic Range 218
Equalizer Phase Shift 218

Chapter 8 Amplitude Domain Processing — 221
Compressors and Limiters 222
Expanders 234
Multi-Function Processors 242

Chapter 9 Analog Noise Reduction — 245
White Noise 246
Residual Noise Level 247
Signal-to-Noise Ratio 248
Noise Reduction Systems 249
Summary 265

Section IV: Analog Audio Recording Systems — 269

Chapter 10 Magnetic Tape — 271
The Magnetic Recording System 273
The Recording Process 276
Tape Differences 288
High Output Tapes 288
The Print-Through Phenomenon 289

Chapter 11 The Analog Tape Recorder — 291
Track Format Standards 291
Tape Speed 294
The Recording Process 294
The Playback Process 296
Other Features and Characteristics 304

■ Chapter 12 Analog Tape Recorder Maintenance and Alignment   313

Cleaning and Demagnetizing 313
Test Tapes 314
Operating Levels 314
Electronic Adjustments 315
Mechanical Adjustments 323

Section V: Digital Audio Systems   329

■ Chapter 13 Digital Audio   331

Digital Design Basics 333
Digital Encoding: The Binary System 336
The Record Process 342
The Playback Process 348
The Digital Advantage 352

■ Chapter 14 Digital Audio Recording Systems   357

Tracks Versus Channels 357
Sampling-Rate Converters 358
Fixed-Head Digital Recorders 360
Rotary-Head Recording Systems 365
Fixed-Head Versus Rotary-Head Digital Recorders 376
Hard-Disk Recorders 377
Digital Editing Procedures 380
Optical Disk Storage Systems 392

Section VI: Synchronization Systems   403

■ Chapter 15 Time Code   405

Recorded Time Data 406
Society of Motion Picture and Television Engineers (SMPTE) Time Code 407
Vertical Interval Time Code (VITC) 415

Time Code Implementation 416
Musical Instrument Digital Interface (MIDI) 425
The SMPTE/MIDI Interface 429

Section VII: The Recording Console — 431

■ Chapter 16 Recording Studio Consoles — 433
Overview 433
The Input/Output (IO) Module 437
The Master Module 440
The Monitor Module 440
Components of the In-Line Recording Console 442
Summary of Signal Flow Path—Analog Console 458
The Digital Difference 472
Control Surface to Electronic Rack Interfaces 472
System Configuration Files 474
Clocking the Digital Console 474
The Input Pool 475
Patch (Insert) Points 476
Auxiliary Send Buses 476
The Output Pool: Buses and Direct Outs 478

■ Chapter 17 Console Automation Systems — 481
The Voltage-Controlled Amplifier 482
Automation Data Storage 483
Primary Automation Functions 487
An Automation Session 497

Section VIII: The Recording Process — 501

■ Chapter 18 The Recording Session — 503
Overdubbing 503
Headphone Monitoring 514
Track Assignment and Disk Allocation 515

Preparing for the Multi-Track Session   517
Recording   518
End of Session   521

Chapter 19 The Mixdown Session   523
Editing   523
Track Assignment and Panning   530
Preparing for Mixdown   531
Assistance during Mixdown   532
Recording and Monitor Levels   533
Monitor Loudspeakers for the Mixdown Session   534
The Master   535
Mastering   538
Mixdown Summary   540

■ Glossary   543

■ Suggested Reading   593

■ Index   595

Preface

Digital audio is new and not new. The theories behind digital audio have been with us for some time. Charles Babbage, in 1842, proposed a system for storing and implementing calculations, and this probably would have been the first computer had it been completed. Thomas Edison invented a repeater for the telegraph of the day in 1877, and in 1928 H. Nyquist published *Certain Topics in Telegraph Transmission Theory,* which spelled out the sampling theorem much as we know it today. In 1943 the first computer, ENIAC, was developed by the U.S. Army, and the 1950s saw the publication of papers on error correction from Hamming and a redundancy coding scheme from Huffman. The 1960s were a time of development in Electronic Music technology at both Bell Labs and Princeton University, and Reed and Solomon developed a multiple error correction code that is still in use today. From then through the 1970s, work on digital audio systems escalated with the introduction of the Soundstream Digital System from Thomas Stockam and the introduction of the first in a series of digital audio video-based processors from Sony, known as the PCM-1600.

Today, digital audio is a fundamental part of our lives, and although everyone is not "sold" on the new technology, it is surely here to stay. These are the twilight years for analog audiotape, and as surely as the compact disc (CD) replaced the LP of yesteryear, digital audio workstations, CD recorders, and Digital Versatile Disc (DVD) machines are replacing and will continue to replace the analog reel-to-reel tape recorders that have been the workhorse of the professional audio industry for so long. In 1988 it was estimated that there were 100 million to 150 million CDs. In 1998 the estimate is that there are 500 million CD players and over 10 billion existing CDs.

Some time ago, a good friend sent me a copy of an article from a magazine called *The Gramophone.* The editor stated in the article that "the exaggeration of sibilants by the new method is abominable, and there is often a harshness which recalls the worst excesses of the past." As I read, I thought to myself that this was yet another of the digital naysayers who were prevalent during the earlier days of digital audio. Much to my surprise, when I reached the end of the piece, I found that this editorial was from 1925 and that the process the editor was referring to was the then-new electrical recording process introduced by RCA in 1924. Recent editorials have said similar things about the visual quality of the DVD. Seventy-four years from now, who knows what media we will be using for listening to music or watching moving pictures.

SECTION I

Measuring Sound

■ INTRODUCTION

Simply stated, sounds are produced by mechanical vibrations. A vibrating object disturbs the molecules of air surrounding it, causing periodic vibrations in the air pressure. As the object vibrates back and forth, the pressure becomes alternately more, and then less, dense. These pressure compressions and rarefactions radiate away from the object as waves, eventually reaching the listener's ear, creating the sensation that we know as sound.

The classic example of a vibrating object is a taut length of string, which may be set into vibratory motion by striking it (as in the piano), plucking it (guitar), or by drawing a bow across it (violin). The rate of vibration is a function of the tension applied to the string as well as of its length. In wind or brass instruments, the vibration is that of a column of air, and the vibration rate is regulated by changing the length of the air column. This is accomplished by opening and closing valves that change the length of the pipe (trumpet), or by keys that shorten the effective length of the column of air (clarinet).

The dynamic range of an instrument is a measure of the span between the quietest and loudest sounds it is capable of producing. Musically subjective terms such as pianissimo (very soft), mezzo-forte (moderately loud), forte (loud), and fortissimo (very loud) are used to describe the relative intensity of the note produced.

The dynamic range of the human voice is quite broad, from a pianissimo whisper to a fortissimo shout. In contrast, the range of the harpsichord is practically zero, since each of its strings is plucked with the same amount of force, regardless of the strength with which the keys are struck. Interestingly, the harpsichord's successor, the pianoforte, got its name in recognition of its wide dynamic range capability, as compared to its predecessor. The harder the key on a piano is struck, the more force is applied to the string and the greater the amplitude of the vibration. The striking force, however, has no effect on the rate of vibration, or frequency, of the note struck.

These two components of sound—power and frequency—are the raw materials of the audio recording process. The skilled engineer at the console uses various tools to measure and affect these forces in ways that will optimize the quality of the audio product. The reader should have a basic understanding of these raw materials before exploring the many devices, tools and techniques used in the recording studio. Section I was written with this goal in mind. The mathematics may, at times, be daunting for those who are not so inclined. The reader, however, should press on, for familiarity with these principles will bring a deeper understanding of the devices themselves, and in turn, greater skill in how to use them effectively.

ONE

The Raw Materials of Sound

■ THE DECIBEL

The decibel (dB) is the commonly used unit for the measurement of sound levels. It is always abbreviated small "d" capital "B" since it stands for one-tenth (deci) of a Bel. The Bel is a rather large unit of measure to work with, hence the decibel. Sound, such as a jet aircraft taking off or a quiet sail on a placid lake, can be measured with a sound level meter. We think of a jet taking off, which creates a noise level of around 130dB, as very loud, and a peaceful sail, about 35dB, as very quiet. The sound level meter usually measures sound levels between 0dB and 140dB or more. Interestingly, 0dB is not the total absence of sound but is equated to the threshold of hearing: that is, the lowest sound pressure level that an average listener with good hearing can detect. The mathematical formula for the decibel is:

$$dB = 10 \log \frac{P}{P_R}$$

where: P = the power to be measured
P_R = the reference power

This zero reference level corresponds to a sound pressure of 0.00002 dynes/cm^2. In terms of intensity, this is equivalent to 0.000000000001 watts/meter^2.

Even a place as quiet as an acoustically correct recording studio has a certain amount of background noise. The simple movement of air may be around 25dB at some frequencies. Therefore, we can say that the ambient noise level is 25dB greater than our reference level of 0dB.

Anyone who has ever spent time with a tape recorder has surely noted that the readings on the VU meter are not necessarily related to the actual volume heard in the room. Even with the playback loudspeakers turned off, the meters will still register if a signal is applied to the tape recorder. Changing the listening level in the room will have no effect whatever on the meter readings. Typical meter readings of -10 to +3 suggest that a different zero reference value is being used.

■ THE LOGARITHM

From the above formula, it should be clear that the decibel is defined in terms of logarithms (abbreviated log). Therefore, some comprehension of the mathematical significance of the log is essential to an understanding of the decibel. The reader who is completely familiar with logarithms may wish to skip ahead. Others may also wish to do so, but should avoid the temptation.

Below, several very simple multiplication problems are solved. To the right of each answer, there appears a shorthand notation, consisting of the number 10 followed by a superscript. In mathematics, this superscript is known as an exponent, and to help realize its significance, we may say that an exponent indicates how many times the number 1 is to be multiplied by 10. In the last problem, the exponent, x, is equal to 5, and is read as "ten, raised to the fifth power," or simply, "ten to the fifth." The fact that $10^0 = 1$ may be difficult to understand. It may help to understand that 10^0 is not 1 multiplied by 10 at all, and so it remains simply 1. This explanation may not please the mathematicians, but it will enable us to get on with our introduction to the decibel with a minimum of pain.

$$1 = 1 \text{ or } 10^0$$
$$10 = 10 \text{ or } 10^1$$
$$10 \times 10 = 100 \text{ or } 10^2$$
$$10 \times 10 \times 10 = 1{,}000 \text{ or } 10^3$$
$$10 \times 10 \times 10 \times 10 = 10{,}000 \text{ or } 10^4$$
$$10 \times 10 \times 10 \times 10 \times 10 = 100{,}000 \text{ or } 10^x$$
$$(x = 5)$$

We should also note that numbers less than 1 can be represented by powers of ten. For example:

$$0.1 = 10^{-1}$$
$$0.01 = 10^{-2}$$
$$0.001 = 10^{-3}$$

In fractional form, $0.001 = 1/1,000$ or $1/10^3$. So, we may say that $0.001 = 1/10^3 = 10^{-3}$. In other words, a power of ten in the denominator of a fraction may be moved to the numerator, simply changing the sign, in this case from plus to minus.

Often, a difficult fraction may be simplified by following this procedure. For example:

$$\frac{3,200,000}{0.004} = \frac{32 \times 10^5}{4 \times 10^{-3}} = \frac{32 \times 10^5 \times 10^3}{4} = \frac{32 \times 10^{(5+3)}}{4} =$$

$$\frac{32 \times 10^8}{4} = 8 \times 10^8$$

Note that when powers of ten are to be multiplied, the exponents are merely added to find the product ($10^5 \times 10^3 = 10^{(5+3)} = 10^8$). Likewise, division may be accomplished by subtracting the exponents (10^5 divided by $10^3 = 10^{(5-3)} = 10^2$). The significance of this operation will be appreciated later in solving power and voltage equations in terms of decibels. For now, we may realize that cumbersome numbers become less so when converted to powers of ten. For example, the sound intensity at the threshold of hearing, stated earlier as 0.000000000001 watts/m², becomes simply 1×10^{-12} watts/m². A number such as 0.0000025 may be rewritten as 2.5×10^6, or, if it is more convenient, 25×10^{-7}.

Now, if $1 = 10^0$, and $10 = 10^1$, it stands to reason that we should be able to represent any number between 1 and 10 by 10^x, with x equal to some value between 0 and 1. To find the value for x in a problem such as $3 = 10^x$, we will call this exponent a logarithm, and offer the following explanation.

The logarithm of a number is that power to which 10 must be raised (not multiplied) to equal the number.

The statement may appear to contradict what was just said about multiplying 1 by 10 a number of times. To help resolve the apparent contradiction, we may note that although the number 10 may indeed be simply multiplied by itself, say, four times for example,

($10^4 = 1 \times 10 \times 10 \times 10 \times 10 = 10{,}000$), it is no simple matter to do the same thing 4.7 times. Yet, if $10^4 = 10{,}000$, and $10^5 = 100{,}000$, surely there must be a value for $10^{4.7}$. To find the answer, something beyond simple multiplication is required. Returning to our sample problem, $3 = 10^x$, we may say that the log of 3 is x. The actual calculation of this, or any other log, is an involved calculation appreciated only by mathematicians. The working recording engineer, however, need not be concerned with the process, since logarithms are readily available with the push of a button on most serious calculators. A simplified log table is given below. It is important to understand the significance of the log, so that calculation of decibels will be clearly understood.

By studying the table of logs, we may notice several important characteristics of logs, which are really extensions of what was just pointed out about multiplying and dividing exponents. Since a log is an exponent:

Number	Log	Number	Log	Number	Log
1	0.000	10	1.000	100	2.000
2	0.301	20	1.301	200	2.301
3	0.477	30	1.477	300	2.477
4	0.602	40	1.602	400	2.602
5	0.699	50	1.699	500	2.699
6	0.778	60	1.778	600	2.778
7	0.845	70	1.845	700	2.845
8	0.903	80	1.903	800	2.903
9	0.954	90	1.954	900	2.954
10	1.000	100	2.000	1000	3.000

When numbers are multiplied, the log of the product is equal to the sum of the logs of the numbers.

$$\log (4 \times 5) = \log 4 + \log 5$$
$$\log 20 = \log 4 + \log 5$$
$$1.301 = 0.602 + 0.699$$
$$1.301 = 1.301$$

When numbers are divided, the log of the quotient is equal to the difference of the logs of the numbers.

$$\log (800/20) = \log 800 - \log 20$$
$$\log 40 = \log 800 - \log 20$$

$$1.602 = 2.903 - 1.301$$
$$1.602 = 1.602$$

A final point: note that when any number is doubled, the log increases by a constant amount (0.301). When a number is multiplied by 10, its log increases by 1, and when a number is squared, its log is doubled.

■ POWER

Having bravely endured the fundamental concepts of the decibel and logarithm, we now move on to look at the ways in which these principles serve the audio engineer. There are two measuring systems in which the decibel is commonly used: acoustical and electrical.

Acoustical Power Measurements

The first measurement system, as in our jet plane example from earlier in the chapter, is a measure of acoustic power. Here the decibel tells us how far above the threshold of hearing these sounds are. Figure 1-1 shows the typical sound levels found in our everyday environment. The second measurement—our VU meter—tells us something about the electrical power flowing through the meter circuit. Whenever power measurements are made, whether acoustical or electrical, the formula for the decibel is the same. Let's look again at the formula:

$$dB = 10 \log \frac{P}{P_R}$$

where:
P = the power to be measured
P_R = the reference power

When making acoustical measurements, the reference power P_R corresponds to the pressure produced by a sound at the threshold of hearing. This pressure level may be expressed in microbars, dynes/cm^2, watts/meter^2 or newtons/meter^2. The relationship between these different systems can be expressed as follows:

$$0 \text{ dB} = 0.0002 \text{ microbars} = 0.0002 \text{ dynes/cm}^2 = 0.00002 \text{ newtons/m}^2 = 0.000000000001 \text{ watts/m}^2$$

Therefore,
$$1 \text{ microbar} = 1 \text{ dyne/cm}^2 = 0.1 \text{ newton/m}^2$$

or:
$$10 \text{ microbars} = 10 \text{ dynes/cm}^2 = 1 \text{ newton/m}^2.$$

Electrical measurements require a different reference level which we will see shortly. If we know, then, that our jet aircraft produces a sound intensity of 10 watts/m², the decibel level can be calculated from the aforementioned formula:

$$\text{dB} = 10 \log \frac{P}{P_R} = 10 \log \frac{10^1}{10^{-12}} = 10 \log 10^{(1+12)} = 10 \log 10^{13}$$

$$10 \times 13 = 130 \text{dB}$$

where:

$$P = 10 \text{ watts / m}^2 \text{ (the jet)}$$
$$P_R = 10^{-2} \text{ watts / m}^2 \text{ (the threshold of hearing)}$$

The jet noise is therefore 130dB above the threshold of hearing, or 10^{13} times as intense. Now although this certainly is not anyone's idea of soft, we probably don't think of a jet aircraft being 10,000,000,000,000 times the level of the threshold of hearing. Indeed, we may be surprised to realize that our ears are capable of responding to such an incredibly wide range of intensities.

The decibel notational system makes these large intensity ranges more manageable. As a matter of fact, the human ear responds psycho-acoustically about the same way. For example, if one jet aircraft produces a sound level of 130dB, what sort of acoustic horror awaits us in the presence of two such noises? Two jets will definitely sound louder than one, but not by as much as you might expect. The listener will probably note that the noise level has increased, but certainly it is nowhere near twice as loud as before.

Note that if one jet produces 10 watts/m², two would produce 20 watts/m². Or,

$$\text{NdB} = 10 \log 20 = \frac{10 \log 2 \times 10^1}{10^{-12}} = 10 \log 2(10^{13}) =$$

$$10(\log 2 + \log 10^{13}) = 10(0.301 + 13) = 10(13.301) = 133.01 \text{dB}$$

Figure 1.1 Typical sound pressure levels.

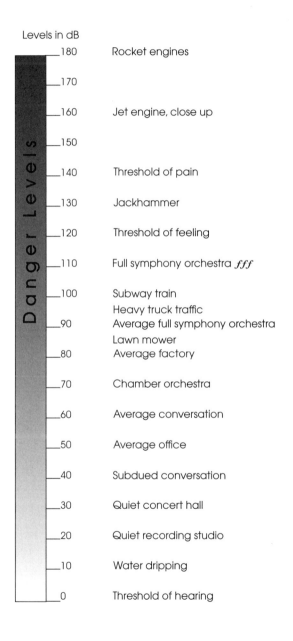

In other words, the two aircraft produce a sound that is only 3dB above the noise of one of them. This dB value rather accurately parallels the listener's subjective impression of the increase in level. In fact, if we returned to a more normal monitoring environment and listened to a musical program, and were then instructed to turn up the volume control until the program was "twice as loud," the measured sound pressure level would likely be about ten times greater than before. Of course, this is merely an approximation, since the conception of change in listening level will vary somewhat from one person to another.

■ Inverse Square Law

As we have just seen, two jets are 3dB louder than one jet. Now consider how the level in decibels would change if we doubled the distance that we are from the jet(s). One would suppose that the sound level would halve and that the level would be decreased by 3dB. However, this is not the case. Let's see what happens as we move from, say, 100 feet in distance from our noise source to 200 feet away. At our original distance, the sound radiates equally into a sphere with a radius of 100 feet. This means that the sound is dispersed over an area of 125,664 square feet ($4\pi r^2$). When we move back to a distance of 200 feet, the sound is now radiating into an area of 502,655 square feet, or 4 times the original area. Since we know that doubling the sound power gives us a 3dB increase in level, it makes sense that half the power level will cause a decrease of 3dB, and that one-fourth the level (another halving) will give us a total decrease of 6dB.

Since the constant level sound is radiating into a space that is four times as large as the original space, the sound power level will diminish by 6dB. Simply stated, this means that every time we double the distance from the source of a sound in an open space, the level of the sound will drop 6dB. And conversely, every time we halve the distance to the source of the sound, the sound level will rise 6dB.

Electrical Power Measurements

In any electrical circuit, a certain amount of power is dissipated. For example, the power dissipated in a resistor may be found from the formula $P = E^2/R$ or $P = I^2R$. Therefore, if we know the value of the resistance (R), and either the current (I) flowing through it or the voltage (E) across it, we may calculate the power. If the power then increases, we may calculate the difference in terms of decibels by comparing the new power with the old. For example, if the power dissipated in a resistor increases from 0.5 watts to 10 watts, the increase in decibels is found from the log of the ratio of the two powers, with the first value, 0.5 watts, used as the reference level.

$$\text{NdB} = \frac{10 \log P}{P_R} = \frac{10 \log 10}{0.5} = 10 \log 20 = 10 \times 1.301 = 13.01 \text{dB}$$

If we had wished instead to calculate a decrease in dissipated power, the reference power would simply be placed in the numerator, and a negative sign would appear in the answer,to indicate a power loss.

In the recording studio, it is more convenient to work with voltage levels. This presents no problem, since power ratings are usually found only after the voltage has been measured anyway. And, since we invariably make all our measurements across the same resistance value, it may be eliminated from our calculations in the following manner.

Consider two different power values, P_a and P_b. Comparing them in terms of dB, $N_{dB} = 10 \log P_a/P_b$. But, since $P = E^2/R$, we may rewrite the formulas as:

$$10 \log \frac{E_a^2/R_a}{E_b^2/R_b}$$

Since the value of the resistance does not change, $R_a = R_b$, and the formula may be simplified to $10 \log E_a^2/E_b^2$, or to $20 \log E_a/E_b$. Now if we measure the voltage across the resistance at, say, 6 volts, and later increase it to 12 volts, the dB increase is:

$$\text{NdB} = 20 \log \frac{12}{6} = 20 \log 2 = 20 \times 0.31 = 6.02 \text{dB}$$

Note that whereas a doubling of power gave us an increase of 3dB, a voltage doubling yields a 6dB gain.

In the case of acoustic power measurements, there is a standard zero reference level. And although this zero reference level represents an extremely low value (the threshold of hearing), in the studio it is more convenient to use an electrical zero reference level equivalent to the voltage found across a resistance in a typical operating condition. This allows us to compare other voltages and note that they are so many decibels above or below our zero reference standard.

Discussing acoustical power and sound pressure level may be confusing, but it is important to note that sound pressure is a potential (E) analogous to voltage, where acoustical power is in watts (P). Therefore the 20 log E/E_R formula is used for calculating sound pressure level while the 10 log P/P_R is used for acoustical power considerations.

■ The dBm

The voltage drop across a 600-ohm resistor through which 1 milliwatt of power is being dissipated is the standard studio zero reference level. Using the formula $P = E^2/R$, we may discover that the zero reference voltage is 0.775 volts. Actually, the meters in the studio are voltmeters, but they are calibrated in decibels to enable us to make the kind of measurements that are most suited to audio signals. Although the range of voltages is nowhere near as great as the sound intensity variations cited earlier, the dB scale is still a most practical measuring system. Over the years, there have been other standard reference levels used. And so, to clarify the fact that our decibel measurements are made relative to 1 milliwatt across 600 ohms, the notation dBm is used. The m refers to the milliwatt reference power level.

Many regular voltmeters will also show a dB scale (Figure 1-2) and, upon inspection of the meter face, we may verify that 0dB = 0.775 volts. However, it should be remembered that these dB scales are not dBm unless the measurements are being made across the 600 ohm line. Studio meters are usually permanently wired across the proper circuit values, but when using a bench-type voltmeter, there is no reason to suppose that all measurements are being made across 600 ohms. Of course, the voltage scale is accurate regardless of the resistance across which the measurement is being made, but if we were making our measurements across a 1,200-ohm resistor, we would have to use the following formula to determine the actual dBm value:

$$NdBm = 10 \log \frac{E^2/R}{E_{Ref}^2/R_{Ref}}$$

When R_{Ref} is less than R, the formula may be rewritten as:

$$NdBm = 20 \log \frac{E}{E_{Ref}} - 10 \log \frac{R}{R_{Ref}}$$

Figure 1.2 Voltmeter showing the relationship between dBm and volts.

E_{Ref} and R_{Ref} are the values of our standard zero reference level: 0.775 volts and 600 ohms. E is the voltage across the nonstandard resistance, and R is the value of that nonstandard resistance.

If we read an apparent 0dBm (0.775 volts) across 1,200 ohms, the actual dBm value would be:

$$\text{NdBm} = 20 \log \frac{E}{E_{Ref}} - 10 \log \frac{R}{R_{Ref}} = 20 \log \frac{0.775}{0.775} - 10 \log \frac{1200}{600} =$$

$20 \log 1 - 10 \log 2 = 20 \times 0 - 10 \times 0.301 = -3.01 \text{dBm} =$

Actual Value in dBm

We may verify this by comparing the power dissipated across the zero reference standard with that dissipated across the 1,200-ohm resistance.

$$P = \frac{E^2}{R} = \frac{(0.775)^2}{600} = 1 \text{ milliwatt in the zero reference standard.}$$

$$P = \frac{E^2}{R} = \frac{(0.775)^2}{1200} = 0.5 \text{ milliwatt in the 1200 } \Omega \text{ resistance.}$$

Since the power across the 1,200-ohm resistance is half the reference power, the decibel value is down 3dB, which confirms our previous calculation.

■ **The dBV (1-volt reference level)** Decibel measurements are often made under conditions where a 600-ohm/1-milliwatt reference is inconvenient. For example, a certain acoustic pressure in front of a microphone may create an output level of -55dBV. The dBV nomenclature indicates a zero reference level of 1 volt. If the microphone's output voltage is 0.0005 volts, its dBV output is:

$$NdBV = -20 \log(\frac{1}{0.0005}) = -20 \log 2{,}000 =$$

$$-20 \times 3.301 = 66.02 dBV$$

Microphone output levels are discussed in greater detail in Chapter 3.

■ **The dBv (0.775-volt reference level)** Some confusion arises from the fact that dBv measurements are frequently made with respect to a 0.775 volt reference level. Since 0.775 volts corresponds to 0dBm, many voltmeters are calibrated so that 0.775 volts and 0dB read the same on the meter face, as shown in Figure 1-2. Such meter faces should contain the legend "1mW, 600 ohms," reminding the user that the scale measures dBm only across a 600-ohm line. But since the meter is so often used under other circuit conditions, where the line impedance remains constant, though not at 600 ohms, there is little point in referring every reading back to a dBm value. In fact, in some cases the impedance of the circuit being measured may be unknown, making it impossible to calculate the actual dBm values. And so, the meter scale is simply read directly. Although readings thus made are not dBm, the arithmetic difference between any readings will be the same as if they were both converted to dBm via the formula given earlier.

This type of measurement is usually given in dBv, and the zero reference voltage may or may not be clearly specified. Generally, the nature of the measurement gives some clue as to the probable zero reference of 0.775 or 1 volt.

■ **The dBu (0.775-volt reference level)** Since many of today's electronic circuits use what are called "active bridging" circuits, a new reference standard is being adopted to define the relative level without regard to a standard impedance. The reference, once again, is 0.775 volt and the unit is called the dBu (u stands for units). The dBu allows us to use the decibel scale to make measurements in circuits without a 600-ohm reference. It takes 1 milliwatt of power to pump 0.775 volts into 600 ohms. dBu is that voltage, but it's only that voltage. It doesn't have to be across 600 ohms. We know that the output source impedance of modern equipment is very low. So, no matter what the output source impedance is, we know that if it puts out 0.775 volts into an open circuit, it will put out very close to that voltage into whatever we connect it to. As with the dBv, readings thus made are not dBm. The arithmetic difference between any readings will be the same as if they were both (the reading and the reference) converted to dBm. The dBu and the dBv are often used interchangeably. The dBm and dBu are numerically equivalent when they both are measured at a 600-ohm load. However, one is still a voltage measurement and the other is still a power measurement.

■ FREQUENCY

When we say that a tone has a frequency of 440 Hertz (abbr. Hz), we mean that the device producing the tone or its air column is vibrating back and forth 440 times each second. Musical instruments do of course produce a wide range of frequencies, and a chart of the typical frequency ranges for various instruments is given in Figure 1-3.

Wavelength As the vibrations from a musical instrument are transmitted through the air, we may calculate the physical length of one complete alternation of air pressure and call this the wavelength of that particular frequency. If we attach a pen to a vibrating object, and slide a piece of paper past it at a constant rate, the line traced on the paper will be a graphical representation of the vibration. If the vibration produces a pure tone—that is, a single frequency with no harmonics or overtones—the graph will show something called a sine wave. In Figure 1-4a, a sine wave is being produced by a tuning fork. If the paper were moving at the speed of sound, the physical length required for one complete oscillation to be drawn on the paper would correspond to the wavelength of the sine wave being measured. Since this is of course impractical, we may simply calculate the wavelength (lambda or λ) as:

$$\lambda = \frac{V}{F}$$ where: λ = wavelength in feet (per cycle)
V = velocity of sound in air (feet per second)
F = frequency in Hertz (cycles per second)

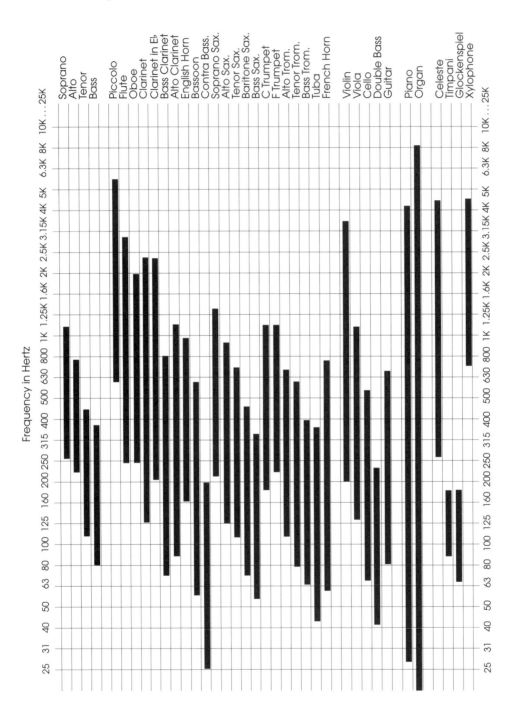

Figure 1.3 The frequency range of various musical instruments.

From the formula, we see that wavelength is dependent not only on frequency, but on the velocity of sound in air. This has been measured to be 1,087 feet/second at a temperature of 32°F, and increases about 1.1 feet/second for each degree in temperature. Therefore, at a temperature of 70°F, the velocity of sound is approximately 1,130 feet/second. Putting this figure into our formula, we find that a frequency of 100Hz has a wavelength of 1,130/100, or 11 feet 4 inches (11.3 ft. = 11 ft. 4 in.). If the temperature changes, the velocity—and therefore the wavelength—changes also. Figure 1-4b shows one complete cycle of a sine wave.

Figure 1.4 a. A Graph of the tuning fork's vibration is traced on the moving paper; b. One cycle of a sine wave.

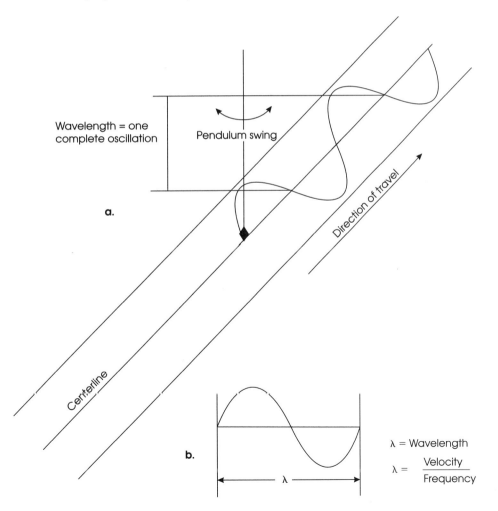

Frequency Response and Dynamic Range of Audio Equipment

Note that the chart in Figure 1-3 gives no information about the relative amplitude of the frequencies produced by the instruments. In most cases, the musician may vary the amplitude of each note played according to his or her taste. Subjective terms such as piano and forte cannot be used to describe audio equipment. Although the equipment must be able to reproduce the dynamic range of whatever musical program is passing through it, its performance is also evaluated in terms of frequency response. The frequency response of a device is a measure of its relative amplitude at various frequencies within its range.

■ Frequency Response and Logarithms

As mentioned earlier, the listener's subjective response to loudness varies logarithmically in proportion to the applied energy. In a similar manner, our perception of the interval between any two frequencies depends on the relative location of those frequencies within the audio bandwidth. To a musician, a 220-Hz interval between the two tones of 220Hz and 440Hz is recognized as an octave (abbr. 8va). However, the same 220-Hz interval between 440Hz and 660Hz would be a perfect fifth in musical terms. An octave above an "A" of 440Hz would be an "A" of 880Hz. Therefore, we may define an octave as:

> The interval between any two frequencies, f_1 and f_2, where $f_2 = 2f_1$.

We can see that as the frequency of f_1 increases, the arithmetic frequency interval within the octave grows larger. Yet to the ear, any octave relationship "sounds" the same as any other. By definition, then, the ear's response to frequency is logarithmic.

■ Log Paper

If frequency response measurements are drawn on standard graph paper, as shown in Figure 1-5, it will, for the reasons described before, be seen that each successive octave requires twice as much space as the one preceding it. A more satisfactory arrangement would allot an equal space to each octave, thus compensating for the ear's logarithmic response to frequency.

Log paper is used for just this purpose, and may be prepared as shown in Figure 1-5. Four equally spaced intervals, a-d, are drawn along the horizontal axis. The beginning of each interval is numerically labeled at twice the one before it (1, 2, 4, 8, 16).

The vertical axis is a simple arithmetic progression (1, 2, 3, 4,...16). Points are placed on the graph (P_1, P_2, P_4, P_8, P_{16}) at the intersections of lines drawn from corresponding numbers on both axes (1-1, 2-2, 4-4, 8-8, 16-16). A curve is drawn connecting the points. Now, lines from points 1 through 10 on the arithmetic axis are drawn to the curve, and then down (dashed lines) to complete the logarithmic scale. The 10 points on the logarithmic scale are labeled 1 through 10, or they may be labeled 10–100, 100–1,000, etc. If the scale is repeated several times, as in Figure 1-5b, it may be seen that every octave now takes the same amount of space, regardless of its frequency interval.

Figure 1.5 a. When standard graph paper is used for frequency response data, each octave uses more space than the one preceding it.
b. A method of producing a logarithmic scale, so that each octave uses an equal amount of space.

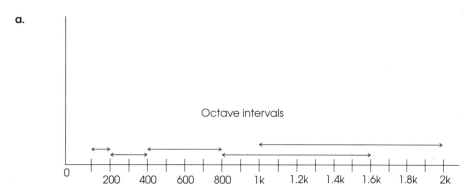

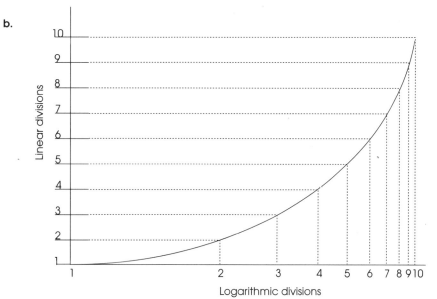

The dB scale (vertical) is not drawn logarithmically since the dB itself takes into account the logarithmic sensitivity of the ear to loudness.

■ **Amplifiers** Presumably, an amplifier will have a "flat" frequency response so that it supplies a gain of, say, 60dB equally to all frequencies within its range, plus or minus whatever dB tolerance from that "flat" response is specified. In addition to a flat frequency response, an amplifier should not alter the full-scale dynamic range of the signal passing through it. However, this is not always possible, since the dynamic range of some program material may exceed the dynamic range capabilities of an audio system or the capacity of the storage medium. [This will be discussed in greater detail in the chapter on compressors, limiters and expanders (chapter 8).]

■ **Bandwidth** The bandwidth of an amplifier or any other audio device is the range between the lowest and highest frequencies that are no more than 3dB down in level. Figure 1-6a shows the frequency response of an amplifier with a flat response between 40Hz and 10,000Hz, and a 20Hz to 20,000Hz bandwidth. Figure 1-6b shows the same amplifier's response on the logarithmic scale. Note that the amplifier also passes frequencies below 20Hz and above 20,000Hz, but that these are considered as being outside the amplifier's bandwidth, since they are attenuated by more than 3dB.

■ **The Human Voice** Like a musical instrument, the human voice may be analyzed in terms of frequency, amplitude and bandwidth. Speech sounds are of course quite complex, and are not generally thought of as having a recognizable musical pitch. Some voices are deeper than others, but people do not speak in musical tones, since every spoken word produces energy at a great many frequencies simultaneously.

When the voice is analyzed, its "frequency response" is usually shown as an energy distribution curve, as in Figure 1-7. The curves show the relative amount of acoustic energy present at each frequency in a typical male and female voice. Sibilant sounds such as s, z, sh, and zh produce a surprisingly large concentration of high level high frequency acoustic energy. Although the listener may not be distracted by an occasional sibilant peak (discussed in Chapter 2), an amplifier may be driven into distortion producing a very unpleasant "spitty" sound if corrective measures are not taken. These measures will also be discussed in greater detail in Chapter 8.

Figure 1.6 a. An amplifier with a "flat" frequency response between 40Hz and 10,000Hz, and a bandwidth from 20Hz to 20,000Hz; b. the frequency response of Figure 1-6a, redrawn on logarithmic scale paper.

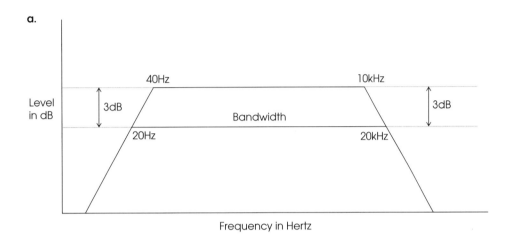

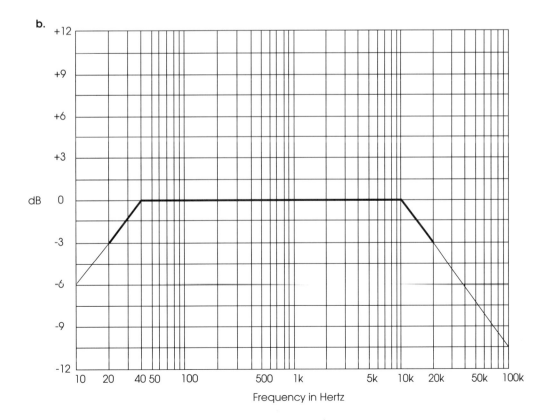

Wavelength on Tape

When any frequency is to be recorded on tape, the velocity is by no means the same as, or as constant as, the speed of sound in air. Most tape recorders are capable of running at several speeds, and so the tape velocity may be halved or doubled, depending on the speed selected. In Figure 1-4, the amount of paper taken up by one complete cycle depends on the speed at which the paper moves. Obviously, the slower the paper travels past the tuning fork, the less distance is used for each cycle and the smaller is the wavelength. And so it is with magnetic tape as it travels past the record head. The time it takes a wave to complete one complete oscillation is called its period and can be represented as:

$$P = \frac{1}{F}$$

where:
    P = period in fractions of a second
    F = frequency in Hertz (cycles/second).

If a 1,000Hz sine wave is recorded and then played back at the same speed, each recorded cycle will have a period of one one-thousandth of a second and will be spread over whatever amount of tape passed the record head during that interval of time.

Later on, there would be no way to determine the frequency of the original sine wave unless we knew the speed at which the recording had been made. On playback, the frequency reproduced will depend on the speed at which the tape is moving. For example, at half speed, each cycle of a sine wave that was originally 1,000Hz will now take two one-thousandths of a second to pass the playback head and the output will appear to be a 500Hz tone. The recorded wavelength remains unchanged, yet the frequency heard has dropped a full octave. Of course, an actual musical program played back at the wrong speed would be painfully obvious, but for the moment we are concerned with sine waves only. A sine wave with a recorded wavelength of one one-thousandth of an inch (1 mil) will produce the following frequencies:

> 7,500Hz at a tape speed of 7.5 inches/second
> 15,000Hz at a tape speed of 15 inches/second
> 30,000Hz at a tape speed of 30 inches/second

In evaluating the performance of an analog tape recorder, references to playback frequency response are invariably made, since our primary concern is: how well does the machine reproduce the audio

Figure 1.7 Energy distribution curves for typical speaking voices.

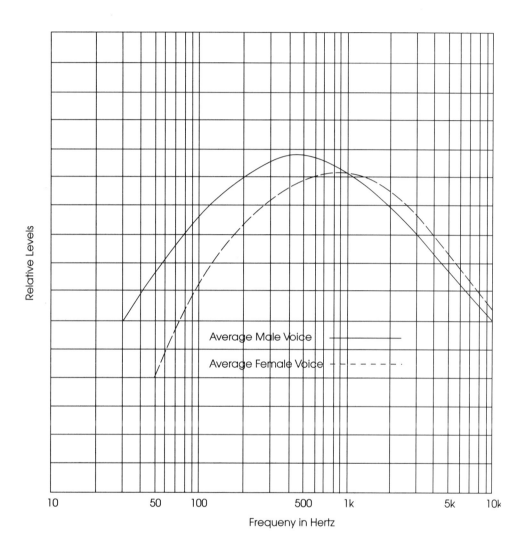

bandwidth? However, the playback head is actually responding to the wavelength on tape, and the frequency response depends on the "wavelength response" of the playback head, and the tape speed (velocity) selected.

Diffraction and Reflection

When a sound wave travels past an obstacle, a portion of the wave bends around the obstacle and moves off at an angle to the original straight line path. This is called diffraction and occurs when the wavelength of the sound is larger than the obstacle in its path. This phenomenon recalls the popular high school physics demonstration in which a beam of white light is made to pass through a prism, where it is dispersed into a rainbow of colors. White light is a composite of energy of many different wavelengths, and as it passes through the prism, each wavelength—with its characteristic color—is refracted at a different angle. Now while energy passing through an object is refracted, when it travels around the object it is diffracted.

When the wavelength of the sound is shorter than the obstacle in its path, it is reflected and bounces off the object at an angle that is equal to the angle of approach—much like a mirror reflects light. In either case, the result is the same: some of the energy is re-routed in a different direction, and the angle of deflection depends on the angle of incidence and the wavelength of the energy wave.

In the recording studio, diffraction and reflection are important considerations. A long wavelength (low frequency) may be bent (diffracted) almost completely around an obstacle in its path, while short wavelengths (high frequencies) will be reflected. Therefore, a listener (or microphone) in the shadow of the obstacle will hear a frequency-distorted version of the original sound wave.

Effect of Frequency on Threshold

In the section on decibels, we learned that the ear regularly encounters sound pressure levels between a very few dB and somewhat more than 110dB. 0dB was defined as the threshold of hearing. It should be pointed out that the actual sound pressure level at the threshold of hearing varies considerably, depending on the frequency of the sound. The ear is highly sensitive to frequencies between 3,000Hz and 4,000Hz, and less sensitive to frequencies outside this bandwidth.

For example, if a 3,500Hz tone is heard at the threshold of hearing, the sound pressure level will be 0dB (10^{-12} watts/m^2) as stated before. But if the frequency of the tone is changed to 350Hz (more than 3 octaves below), the sound pressure level will have to be raised about 17dB above this reference level in order for the average listener to detect the signal. At 35Hz, the sound pressure level may have to be raised more than 50dB in order for the signal to be heard.

Equal Loudness Contours

Therefore, our threshold of hearing is by no means constant over the entire range of audible frequencies. Figure 1-8 is a graph of the equal loudness contours as developed by Robinson and Dadson in 1956. Notice that the various contours are labeled in Phons, ranging from 0 to 130, and that at 1kHz, the Phon rating corresponds to the sound pressure level in decibels. The Phon is a measure of "equal loudness." If we first listen to a 1,000Hz tone measured at 50dB sound pressure level, and then adjust another tone so that it appears to be just as loud, both tones are considered to have a loudness of 50 Phons, even though the measured sound pressure level of the second tone differs from the 50dB level of the 1,000Hz standard. The 0 phon curve indicates the average listener's perception of equal loudness for frequencies determined to be at the threshold of hearing. As noted, low frequencies must be boosted considerably in sound pressure level in order to be heard, and frequencies above the maximum sensitivity area of 3kHz to 4kHz must also be boosted, though not to the same extent.

Notice also that although the equal loudness contours each follow the same general direction, there are significant differences in the ear's relative sensitivity at different listening levels. The sensitivity of the human ear is always greatest in that 3kHz to 4kHz region, and falls off as the frequency is lowered. However, in the 30 to 100 Phon range, there is again some increase in sensitivity as the frequency falls from 1,000Hz to about 450Hz, and then sensitivity again decreases until, at the low end of the frequency spectrum, the ear is at its least sensitive point.

The equal loudness contours are often referred to as the Fletcher-Munson curves, named after the two men who did early research in this area. The current version in Figure 1-8 shows the recent revision of the curves by Robinson-Dadson and also shows the minimum audible field (MAF) curve. This curve represents, at all frequencies, the threshold of audible perception.

The implications of the equal loudness contours should be clearly understood by every recording engineer. As the overall loudness level is changed, the ear's frequency response is significantly altered. Therefore, a frequency balance that is satisfactory at one listening level may not be so at another. Many engineers prefer a loud monitoring level, claiming that it allows them to hear every sound and nuance clearly. Later, when the level is reduced to a more normal loudness, there is inevitably some disappointment with the apparent lack of bass. Here, the equal loudness contours are working against the engineer.

Figure 1.8 Equal Loudness Contours as developed by Robinson and Dadson. These countours demonstrate the ear's sensitivity at various sound pressure levels with respect to frequency. The MAF (Minimum Audible Frequency) curve defines the ear's minimum sensivitity.

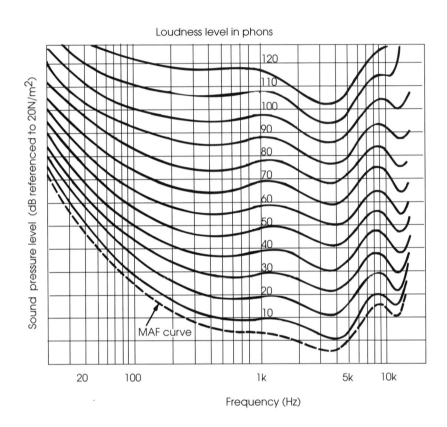

For example, from Figure 1-8, we see that at 120 Phons, the ear is 15dB more sensitive to 3,500Hz than it is to 1,000Hz. Below 1,000Hz, the ear's sensitivity is uniform to about 200Hz and then sensitivity falls off gradually. At 40 Phons, there is only a 7dB sensitivity difference between 1,000Hz and 3,500Hz, and below 300Hz the sensitivity decreases rapidly. So, if a recording is balanced at 120 Phons, and then played back at 40 Phons, the bass will invariably sound weaker, and there will be a lack of "presence" in the 3,500Hz area if these frequencies were attenuated earlier as a reaction to the ear's extreme sensitivity to them at 120 Phons.

The situation may be clarified somewhat if we invert the 120 and 40 Phon contours, as shown in Figure 1-9. The curves now depict a graph of the ear's relative frequency response at 40 and 120 Phons. Now imagine that we added some equalization, while listening at 120 Phons, in order to create the apparently flat frequency response shown by the dashed line. If we now observe the effect of this equalization on the 40 Phon contour, we arrive at the unsatisfactory frequency response shown in curve "D." The low-frequency response has apparently fallen off, and there is also a pronounced fall-off in the 3kHz to 4kHz range.

It is an oversimplification to state that recordings should be monitored at the same loudness level at which they will be heard later on. Obviously, the engineer has no control over the record or CD buyer's listening habits, and cannot predict the level at which he will choose to listen to the finished product. However, the equal loudness contours do suggest that if studio monitoring levels are kept on the conservative side, there will be less disappointment later on. If the proper amount of bass is heard at lower listening levels, any subsequent level increase will bring with it an apparent

Figure 1.9 The 120- and 40-Phon contours, inverted to compare the ear's relative sensitivity at these two listening levels. A. The 120-Phon contour; Note the ear's extreme sensitivity in the 3,500Hz area; B. Equalization required to achieve an apparently flat frequency response (dashed line) at a 120-Phon listening level; C. The 40-Phon contour; Note the sharper fall-off in sensitivity at low frequencies; D. The apparent frequency response at 40 Phons, as a result of the equalization that was added when listening at 120 Phons; note the deficient bass response and lack of "presence" at 3,500Hz.

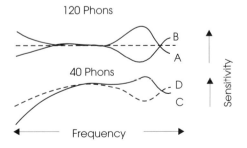

bass boost. Generally, this is to be preferred over the opposite condition, where the bass apparently decreases from the ideal point as the level is dropped. Today, many studios have adopted the Dolby Level Standard used in the motion picture industry, where material for the home is monitored at an average level of 79dB, while sound for the theater is monitored at a level of 85dB.

Perception of Direction

The ear is an extremely sensitive and complex device and, as we have already mentioned, is capable of wide band frequency response and extended dynamic range. However, as responsive as the ear may be, one ear is limited in its ability to perceive from what direction a sound is coming. A single ear can tell us pitch, loudness, and even the character or timbre of a sound, but cannot tell us the location of that sound. For us to be able to localize a sound, we need two ears. It is the difference in the information received by these two ears that lets the ear/brain combination tell us where that sound source has originated. This binaural hearing uses the time-of-arrival differences and loudness differences between the left and right ear to determine the location of the sound.

If a single sound source is positioned directly in front of us, as in Figure 1-10a, the intensity or loudness will be the same at each ear, and the sound will arrive at both ears simultaneously. However, if a sound source is moved to a position that is to the left of center (Figure 1-10b), because of the diffraction around the head and the distance between our ears, the sound will arrive at the left ear before the right one. The sound will also be perceptibly louder at the left ear than at the right. These binaural cues are extremely important for our perception of direction and contribute to a good stereophonic recording as will be discussed in Chapter 4.

Phase and Coherence

In Figure 1-11a, various points along a sine wave are marked off in degrees. A second sine wave is drawn in Figure 1-11b. Although the two are of course identical in wave shape, amplitude and period, we may say that the second one has been shifted by 180° due to its relative position with respect to the first sine wave. Notice that as one sine wave reaches positive maximum amplitude (points "a"), the other reaches negative maximum amplitude. Both sine waves pass through zero amplitude at the same time (points "b"). If the two waves are combined graphically, as in Figure 1-11c, the resultant

Figure 1.10 Sound originating from the center (a) will arrive at both ears simultaneously, while sound from off-center (b) will cause amplitude and time differences.

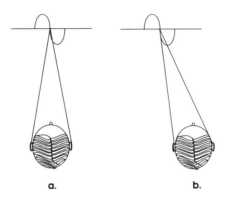

wave form will be a straight line, indicating zero amplitude. Since the waves are at all times equal in amplitude and opposite in polarity, they have canceled each other out.

A similar condition will occur electrically if audio signals of equal amplitude and opposite polarity are combined. This might happen to an audio signal that is passing through two different signal paths if there is a wiring reversal in one of the paths as shown in Figure 1-12. When the two are eventually combined, any signal common to both of them is canceled out. The signals are said to be electrically out of phase.

An acoustic phase cancellation might occur if two microphones are positioned in such a way that one receives a positive pressure center at the same time that the other receives a negative one, as shown in Figure 1-13. This is a very real problem, which will be discussed in greater detail in the chapter on microphone applications.

Of course, actual program waveforms are considerably more complex than sine waves, and the relationship of one to another cannot really be measured as precisely as two sine waves of the same frequency. Phase relationships then become meaningless, and about all that can be said for the two waveforms shown in Figure 1-14

Figure 1.11 The combination of two sine waves.

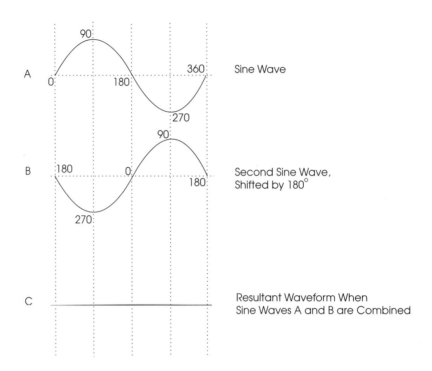

is that sometimes they are of the same polarity (+,+ or -,-) and other times they are not (+,- or -,+). Even when the polarities are opposite, the wave forms do not cancel out, since their amplitudes are usually unequal. During those intervals when the waveforms are of the same polarity, they are said to be "coherent" and they interfere constructively; when the polarities are opposite, the waveforms are incoherent and interfere destructively.

When the same signal is applied to two signal paths, they are totally coherent since, regardless of the complexity of the waveforms, they are of course identical. But if one of the paths has an electrical phase reversal, the two signals become totally incoherent since they are at all times identical in amplitude and frequency yet opposite in polarity. As with the sine waves shifted by 180°, the

Figure 1.12 The effect of an electrical phase reversal.

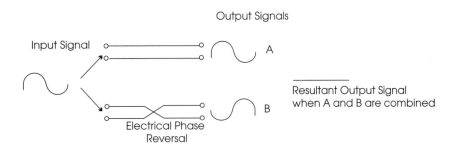

Figure 1.13 The acoustic phase reversal between the two microphones will cause a cancellation if their outputs are combined.

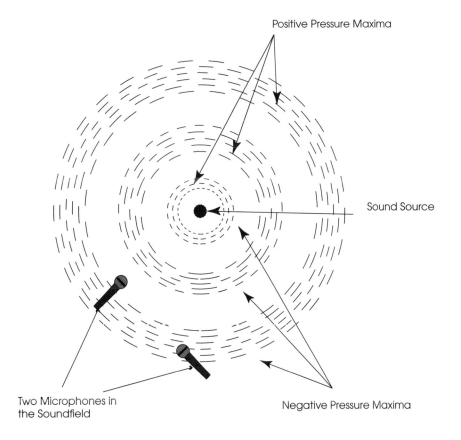

Figure 1.14 Two complex audio waveforms.

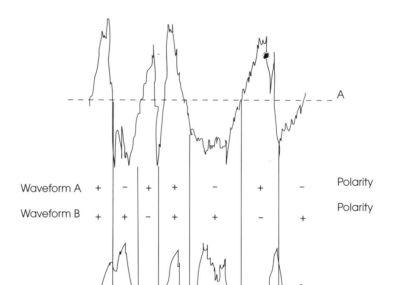

totally incoherent wave forms will cancel out if combined. Of course, if two totally coherent waveforms are combined they will sum and show an increase of 3dB in power.

Signals from two different sources such as a guitar and a piano will display a random coherence, since there is no signal component common to both signals. An electrical phase reversal will have no effect on an eventual combination of the signals. However, if a signal from a centrally-placed soloist is common to both sources, the phase reversal will cause it to cancel out when the signals are combined.

TWO
Sound Measuring Devices

■ **THE VOLUME INDICATOR (VU METER)**

So far, we have talked about studio meters in terms of decibels. However, when audio signals are being measured, the proper unit of measurement is usually the volume unit, or VU. The decibel has been defined as a ratio of powers, or perhaps voltages. As long as we are measuring a constant level, such as a 1,000Hz reference tone, we may express the reading in dBm. On the other hand, an audio signal has a constantly changing level. Occasional peaks may be far above the average signal level, yet they do not contribute much to the listener's sensation of program loudness, for the ear has a tendency to average the fluctuating program level in evaluating its subjective loudness. Thus, a program with a fairly constant, though moderate, level will be considered much louder than a low-level program with some recurring high-level peaks.

Equating this to the meter, we may note that a regular bench type voltmeter does a poor job of responding to the rapid level changes of an audio program. The meter ballistics are such that it is accurate only on steady-state tones. For audio measurements, a different meter movement has been designed. This is known as the volume indicator, and it is calibrated in volume units, or VU. When measuring steady-state tones, the volume indicator readings will correspond to those seen on a regular dBm-calibrated voltmeter, yet on program measurements, the volume indicator will read somewhere between the average and peak values of the complex audio waveform. Volume indicator ballistics are designed so that the meter scale movement approximates the response of the ear; thus the meter does not register instantaneous peaks above the average level. In fact, the transient response of the human ear is limited by the reaction of the Staepedius muscle, which is connected to the ear drum and the three small bones of the middle ear. The typical reaction time of this small muscle is about 300 milliseconds. Since our VU meter is calibrated to have similar reaction times, this means that any signal having a peak value shorter than 300ms in duration will not be accurately metered. If the peaks are frequent, the meter will read somewhat higher than the average level, yet still below the actual peak level. The ballistics of the ASA (American Standards Association) standard volume indicator cause the meter to read 99% of its 100% deflection in 300 milliseconds. Its fall rate is the same as its rise time.

The fact that the volume indicator's reading corresponds with the listener's subjective impression of loudness makes it a valuable recording tool. However, it should be clearly understood that actual recorded levels are consistently higher than the meter readings. For example, the maximum-to-average ratio of program levels may be such that the volume indicator reads, say, 0VU. But the instantaneous level that causes that 0VU reading may be +10dBm. Peak levels are often 9dB to 10dB above the average level seen on a standard VI meter. Both in practice and in the literature on recording there is a tendency to refer to audio program measurements in decibels. Although there is surely little real danger in honoring this custom, the recording engineer should understand the difference between the decibel and the volume unit. The difference is particularly significant when high-level peaks are causing distortion, while the meter reading displays an apparently conservative recording level.

The Decibal or Volume Unit in Equipment Specifications

In reading other chapters of this book, continuing reference is made to the decibel (dBm), with little or no further mention of the volume unit (VU). For example, we may read that a tape saturates at +10dBm, or that an equalizer supplies a 6dBm cut at 3,000Hz. Since the effects of these variables are so often observed on volume indicators, it may seem contradictory to discuss them in terms of decibels. Although the volume unit is a measure of a complex audio waveform, for ease of measurement a steady tone is invariably used to determine equalizer performance or tape saturation. Therefore, the dBm becomes the correct unit of measurement. As an example, consider an equalizer set to give a boost at 1,000Hz of +3. When the equalizer is inserted, it might be expected that 1,000Hz tones will now be higher by three volume units. However, if the 1,000Hz tone is in the form of a sharp transient, the volume indicator may show little or no increase in reading, for the reasons discussed earlier. On the other hand, a sustained 1,000Hz tone will be three volume units higher than before. Therefore, since the meter's response to the equalization change remains dependent on the nature of the program, it cannot be stated with certainty that a boost of +3 will always produce an increase of 3VU. Consequently, these variables are specified in dBm, taking into account the steady-state conditions under which the measurements or calibrations were made.

■ THE SOUND LEVEL METER

Although the recording engineer is concerned with the decibel primarily in terms of electrical power or voltage ratios, some understanding of the sound level meter is essential to good engineering practice.

Basically, a sound level meter consists of a microphone, an amplifier, and a meter calibrated in decibels. However, such a device may not be created by simply connecting any available microphone, amplifier and meter. In addition to a specially calibrated microphone, the sound level meter will contain several filters and "weighting" networks. The filter networks allow the meter to respond to sound energy within various narrow bandwidths. For example, if the filter network was set at 1kHz, sound (or noise) containing 1kHz components could be measured, while the meter would be comparatively insensitive over the rest of the audio spectrum. This tunable feature allows the engineer to determine the frequency band which is contributing the greatest energy to the overall sound or noise level.

Weighting networks are filters that create a response in the meter corresponding to the ear's varying sensitivity at different loudness levels. As described and shown by the Robinson-Dadson curves in Figure 1-8, the ear is less sensitive to extreme high and low frequencies at lower listening levels. Therefore, at lower listening levels, noise in the low and high frequency regions would of course be less objectionable than the same amount of noise in the mid frequency area.

The "A," "B" and "C" weighting networks are filters which correspond to the sensitivity of the ear at listening levels of 40, 70 and

100 Phons, respectively. When making measurements at these levels, the insertion of the appropriate weighting network will give a meter reading that is pretty much in accordance with what the listener subjectively hears.

Although noise level measurements of recording studio equipment are commonly made on a standard voltmeter, an "A" weighting network is often inserted just before the meter input so that, as in sound level measurements, the reading conforms to the subjective impression that the noise would make at low listening levels. Accordingly, noise specifications are frequently quoted at so many decibels, dBA. Or the level may be quoted in dBm, with the notation "A Weighting." An "A" weighting network is shown in Figure 2-1a, and the "A," "B" and "C" weighting curves are drawn in Figure 2-1b. Figure 2-2 is a photo of a high-quality sound level meter.

■ THE PEAK READING METER

The peak reading meter is more responsive to actual program peaks. Its ballistics are such that it will more accurately track the attacks of sudden high-level transient peaks. But since these peaks may be over just as suddenly as they occur, the movement is usually tailored to provide a more gradual fall-off, lest the rapid up and down movement prove to be too fast for the eye to follow. The standard peak reading meter has a rise time of 2.5 milliseconds followed by an exponential fall. Although many engineers use the loudness-related volume indicator, the peak reading meter gives a more accurate indication of what is actually being recorded. Since magnetic tape, as well as amplifiers, may be overloaded (and therefore distorted) by sudden bursts of high-level energy, some engineers prefer the peak reading meter, since it is more reliable in alerting the operator to potentially troublesome peaks that may escape the notice of the ear. Of course, a distorted amplifier will certainly be heard, but if the distortion takes the form of tape overload, it may not be noticed until the tape is played back later.

As noted earlier, the volume indicator and peak reading meter will read substantially the same on a sustained level program. This may cause the engineer to be too conservative with the record levels on program material such as organ music, which has very little peak information. However, with a program containing a significant number of high-level transients (for example, drums, tambourines,

Figure 2.1 a. A schematic. The network produces a 4dB insertion loss; b. The "A," "B" and "C" weighting curves.

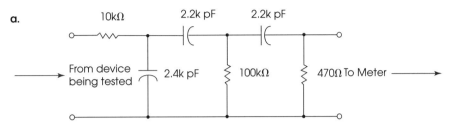

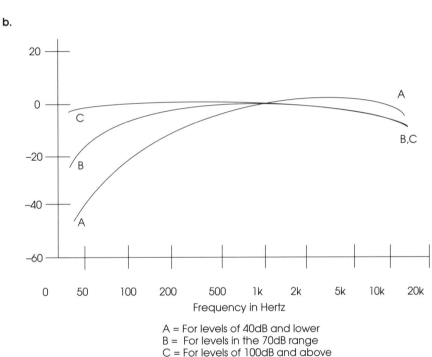

A = For levels of 40dB and lower
B = For levels in the 70dB range
C = For levels of 100dB and above

etc.), the peak reading meter may indicate levels 10dBm or more above the volume indicator and can prevent the engineer from overloading the tape or amplifier. Since the peak reading meter is not a volume indicator, it is correct to read it in decibels, rather than volume units.

Figure 2.2 A sound level meter (Bruel and Kjaer)(B&K Photo).

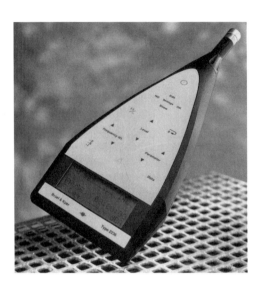

■ STUDIO AND BROADCAST REFERENCE LEVELS

Up until now, we have been talking about a zero reference level equal to 1 mW of power dissipated across a 600-ohm resistor. This value has no inherent significance—it is merely a mathematically convenient reference point. However, the assignment of handy reference points is one thing, and the design of efficient meter movements is quite another. Early on, it was found that, due to the vagaries of meter construction, it was difficult to build a meter movement that would satisfactorily register 0dBm (or 0VU). Such a meter would have an impedance of less than 4,000 ohms, and would therefore load the circuit it was supposed to be measuring, causing misleading readings.

To negate the loading effect of the meter movement, a series resistor (empirically valued at 3,600 ohms) was inserted in the meter circuit. But now, although the meter no longer affected the circuit being measured, the presence of the series resistor caused the meter to read about 4dBm lower, when measuring an actual 0dBm. With the 3,600-ohm resistor in the meter circuit, when the meter did read 0VU, the level would actually be +4dBm. Rather than re-define the zero reference level, it was considered expedient to leave 0dBm at 1mW across 600 ohms, with the understanding that 4dBm above this value would correspond to a zero meter reading. Most, if not all, recording studio equipment now adheres to this 0VU = +4dBm convention. However, the broadcast and telephone industries sometimes go one step further and define a zero reference of +8dBm. This takes into account the higher output levels required for telephone lines, which play such an important part in broadcasting. This is not a consistent policy, and the prudent engineer should be aware of which reference is in use. As may be noted in Figure 2-3, a standard volume indicator's meter face reads from -20 to $+3$. Therefore, if 0VU = +4dBm, the meter is capable of reading levels up to +3VU = +7dBm. For broadcast applications, an additional 4dBm of attenuation is placed in the meter circuit, so that 0VU = +8dBm. Now, a meter reading of +3VU signifies +11dBm.

Many of the so-called semi-professional and consumer tape recorders use an output level of -10dBm to correspond to a reading of 0VU. Because of these conflicting zero reference levels, there is often some confusion about differences—if any—in recorded levels between studio, consumer, and broadcast tape recorders. It often comes as a surprise to learn that there is actually no difference in recorded level, regardless of what level (-10, $+4$, $+8$, or whatever) has been defined as a zero reference.

For example, a standard test tape is used to align all machines, whether used in broadcast or recording studio work. For standard operating level, the output level control is rotated until the meter reads zero. Depending on the machine's meter circuit, this zero will produce a level of either -10dBm, +4dBm or +8dBm at the machines's output. Nevertheless, the input level—in this case from the test tape—remains the same. Consequently, when the machines are placed in the record mode, the same input level applied to both machines will cause identical meter readings. Of course, if the actual output levels are compared, the studio and broadcast machines will

Figure 2.3 A typical volume indicator, calibrated in volume units (VU).

be louder. But this is because their playback amplifiers have been turned up to +4dBm or +8dBm respectively. In any case, the input level, and therefore the level recorded on the tape, remains the same.

■ THE SPECTRUM ANALYZER

The spectrum analyzer is a specialized type of sound level meter. It allows the engineer to examine the amplitude versus the frequency spectrum of a sound. This is accomplished using multiple sets of filter networks and the information is displayed in vertical columns arranged from low to high frequencies, left to right. These analyzers are usually found with third-octave spacings between the frequencies within a range between 20Hz and 20kHz. Occasionally they are found with octave bands, that is, with a doubling of frequency above each preceding frequency column. The third-octave bands are determined by using the cube root of 2. For instance, the next band above 1Hz would be the cube root of 2 (1.25), followed by the cube root of 2 squared (1.58), the cube root of 2 to the third power (2), the cube root of 2 to the fourth power (2.5), etc. Each succeding octave is multiplied by a factor of 10. These ISO (International Standards Organization)-preferred frequencies are used for equalization centers as well, and will be discussed further in Chapter 7. The frequencies used are as follows:

1/3 8va bands

1, 1.25, 1.63, 2, 2.5, 3.15, 4, 5, 6.3, 8, 10, 12.5, 16.3, 20, 25, 31.5, 40, 50, 63, 80, 100, 125, 163, 200, 250, 315, 400, 500, 630, 800, 1000, 1250, 1630, 2k, 2500, 3150, 4k, 5k, 6300, 8k, 10k, 12.5k, 16k, 20k etc., as necessary;

8va bands

31.5, 63, 125, 250, 500, 1000, 2k, 4k, 8k, 16k etc., as necessary.

Many portable spectrum analyzers also allow the user to select the "A," "B" or "C" weighting scale as well. And, as will be discussed in Chapter 16, spectrum analyzers may often be found as part of the metering circuit on larger audio recording consoles.

■ THE OSCILLOSCOPE

The oscilloscope is an effective monitor of phase and coherency information. If a sine wave is applied to the scope's X Axis input, a vertical line will be seen, as in Figure 2-4a. If the signal is instead applied to the Y Axis input, a horizontal line will be seen, as in Figure 2-4b. In both cases, the length of the line is an indication of the amplitude of the applied signal.

If the sine wave is applied in phase to both signal inputs, a diagonal line will be seen as shown in Figure 2-4c. The angle of the line will indicate the relative amplitudes of the signal at both inputs. If the amplitudes are equal, the line will appear at a 45° angle, as seen in Figure 2-4c. If the amplitudes are unequal, the angle will change towards the vertical or the horizontal axis, depending on which amplitude is greater. These displays are called lissajous patterns and, if there is an electrical phase reversal of 180° in one of the inputs, the slope of the diagonal line will be reversed (Figure 2-4d). If there is a phase incoherence of 90° between the two input signals, a perfect circle will be displayed, as seen in Figure 2-4e. These visual displays are a clear indication of any phase incoherence which might otherwise escape notice.

If complex audio waveforms are applied to both oscilloscope inputs, the coherent component will tend to produce a diagonal display in the in-phase direction, while the incoherent component will tend toward an out-of-phase display. The net resultant pattern will resemble those patterns shown in Figures 2-4f through h. The general diagonal orientation of the display indicates the amount of coherency.

An accidental phase reversal, whether due to an improperly wired cable or an acoustical condition in the studio, may be difficult or impossible to detect by ear alone. In fact, it may actually enhance the sound by giving it a more spacious feeling. However, later on when the signals are combined during the mixdown process, the signals may cancel out almost completely. Therefore, the use of an oscilloscope as a visual monitor while recording cannot be overemphasized.

Most stereo programs contain a signal component that is common to both left and right. This is the phantom center channel information which is heard equally from both speakers. If the relative level of this center channel component is high, the oscilloscope pattern will take on an in-phase orientation, as in Figure 2-4f. However, if an accidental phase reversal has occurred, the general orientation

Figure 2.4 Cathode ray oscilloscope patterns, indicating phase and coherence between two signals. a. Signal applied to Y axis input only (generally signifies left track of a stereo program); b. Signal applied to X axis input only (generally signifies right track); c. Same signal applied to both inputs; d. Same signal applied to both inputs, but with a phase reversal in one input line; e. Same signal applied to both inputs with a 90 degree phase relationship between them; f. Stereo signal with a strong coherent component, resulting in an "in-phase" orientation; g. Stereo signal with random coherency, resulting in a circular pattern with no noticeable diagonal orientation; h. Stereo signal with a strong incoherent component, resulting in an "out-of-phase" orientation.

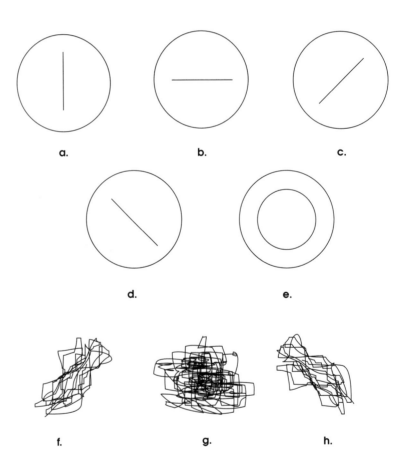

will be reversed, as in Figure 2-4h. This visual display will warn the engineer of a potentially troublesome condition which should be traced and corrected.

■ OTHER METERING STANDARDS

The volume indicator is the most commonly used style of metering in the United States today; however, some form of peak metering is often found in conjunction. Peak metering, often called IEC (International Electrotechnical Commission) metering, has several different standards in various parts of the world (Figure 2-5). The IEC Type I is often called DIN metering, while the IEC Type II is often referred to as Nordic metering. The Type III IEC metering is commonly called BBC (British Broadcasting Corporation) metering. The ballistics and standards for the VI (or VU) meter are shown for comparison. Note that not only do the ballistics (rise and fall response times) vary, but the O reference value (in dBm) varies as well.

The last column shows the ballistics and characteristics for a typical meter used to measure signals in the digital domain. As will be noted later in Chapter 14 on digital audio, a meter of this type uses zero on the scale to represent full digital level. This may vary relative to the analog equivalent depending on the number of bits used by the system. Therefore, readings on a meter designed for digital audio will read in dB full scale or simply dBfs. The standards for relating digital audio levels to the more conventional analog power levels is still in a state of flux. Although the 1kHz tone at +4dBm is shown as equal to -20dBfs, it may just as often be found to equal -18dBfs depending on the manufacturer of the equipment.

Figure 2.5 Peak metering standards in various parts of the world.

	VU	BBC	IEC Type I	IEC Type II	Digital
Scale	-20VU to +3 VU	MK 1 to MK 7	-50dB to +5dB	-36dB to +9dB	-50dBfs to 0dBfs
Rise Time	300ms to 0VU	100ms to MK 6	10ms to -1dB	10ms to -1dB	10ms to -1dBfs
Fall Time	300ms to -20VU	2.85s to MK1	13.3dB per sec.	13.3dB per sec.	13.3dB per sec.
Reference	0VU = +4dBm	MK 4 = 0dBm	0dB = +6dBm	0dB = 0dBm	0dBfs = +24dBm
1kHz tone @+4dBm	= 0VU	= MK 5	= -2dB	= +4dB	= -20dBfs

44   THE AUDIO RECORDING HANDBOOK

It is important for the engineer to understand thoroughly the type of meter in use. Only then can he be sure that what he is seeing is actually the signal flowing in the circuit in question. Any of the above standards can be used effectively, and many engineers have a favorite type of metering that they prefer to use if given the choice. Figure 2-6a shows a typical peak meter while Figure 2-6b pictures a sophisticated digital meter.

Figure 2.6   a. A vacuum fluorescent peak program meter (PPM); b. A modern digital meter that reads in dBfs.

SECTION II
Transducers: Microphones and Loudspeakers

■ INTRODUCTION

Transducer:

    1. A device actuated by power from one system and supplying power in the same or any other form to a second system;

    2. A microphone or loudspeaker.

Modern recording studio practice is a curious mixture of art and science. The engineer who may describe the operating parameters of every Op-amp in the studio may be content to describe his favorite microphone as "warm." While demanding less than a fraction of a percent distortion from an equalizer, he may not care about the sensitivity rating of a new microphone, providing it is "clean" and sounds "right" to him.

As for monitor speakers, he's apt to be more concerned with similar subjective value judgments than with the rated efficiency, radiation angles, and so forth.

Needless to say, these subjective evaluations are not much help to the beginning engineer who is trying hard to learn more about the science of recording. Yet it is nearly impossible to avoid such descriptions of transducers in the recording studio.

Perhaps this avoidance of a more precise terminology is a necessary function of the transducing processes in the signal path from the musical instrument to the listener. Along the way, many subtle changes take place. The microphone converts acoustic energy into mechanical, then electrical energy. The loudspeaker reverses the process, and at the end of the chain, the listener's ear converts acoustic energy into brain waves that are subjectively interpreted by the listener. Even in the control room, with the instrumentalists only a few feet away, the listener is several generations away from the actual musical event, and is actually evaluating the performance of the transducing system as well as the musical performance.

Personal taste, upbringing, and age are very few of the factors that influence our musical taste and perception, and all these variables come into play when we evaluate what we hear in the control room. Another engineer will have another opinion, just as subjective, about the reproduced sound. One may prefer microphone "A" while another likes microphone "B." If the monitor system is changed, opinions may change with it, and under no circumstances will there be unanimous agreement as to what sounds "right." Since microphones are easily changed and speakers are not, the listener tends sometimes to ignore the role of the speakers in his evaluation of the microphone. Yet a change in speakers will certainly influence one's preference in microphones.

Given the subjectivity of music, and the inter-dependence of one transducer on the other, definitive statements—especially on microphone technique—cannot easily be made. Nevertheless, a basic understanding of the more scientific aspects of both transducing systems will help the engineer develop his or her own personal recording technique.

Chapter 3 covers the basics of microphone theory, while Chapter 4 discusses some of the more general microphone techniques, which may be adapted or modified to suit the demands of everyday studio practice. Chapter 5 concludes the section with a description of the loudspeaker and the enclosure in which it is placed, and also considers the interface between the speaker and the listening room.

THREE

Microphone Design

If a low mass membrane is exposed to changes in acoustical energy, it will move relative to the energy changes. At the heart of any microphone is its diaphragm, where acoustical energy is converted into an electrical signal as the membrane vibrates in response to the impinging sound wave. Microphones are commonly classified according to the manner in which this energy conversion takes place, and a description of the most popular classifications of studio microphones, dynamic and condenser, follows.

■ DYNAMIC MICROPHONES

Microphones in which an electrical signal is produced by the motion of a conductor within a magnetic field are classified as dynamic microphones. As you may remember from your physics classes, when a magnet is moved up and down within a coil of wire a voltage is induced in the coil. If we were to picture the output of the coil in Figure 3-1, it would appear as a sine wave, with the positive portion occurring as the magnet is plunged into the coil, and the negative portion registering as the magnet is withdrawn. The amplitude depends on the strength of the magnet and the number of turns in the coil of wire, while the frequency of the signal is controlled by how slowly or rapidly the magnet is moved in and out of the coil. Obviously, the same is true if the coil moves and the magnet is stationary.

Figure 3.1 A magnet, moved up and down within a coil of wire, will produce a signal with both positive and negative components at the output of the coil.

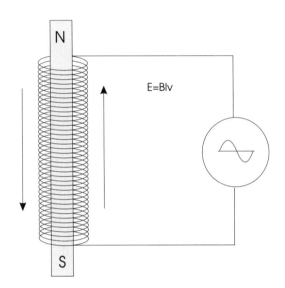

Dynamic Moving-Coil Microphones

In this type of microphone, a coil of wire is attached to the rear of the diaphragm, and the coil is suspended in a magnetic field as shown in Figure 3-2. As the diaphragm vibrates, so does the coil, and the magnetic field induces a voltage within the coil. The voltage is the electrical equivalent of the acoustical energy that caused the diaphragm to vibrate and is based on the following formula:

$$E = Blv$$

where:

E = the output signal, in volts

B = the fluxivity of the magnet, in gauss

l = the length of the coil, in feet

v = the velocity of the movement, in feet per second

Figure 3-3 shows two typical dynamic moving-coil microphones.

Figure 3.2 Cross-sectional sketch of a dynamic moving-coil microphone.

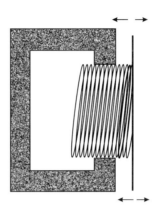

Figure 3.3 Two typical dynamic moving-coil microphones; a. Sennheiser MD-441 (Sennheiser Photo); b. Shure SM-58 (Shure Bros. Photo).

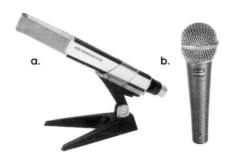

Dynamic Ribbon Microphones

In this type of dynamic microphone, a thin corrugated sheet or "ribbon" of metal foil takes the place of the diaphragm/moving coil combination. The ribbon is suspended within a magnetic field, and a small voltage is induced as the ribbon vibrates to and fro. The corrugations lend some structural strength to the ribbon which in early designs was quite delicate (Figure 3-4). Modern ribbon microphones, however, are well-suited to studio use and may be just as sturdy as their moving-coil counterparts. A ribbon microphone contains a built-in output transformer, since the impedance and the output of the ribbon is quite low (usually on the order of a fraction of an ohm). A transformer is thus required to raise the microphone's output impedance and level to a usable value. A ribbon microphone may also be referred to as a velocity microphone since it responds to the particle velocity which results from the pressure difference, or gradient, between the two sides of the ribbon. A typical ribbon microphone is pictured in Figure 3-5. (Note: a possible source of confusion is the common studio practice of referring to moving-coil microphones as "dynamics" while ribbons, which are also dynamic, are known simply as "ribbons.")

■ CONDENSER MICROPHONES

In the condenser microphone, the diaphragm is actually one of the plates of a capacitor. A capacitor is an electrical device capable of storing an electrical charge. (In earlier times, what we refer to today

Figure 3.4 Close-up of a ribbon microphone with the protective cover removed (Beyer M260, Peabody Recording Studio Photo).

Figure 3.5 A studio ribbon microphone (Beyer M160)(Beyer Photo).

as a capacitor was known as a condenser; hence the nomenclature "condenser microphone.") The condenser microphone's diaphragm is metal-coated film that acts as one plate of a capacitor. The other plate is a stationary plate or backplate. As the condenser microphone's diaphragm vibrates, the spacing between it and a stationary back plate varies, producing a changing capacitance. This change in capacitance causes an inverse proportional change in the voltage potential. This is shown in the following formula:

$$C = \frac{Q}{V}$$

where:

C = capacitance in farads

Q = charge in Coulombs

V = applied potential in Volts

A usable signal voltage is thus derived, in conjunction with a pre-amplifier built into the microphone's case. A pre-amplifier is required since the capacitor's output and impedance must be converted to microphone level values. The pre-amplifier requires a voltage supply, which is usually provided externally. The power supply furnishes both a direct-current polarizing voltage to the condenser/diaphragm plates, and the necessary voltages for the transistors or tubes within the microphone's pre-amplifier as well. Figure 3-6 illustrates the built-in pre-amplifiers in several condenser microphones.

Figure 3.6 a. Neumann U-87 (Neumann USA Photo); b. Neumann M-149 (Neumann USA Photo).

Electret Condenser Microphones

The Electret condenser microphone has for its diaphragm two capacitor plates that have been permanently polarized by the manufacturer. Consequently, the diaphragm does not require an external power source. A voltage supply is still needed for the transistors in the pre-amplifier, however, which may be supplied by a simple battery.

RF Condenser Microphones

Some transistorized condenser microphones employ a circuit that replaces the polarizing DC voltage with a radio frequency oscillator and a diode. Older versions of this type of condenser microphone were plagued with noise problems; however, recent improvements in this technology have produced some excellent models. This design now provides a very low noise floor and superior resistance to "popping" in high humidity conditions due to its extremely low output impedance. The condenser element is part of a tuned circuit where changes in capacitance modify the tuning of the circuit and produce a frequency-modulated signal. This FM signal is demodulated in a manner similar to that of an FM radio and an audio signal is produced. A power supply is necessary to provide voltages to the oscillator as well as to the built-in pre-amplifier.

Condenser Microphone Power Supplies

Like any pre-amplifier, the one within the microphone requires a power supply, as noted above. In the case of older (and some new) tube-type condenser microphones, the power supply furnishes filament voltages as well as a polarizing voltage to the diaphragm. In the case of these tube-type condenser microphones, a special cable between the microphone and the power supply is required, and contains extra conductors which carry the required voltages separately from the audio signal.

Most modern transistor condenser microphones employ "phantom power" supply circuits, in which both the audio and the direct-current supply voltage travel within the same conductors. The standard power supply in use today provides 48 volts DC to the microphone, and is usually found built into the recording console. The phantom circuit has greatly simplified recording set-ups, since now the engineer need not set up a separate power supply for each condenser microphone. The condenser microphone will automatically draw the required power as soon as it is plugged in, while other non-condenser microphones should not be affected by the phantom power supply voltages.

If the recording console uses center-tapped microphone input transformers, the required 48-volt positive supply voltage may be applied as shown in Figure 3-7a. Since transformers do not pass direct current from one winding to another, the only path for the supply voltage is through both conductors in the microphone cable, back towards the microphone. Within the microphone, another transformer, also center-tapped, passes the supply voltage to the pre-amplifier.

Most consoles today utilize a non-center-tapped transformer configuration, or totally transformerless microphone pre-amplifier designs. It is now common practice to create an artificial center tap using two precision-matched resistors, as shown in Figure 3-7b. Note that, on the return, the DC supply voltage goes to the negative side

Figure 3.7 a. The phantom powering system is wired to the center tap of the console's input transformer; b. An artificial center tap is created with two matched resistors.

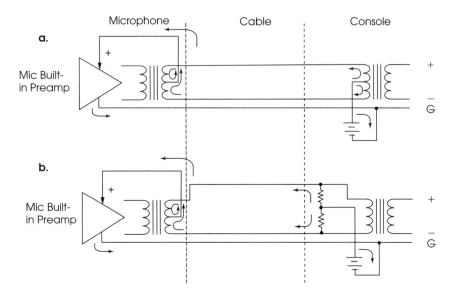

of the power supply leaving the AC audio signal free and clear. Thus it becomes doubly important that all microphone cables have their shields well-connected at both ends of the cable, since the shield must now provide this DC path back to the power supply, as well as fulfilling its primary purpose of shielding the signal leads against noise. Although a slight interruption in the shield may not make the cable unusable with dynamic microphones, it will prevent a phantom powered condenser microphone from properly functioning. Microphone cables are discussed in some detail at the end of this chapter (balanced and unbalanced lines).

Formerly, power requirements often varied from one microphone manufacturer to the next. Today, however, this is seldom the case as most microphone manufacturers adhere to the DIN standard for the powering of condenser (capacitor) microphones. A typical circuit for powering multiple condenser microphones from one voltage source is shown in Figure 3-8. Of course, vacuum tube condenser microphones, whether vintage or modern, will still require filament voltage which is not available through phantom powering.

Figure 3.8 Wiring supply diagram for a phantom power distribution system. Resistors must be 1% precision-type and are typically 6.8k ohms in value when used with a 48 VDC supply.

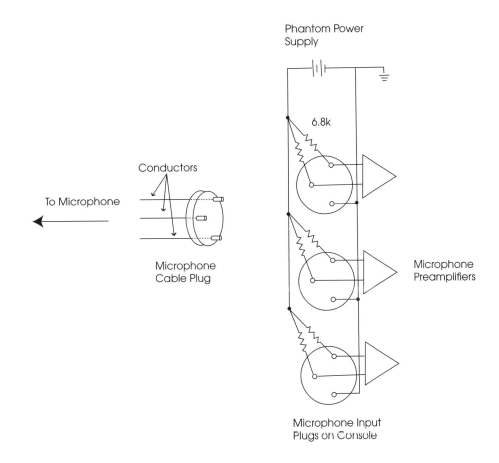

Digital Condenser Microphones

Very recently, several digital microphones have been introduced. This is a little bit of a misnomer since the acoustical principles for deriving an audio signal from a moving diaphragm are most certainly analog. Although from outward appearances the microphone seems identical to its analog cousins, the difference is that the analog-to-digital converter is placed immediately after the microphone capsule. Here, the signal is no longer limited in its dynamic range by the built-in pre-amplifier since it is immediately digitized. Current versions quantize the capsule output to 22 bits, and the signal is sampled at a rate of 48kHz. This provides a frequency response to well over 20kHz and a theoretical dynamic range of nearly 130 decibels. The microphone still requires some form of powering, and several types of systems are under development. At this time there is no standard powering method and each microphone requires a dedicated power supply type. The output of the microphone is sent to the receiving console as an AES/EBU signal. A major advantage of this type of signal is that it can be routed for very long distances, if required, with little or no loss of quality. At very long distances a standard condenser output can be noisy and prone to interference pickup. The reader is directed to Chapter 13 for a discussion of quantization, sampling, and digital signal format.

■ ACOUSTICAL SPECIFICATIONS

Directional Characteristics of Microphones

In any normal listening situation, we generally prefer to face the source of the sound which interests us. In fact, sounds originating from directly in front of us (called on-axis sounds) may often be clearly heard despite the presence of loud distracting sounds in the surrounding but off-axis area. It seems that the brain allows us some flexibility in focusing on what it is we wish to hear, and with concentration we may be able to tune out other distracting sounds. This ability is greatly influenced by the relative direction from which the sound arrives. We find that we invariably attempt to face in the general direction of the sound to which we are listening. For example, consider a concert performance by a large orchestra. A listener in the theater audience might have little trouble concentrating on a particular instrument within the ensemble, and a reasonable amount of distraction (air conditioning noises, outside traffic, and such) would be all but unheard, due to the listener's ability to concentrate on the music.

On the other hand, a single microphone in the audience will hear and pass along every sound occurring within its range, without regard for the musical or non-musical significance of one sound over another. Obviously, the microphone does not enjoy the listener's ability to concentrate on the music alone. If a recording were made in this manner, the microphone would pick up and send to the listener both music and noise with equal facility as shown in Figure 3-9a.

Later on, when the playback lacks the visual cues which help the brain to concentrate, the listener will find it difficult or impossible to sort out the music from the noise, or to concentrate on an instrument which may have been clearly heard during the actual performance. And, obviously, the recording engineer will have no control over anything but the overall recorded level of the total environment.

A dramatic improvement in perspective, if not control, may be realized by employing two microphones in an attempt to more closely simulate the effect of listening with two ears. The brain perceives the placement of sound from this binaural effect. A person with only one ear can perceive pitch, loudness, and timbre, but will be unable to correctly determine the exact location of the source of the sound. Yet two microphones have no more "brain power" than one, and some attempt must be made to separate, for the microphone's benefit, those sounds that are to be recorded from those that are not desired. To accomplish this, we might put the microphones very close to the sound source of interest as in Figure 3-9b, so that all other sounds are, from the microphone's view, drowned out by the sheer volume of the close-up instrument or voice. Even the most unwanted of noises, if very loud and close to the ear, will mask all other sounds regardless of the listener's interest in them. Likewise, a microphone placed quite close to a musical instrument will not hear very much of anything else, wanted or not. This means that several microphones may be required to cover a large group, or even a small ensemble, if each microphone is placed close to an instrument or group of instruments.

In the case of a severe noise problem—and here "noise" may even mean the sound of other musical instruments—we may place certain instruments in isolation booths (Figure 3-9c) to acoustically protect them from being drowned out by other, much louder sound sources.

Another way to deal with this situation is to take into consideration the actual design of the microphone. In any typical recording application, the microphone is no doubt placed so that the wanted

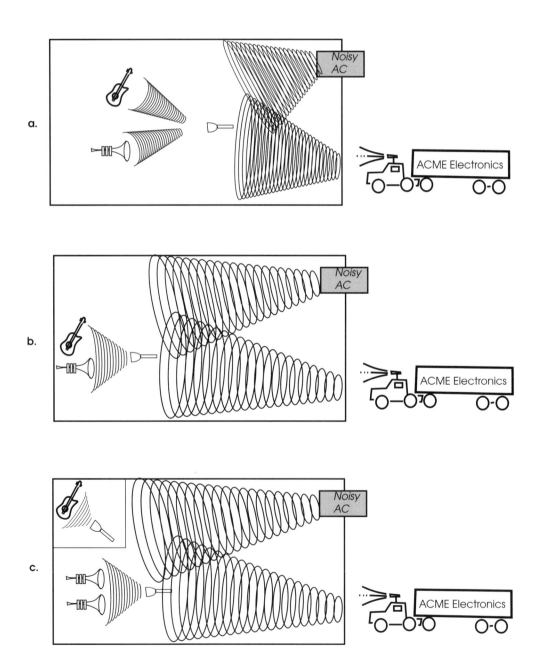

Figure 3.9 a. A microphone in the middle of the room hears both wanted and unwanted sounds; b. A microphone placed close up hears the instrument directly in front of it, and most other sounds are proportionally attenuated; c. In very noisy rooms, the isolation booth helps shield the microphone from unwanted sounds.

sounds arrive on-axis from directly in front of the microphone. Presumably, unwanted sounds are off to the side or rear of the microphone (off-axis). A certain degree of insensitivity to these unwanted sounds may be designed into the microphone. It should be kept in mind that the microphone's insensitivity has nothing to do with whether a sound is wanted or not, but is strictly a function of the angle from which the sound arrives.

Directional Sensitivity: Omni-, Bi-, and Uni-directional Microphones

A microphone may be classified as omni-directional, bi-directional (bi-radial), or uni-directional in nature, with these terms referring to the type of directional sensitivity that is inherent in its design. A brief description of each type of pattern follows.

■ Omni-directional (Pressure) Microphones

A microphone that is equally sensitive to all sound sources, regardless of their relative direction, is known as an omni-directional microphone. Basically, the microphone consists of a diaphragm and a sealed enclosure, as shown in Figure 3-10. It is often also referred to as a pressure microphone, since it responds to the instantaneous variations in air pressure caused by the sound wave. Since the microphone has no way of determining the location of the sound source causing the pressure variation, it responds with equal sensitivity to sounds (pressure variations) coming from all directions; hence the name omni-directional.

The microphone approximates the directional characteristics of the ear, which is likewise an omni-directional transducer. However, in the case of the ear, the head itself is somewhat of an acoustical obstruction to sounds arriving from certain off-axis locations. High frequencies in particular are better heard by the ear closest to the source producing them. In a similar manner, the casing of a physically large microphone may somewhat obstruct rear-originating sounds, causing a slight decrease in sensitivity. Once again, this will be most apparent at the higher frequencies.

■ Bi-directional (Pressure Gradient) Microphones

The bi-directional microphone is equally sensitive to sounds originating from directly in front (0°) and from directly behind it (180°). It is least sensitive to sounds arriving from the sides (90° and 270°). This bi-directional characteristic is generally realized by leaving

Figure 3.10 The omni-directional microphone.

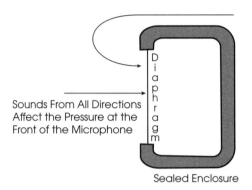

both sides of the diaphragm exposed, as shown in Figure 3-11. Thus, sounds may strike either the front or the rear of the diaphragm. This type of microphone is often referred to as a pressure gradient microphone, since the movement of the diaphragm is in response to the pressure gradient—that is, the difference in acoustic pressure between the front and rear of the diaphragm. Sounds that originate from the sides reach both the front and rear diaphragm at the same time and intensity. This creates a net pressure gradient, or difference, of zero, thus explaining why this type of microphone is so highly insensitive at 90° and 270°. The bi-directional microphone is popularly called the figure-8 microphone since its pattern, as viewed from the top, resembles that number.

An important aspect of the bi-directional microphone is the electrical phase reversal between sounds arriving at its front and rear. To understand the significance of this, consider an instantaneous positive pressure wave striking the diaphragm. If the wave arrives from in front of the microphone, the diaphragm moves toward the rear, and a positive voltage is produced. If the instantaneous positive pressure wave was to simultaneously strike the rear of another bi-directional microphone, its diaphragm would move in the same relative direction (that is, away from the pressure wave). This would be toward the front of this microphone, producing a negative voltage.

Although either voltage would constitute a usable audio signal, their equal and opposite polarities would cause a complete cancellation if their outputs were combined. In like manner, a complex audio wave form that simultaneously reached the front and rear of two bi-directional microphones would be severely attenuated when the outputs of these microphones were combined, either while recording or later on during a mixdown session. Therefore, when two bi-directional microphones are used, it is important to take into consideration this phase reversal and to make sure that no signal source is located between the rear of the bi-directional microphone and the front of another microphone, whether bi-directional or not. In practice, the attenuation is rarely total, especially if the individual frequency responses of the microphones are dissimilar. Characteristically, the combined output signal may sound thin or generally distorted in frequency response, as the two signals combine subtractively.

■ **Cardioid (Phase Shift) Microphones**

A cardioid microphone, sometimes referred to as a uni-directional microphone, is most sensitive to sounds that originate from directly in front of it, that is, at 0°, or what we consider on-axis. If the microphone is slowly rotated through a 360-degree arc, while its

Figure 3.11 The bi-directional microphone.

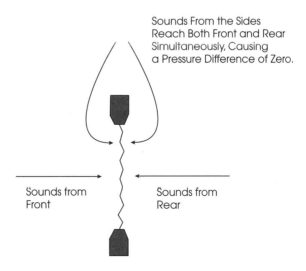

diaphragm remains at a constant distance from the source of the sound, its output level will decrease until the microphone is facing directly away from the source of the sound (180°). As the microphone continues to be rotated through this arc from 180° to 360°, its output level will increase until full level is reached once again at the on-axis point.

The fall-off in sensitivity at the rear can be created by several methods. In single-diaphragm microphones, this effect is generally the result of one or a series of side or rear entry ports, as shown in Figures 3-12a, b and c. The ports are arranged so that sounds originating from the rear can reach the diaphragm by two paths: 1) around the microphone to the front (P1) and 2) through an entry port located at the rear of the diaphragm (P2). If the two paths are of equal length, the pressure on both sides of the diaphragm will be the same, causing no movement of the diaphragm and therefore creating no output voltage. On the other hand, sounds arriving from the front (P3) must travel an extra distance to reach the rear of the diaphragm (P4). The phase shift caused by the difference in path lengths P3 and P4 causes a reinforcement of on-axis signals. In dual diaphragm microphones, this front-to-back sensitivity difference is caused by combining a pressure device with a pressure gradient element. An early derived-cardioid-pattern microphone is shown in Figure 3-13. Here a ribbon (pressure gradient bi-directional pattern) element is combined with a moving coil (pressure gradient omni pattern) element to create a cardioid. This will be discussed in greater detail later in this chapter.

Microphone Polar Patterns

A polar pattern is a graph of a microphone's relative sensitivity to sounds that originate at various locations around it. Polar patterns for omni-directional, bi-directional, and uni-directional microphones are shown in figures 3-14 through 3-16. In each case, the heavy line is the polar pattern itself, while the concentric circles indicate 5dB decreasing increments of sensitivity. In a real polar plot of a specific microphone, there would be several pattern plots drawn on one graph, with each one representing a specific frequency. For simplicity, only one line is shown representing a frequency of 1000Hz. Pattern with respect to frequency will be shown in greater detail later in this chapter.

The omni-directional polar pattern, in an ideal case, is simply a circle. This indicates that the microphone is equally sensitive to sounds originating from any direction. In practice, however, the

MICROPHONE DESIGN   63

Figure 3.12   a. The uni-directional or cardioid microphone; b and c. The side- and rear-entry ports on studio-quality cardioid microphones (Neumann KM183, KM184, KM185 (Neumann USA Photo); KM84 and KM83 capsules (Peabody Recording Studio Photo)).

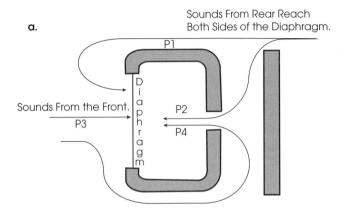

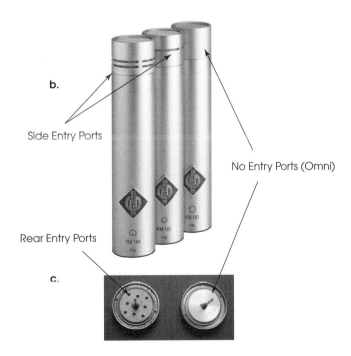

Figure 3.13   Western Electric 639A, circa 1939.

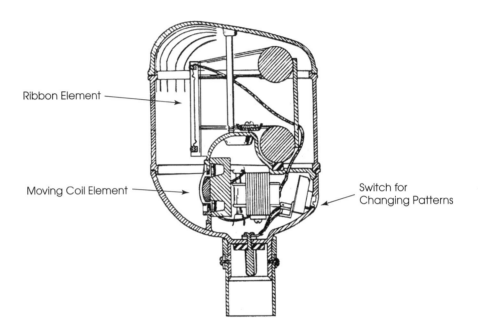

omni-directional microphone will have a tendency to be slightly less sensitive to sounds originating from the rear since, as mentioned earlier, the case may serve as somewhat of an acoustical barrier. This effect is noted by the slight flattening of the polar pattern in the vicinity of 180°, as shown in Figure 3-14. In this area, the pattern almost touches the concentric circle marked 5dB. This means that the microphone being measured is actually almost 5dB less sensitive to sounds originating from that direction.

The bi-directional polar pattern shown in Figure 3-15 illustrates this microphone's equal sensitivity to sounds originating from the front and rear, as well as its insensitivity to sounds originating from 90° and 270° off axis. As mentioned previously, the bi-directional microphone is popularly known as a figure-8, due to the characteristic shape of its polar pattern.

The cardioid polar pattern is shown in Figure 3-16. Notice that the polar pattern crosses the concentric circle at the 6dB point which is

Figure 3.14 The omni-directional polar pattern. The dashed line indicates a slight loss of sensitivity in the rear.

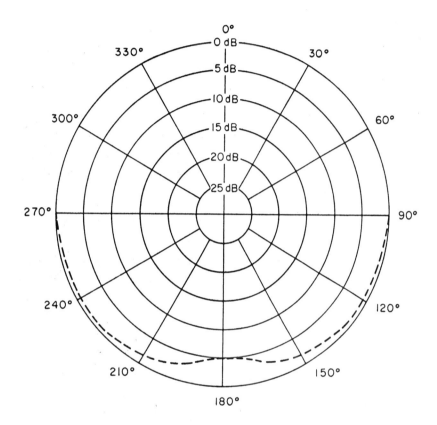

approximately 90°, and that at the rear (180°) it just touches the 25dB circle. This means that sounds originating at these points will be attenuated by 6dB and 25dB respectively as compared to the same sound from on-axis or 0°. It is important to realize that in reality the practical cardioid microphone is certainly not totally deaf to off-axis sounds, but that these sounds are attenuated by a certain number of dB as the polar pattern indicates.

Somewhere between the polar pattern of the bi-directional or figure-8 microphone and the polar pattern of the cardioid or uni-directional microphone, a series of intermediate patterns may be derived in which the rear lobe of the figure-8 pattern becomes progressively

Figure 3.15 The bi-directional polar pattern.

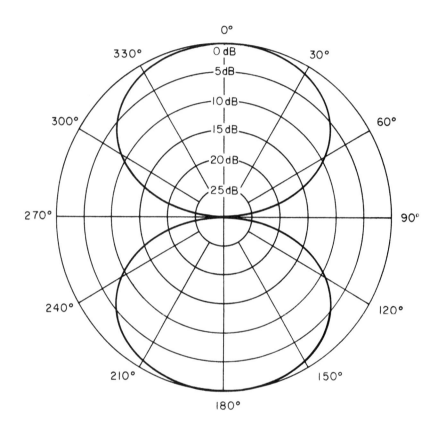

smaller while the front lobe takes on a more cardioid or wider shape. Although many of these patterns are purely mathematical models, several of them may be found in studio quality microphones. The super-cardioid pattern, as shown in Figure 3-17, is often used as a viable alternative to the cardioid pattern. In Figure 3-18, the super-cardioid is overlaid on a regular cardioid pattern. It will be seen that the super-cardioid pattern is more sensitive to sounds originating from the rear, but somewhat less sensitive to side-originating sounds than the cardioid. The super-cardioid has two areas of minimum sensitivity: one at 126°, and the other at an angle of 234°. Another polar pattern that may be found is the hyper-cardioid. It is slightly less sensitive at the sides than the super-cardioid, yet more sensitive to sounds originating from 180°, as shown in Figure 3-19. Between the omni-directional polar pattern and the cardioid polar pattern is another variation that is becoming more widely available, particularly on multi-pattern microphones. This so-called hypo-cardioid or wide-angle cardioid pattern (also sometimes mistakenly referred to as a forward-facing omni) is less sensitive to rear-originating sounds than the omni, but more sensitive to side-originating sounds than the cardioid pattern. This polar pattern is shown in Figure 3-20.

The polar pattern for an interference or so-called *shotgun* microphone generally resembles a flattened-out front lobe with a series of very small rear lobes, as shown in Figure 3-21. The microphone consists of a long tube with many open ports along the full length.

Figure 3.16 The uni-directional (cardioid) polar pattern.

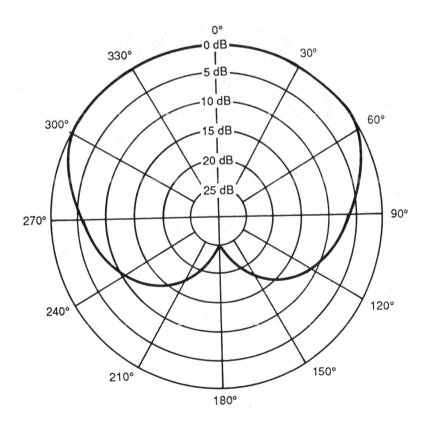

Sounds originating from the front enter the microphone unobstructed, while sounds arriving from the sides are "interfered with" and phase cancelled. The more off-axis the sounds, the greater the cancellation. This microphone is not used very often in recording studios, but may be a valuable tool on remote sessions when it is impossible to place a cardioid or other microphone close to the source of the sound. It is widely used in television studios and on movie sets, where it is important that the microphone not be seen in the picture. A "shotgun" microphone is shown in figure 3-22.

Figure 3.17 The super-cardioid polar pattern.

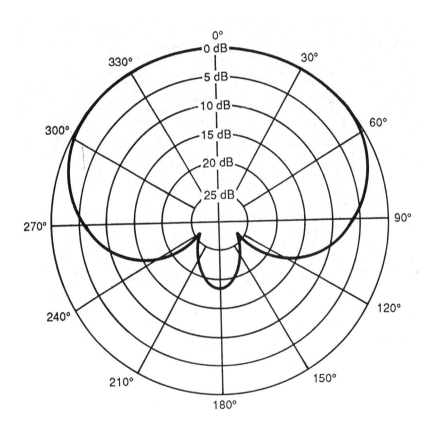

Multi-Pattern Microphones

Many studio-grade microphones have the ability to produce more than one polar pattern. This flexibility allows a microphone to perform well in a multitude of conditions. The most common method of accomplishing this is to build a microphone with two diaphragms.

One of the methods used in this dual-diaphragm design takes advantage of the fact that two cardioid patterns may be combined to produce either a bi-directional pattern or an omni-directional one. Figure 3-23 is a simplified illustration of this principle, where two diaphragms D1 and D2 are on either side of two backplates charged by a common voltage. Each diaphragm used alone will produce a cardioid pattern. With the pattern selector switch in position 1, only one diaphragm, D1, is in use, and the microphone functions

Figure 3.18 A comparison of cardioid and super-cardioid polar patterns.

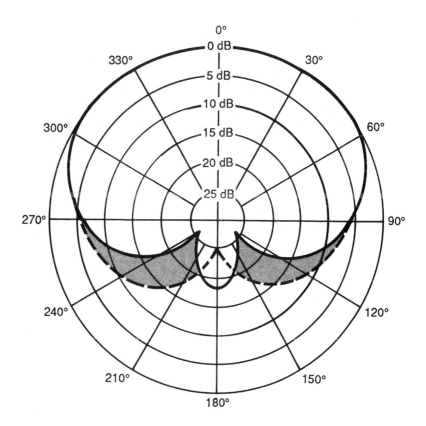

as a simple cardioid microphone. In position 2, the second diaphragm, D2, is also energized, and the two cardioid patterns combine to produce the omni-directional polar pattern as shown in Figure 3-24a. With the switch in position 3, the polarity of diaphragm D2 is reversed, and the opposing polarities cause a cancellation in the areas where the two patterns overlap. The result is a bi-directional pattern as shown in Figure 3-24b. When the pattern switch is in position 4, the second diaphragm is still negatively polarized, yet the resistor, R, drops the polarizing voltage somewhat. Consequently, although D2 still yields a cardioid pattern, it is somewhat smaller than the D1 pattern, and their combination

Figure 3.19 The hyper-cardioid polar pattern.

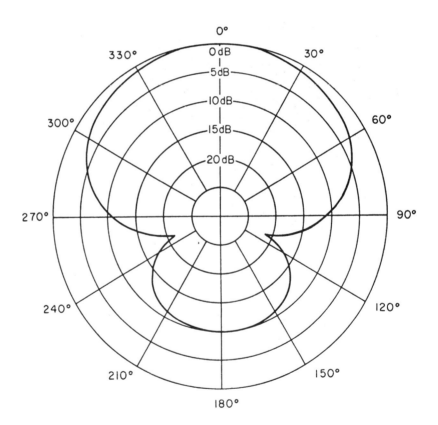

yields the intermediate pattern shown in Figure 3-24c. The size of the rear lobe depends on the actual value of R. If R=0, the pattern would be fully bi-directional, the equivalent of the pattern switch being in position 3. If R=& (open circuit), the pattern becomes cardioid again, since the second diaphragm is out of the circuit.

Another type of dual-diaphragm multi-pattern microphone uses a fixed back plate and two diaphragms. However, in this capsule, the back plate common to both diaphragms is perforated. By varying the electrical potential on the two diaphragms, a variety of patterns can be produced. If the polarity on both diaphragms is positive, a pressure device is formed, responding equally to sounds from all directions. When the polarity of the rear diaphragm is changed to

Figure 3.20 The hypo-cardioid or near-cardioid polar pattern.

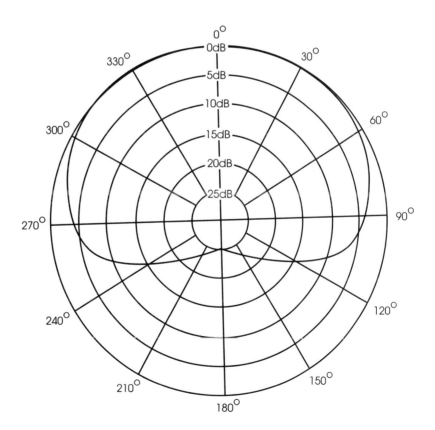

Figure 3.21 A typical "shotgun" polar pattern.

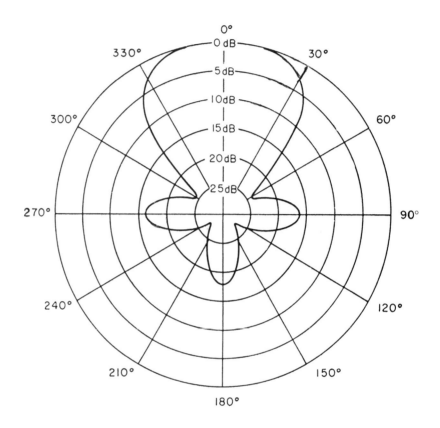

Figure 3.22 A modern ultra-directional or "shotgun" microphone (Neumann USA Photo).

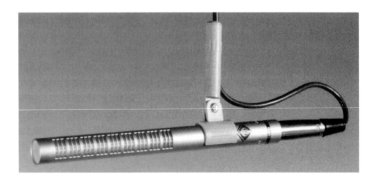

negative, a velocity or pressure gradient can be produced. And by combining the two, the cardioid pattern can be derived. In a condenser microphone, this is referred to as a *Braunmühl-Weber* microphone capsule. The perforated back plate controls the microphone output with regard to the sound source direction. When the pressure is the same on both sides and the polarity on both plates is positive, the opposite but equal movement of the two diaphragms produces a pressure response, while parallel diaphragm movement produces no output. As the diaphragms move, the air between them is alternately compressed and rarified and this dampens the diaphragmatic movement. When the diaphragms are polarized opposite with regard to each other, equal pressure on both diaphragms causes a signal cancellation while parallel motion produces a signal output, thus acting as a pressure gradient or bi-directional pattern. Figure 3-25a shows the patterns resulting from combining various amounts of the omni-directional and figure-8 patterns, while Figure 3-25b shows the circuit diagram for such a capsule. Figure 3-25c shows a Braunmühl-Weber capsule disassembled.

Figure 3.23   A dual-diaphragm condenser microphone. 1. Cardioid pattern; 2. Omni-directional pattern; 3. Bi-directional pattern; 4. Intermediate (hyper- or super-cardioid) pattern.

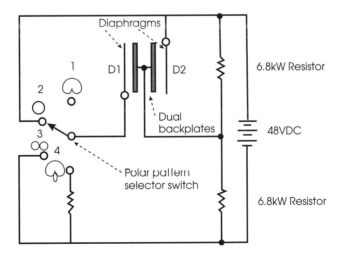

Figure 3-26 is a table showing the characteristics of various microphone patterns. Note that the pickup arcs for both the 3dB down point and the 6dB down point are listed. The null point or point of maximum rejection is also noted.

In some multi-pattern condenser microphone systems, the pattern selector is replaced by a potentiometer, allowing a variable polarizing voltage to be supplied to the diaphragms. In this way, the microphone's polar pattern is continuously variable from bi-directional to omni-directional, with infinite intermediate patterns. Often, in the case of stereophonic microphones (discussed in Chapter 4), this continuously variable potentiometer is remotely located. This allows the polar pattern of the microphone capsules to be changed without having to go to the microphone itself to do so. This is a great convenience when the microphone is located high in the air, or at the end of a long boom. Figure 3-27a shows a microphone with a built-in four-position pattern-selecting switch, while Figure 3-27b pictures a continuously variable pattern control for remote locations.

Figure 3.24 a. Two cardioid patterns, added to create an omni-directional polar pattern; b. Two cardioid patterns, subtracted to create a bi-directional polar pattern; c. Two unequal cardioid patterns, subtracted to create a super-cardioid pattern.

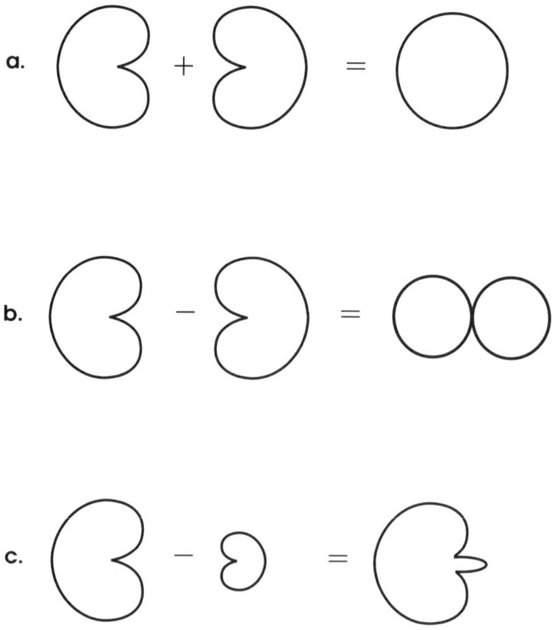

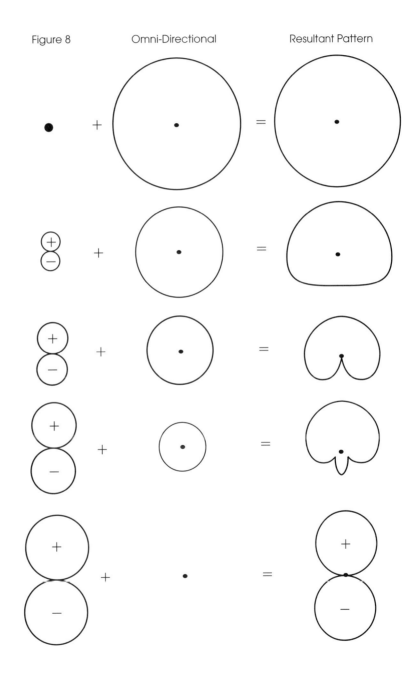

Figure 3.25 Patterns resulting from combining a pressure device (omni) with a pressure-gradient (figure-8) device.

Figure 3.25B A circuit diagram for a multiple-pattern condenser microphone using a Braunmühl-Weber capsule.

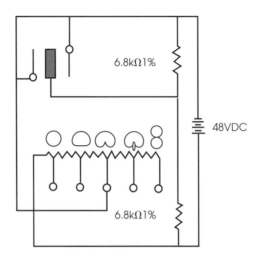

Figure 3.25C Detailed photo of a Braunmühl-Weber microphone capsule (Peabody Recording Studio Photo).

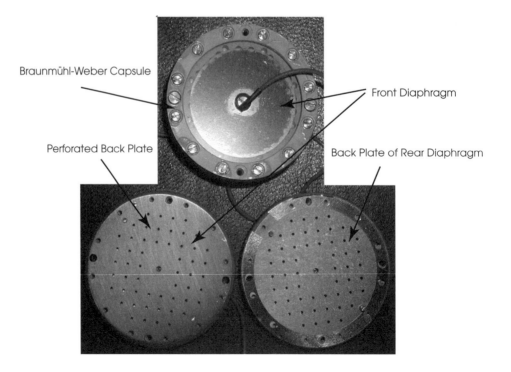

MICROPHONE DESIGN

In another version of the multi-pattern microphone, variations in the polar pattern are realized by changing the diaphragm enclosure. These multi-pattern single-diaphragm microphones have a variety of different capsules available for use on a common pre-amplifier body. A typical system would comprise an omni-directional capsule, a cardioid capsule, a figure-8 capsule and a pre-amplifier body. A capsule for a microphone system of this type is shown in Figure 3-28. In Figure 3-28a, an exploded view of a removable capsule is shown. Figure

Figure 3.26 Table of microphone pattern characteristics.

Characteristics	Omni-Directional	Cardioid	Super Cardioid	Hyper Cardioid	Bi-Directional	Hypo Cardioid
Polar Equation	1	$.5+.5\cos\theta$	$.375+.625\cos\theta$	$.25+.75\cos\theta$	$\cos\theta$	$.7+.3\cos\theta$
Pickup ARC 3dB down (1)	—	131°	115°	105°	90°	180°
Pickup ARC 6dB down	—	180°	156°	141°	120°	—
Relative Output @90° in dB	0	−6	−8.6	−12	−∞	−3
Relative Output @180° in dB	0	−∞	−11.7	−6	0	−12
Angle at which output = 0	—	180°	126°	110°	90°	—
Random Energy Efficiency (REE)	1 0 dB	.333 −4.8dB	.268 −5.7dB (2)	.250 −6.0dB (3)	.333 −4.8dB	.666 −2.4dB
Distance Factor (DF)	1	1.7	1.9	2	1.7	1.3

NOTE:
 1 = Drawn Shaded on Polar Pattern
 2 = Maximum Front-to-Total Random Energy Efficiency for a First Order Cardioid
 3 = Minimum Random Energy Efficiency for a First Order Cardioid

Figure 3.27 a. A dual-diaphragm condenser microphone with a built-in four-position pattern selector (AKG C414EB P48); b. Potentiometers mounted in a separate enclosure provide remote and variable pattern selection (AKGS42).

3-28b shows the same cardioid capsule; note that it is identical to the omni of Figure 3-28c with the exception of the perforated back plate.

Some microphones use two diaphragms for a different reason. These single-pattern dual-diaphragm microphones use one single-diaphragm capsule for low frequency pickup and the other for high frequencies, just as many loudspeakers utilize two drivers—a woofer and a tweeter—to cover the complete audio spectrum. This type of microphone is illustrated in Figure 3-29. The high-frequency system is physically small, while the low-frequency system is approximately twice its size. A crossover frequency is determined that will allow each diaphragm to deliver signal from its optimum frequency range only. The two signals are then combined, producing a single wide-band output.

MICROPHONE DESIGN 79

Figure 3.28 a. Changeable single-diaphragm microphone capsule (Neumann KM-84)(Gotham Audio Illustration); b. Cardioid removable capsule; c. Omni-directional movable capsule (Peabody Recording Studio Photo).

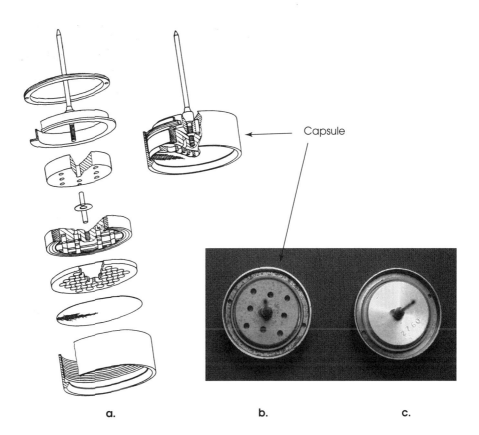

■ **Measuring Microphone Polar Response**

A microphone's polar response is measured by rotating the microphone through a 360° arc while keeping the sound source fixed, in both level and location. The measurements are made within an anechoic chamber as shown in Figure 3-30, so that there will be no reflections from nearby surfaces to influence the off-axis response. The microphone is set up so that as it rotates, its diaphragm remains at a constant distance from the sound source. The many reflections within any audio studio will alter the microphone's apparent polar response with respect to frequency. But there is no way to predict the net effect of these reflections, which will vary from one studio to another and at different locations within the same studio. In fact, the engineer will have to take into account the varying reflective conditions within the studio when determining the best locations for certain instruments, or the exact placement of a particular microphone in front of an instrument or ensemble.

In any case, the reflection-free anechoic chamber serves as a repeatable standard measuring condition, enabling the engineer to compare the ideal polar response patterns of many different microphones. The recording engineer can, and should, study these polar responses so as to be able to predict, with reasonable accuracy, the performance he or she may expect within the studio.

Although polar patterns are drawn only in two dimensions as an artistic and measuring convenience, it should be understood that the patterns are actually three-dimensional, as shown in the five examples of Figure 3-31.

Figure 3.29 A dual-capsule single-pattern condenser microphone (Sanken CU-41)(Sanken Photo).

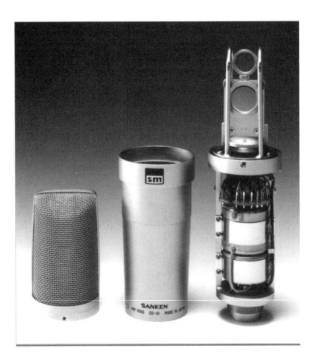

Off-Axis Frequency Response

So far, our examples of polar responses have shown single line patterns and, as mentioned earlier, generally represent the best-case response at around 1000Hz. This would seem to imply that at off-axis positions the microphone is equally sensitive or insensitive to all sounds, regardless of frequency. In practice, unfortunately, this is not the case. For a more accurate indication of the microphone's off-axis response, a polar pattern at several frequencies, equally representing high and low, should be drawn. A typical polar pattern response is shown in Figure 3-32a. From these response curves, we

Figure 3.30 An anechoic chamber. The wedges on all surfaces eliminate virtually all reflections within the room. NIST's anechoic chamber supports research and services that address a growing range of industrial, safety and health needs for high-quality acoustic measurement (National Institute of Standards and Technology Photo).

may draw graphs of the microphone's frequency response at various off-axis angles, as shown in Figure 3-32b.

These response curves indicate that although the microphone performs satisfactorily with respect to on-axis sound, its off-axis response is quite irregular. This condition is known as off-axis coloration, so called because of the distorted, or colored, frequency responses. Off-axis coloration may take the form of an unpleasant muddy sound, since the microphone's off-axis high-frequency response usually falls off at a much greater rate than the lower frequencies. This condition particularly affects the sound of the surrounding ambient field. Note that a cardioid is not a cardioid at all frequencies, and that the omni-directional pattern does not remain so throughout the audio band. It is interesting to note the differences in polar responses between the patterns of a multi-pattern Braunmühl-Weber capsule microphone and the corresponding polar responses of a multi-pattern back-to-back cardioid type of design. Neither is perfect, but the prudent engineer may wish to select one over the other depending on the situation at hand.

■ **Proximity Effect**   Another characteristic of many cardioid microphones is called the *proximity effect*. This is an increase in bass response, in proportion to the high frequencies, as the microphone is moved closer to the sound source. The condition is illustrated in Figure 3-33a, where on-axis response at various microphone-to-source distances is shown. Notice that as the distance decreases, the bass response rises considerably. This rising bass response may be beneficial in some cases, helping to achieve a more robust sound on a voice. However, the slightest movement of the singer, or announcer, towards or away from the microphone will change the overall frequency response noticeably. Especially in hand-held applications, the working distance is continually changing, and a microphone without proximity effect, such as an omni-directional type, may be required. Even when the microphone is stand-mounted close to, for instance, an acoustic guitar, the variations in bass response may be noticeable as the guitarist moves about while playing. On the other hand, when the microphone is placed in front of a stationary guitar amplifier, the working distance may be varied to achieve the desired bass response.

Most large-diaphragm cardioid microphones have a built-in switchable high-pass filter to counteract the proximity effect, such as pictured in Figure 3-33b. The filter restores the low-end response to normal at some specified close working distance. At greater distances,

MICROPHONE DESIGN   83

Figure 3.31   Polar patterns in three dimensions. a. omni-directional; b. bi-directional; c. uni-directional (cardioid); d. uni-directional (super-cardioid); e. ultra-directional "shotgun" (Jon Cresci drawings).

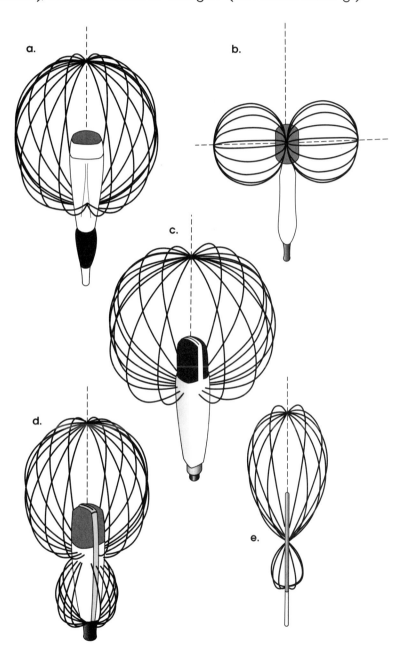

Figure 3.32a Complete polar pattern for an inexpensive cardioid microphone.

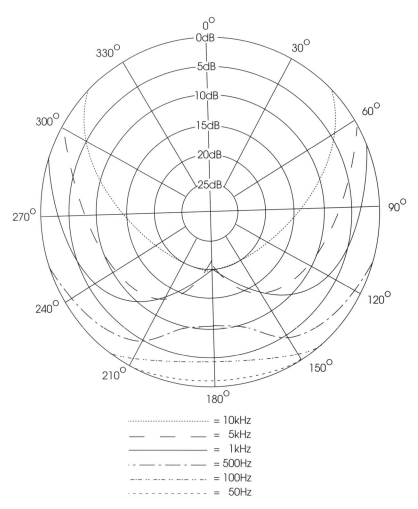

the filter should be switched out, since its fixed low-end attenuation is now unnecessarily rolling off the microphone's normal low-end response. Microphones with omni-directional polar patterns are usually free of this effect.

Impedance

For professional studio applications, low-impedance microphones are an industry standard. Although high-impedance

Figure 3.32b Frequency response at various angles for an inexpensive cardioid microphone.

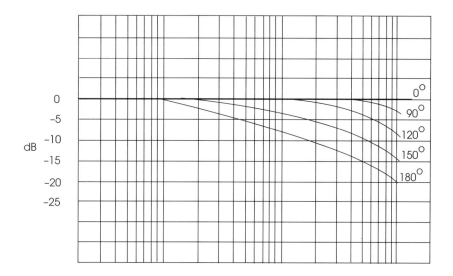

Figure 3.33a Typical proximity effects at various working distances.

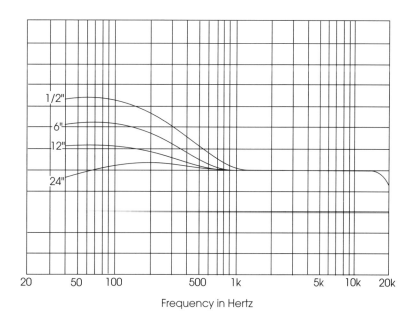

microphones are not necessarily inferior in themselves, the long cable lengths required in most studio installations prohibit their use.

To illustrate why, an equivalent circuit for a microphone that is connected by a length of cable to a console microphone pre-amplifier is drawn in Figure 3-34. Note that R_S and R_L are in series with the source voltage, E. Therefore, if R_S is much smaller then R_L, most of the generated voltage will appear across the console input, R_L. C represents the typical capacitance of any microphone cable; the longer the cable, the greater the capacitance. Now, as frequency increases, this capacitance lowers the impedance of the cable/console combination. Since the microphone's impedance remains constant, the voltage across C (and consequently across R_L), falls off as the frequency rises, since more and more of the total voltage drop is across R_S. This phenomenon can be expressed by the following equation:

$$F = \frac{1}{2\pi R_T C_T}$$

where:

F = the frequency at which the highs begin to roll off at a rate of 6dB/8va
R_T = the total resistance in the circuit expressed as

$$\frac{1}{R_T} = \frac{1}{R_S + R_L}$$

C_T = the total capacitance in the circuit expressed by the number of picofarads/foot, multiplied by the total cable length.

Figure 3.33b A switchable bass roll-off filter to minimize the proximity effect.

It can be seen that, with a high-impedance microphone, this fall-off of high frequencies will be quite noticeable within the audio bandwidth, unless cable length, and therefore total capacitance, is kept to an absolute minimum. On the other hand, a low-impedance microphone will allow the use of long cable runs up to several hundred feet with no adverse effect on the frequency response. Although the cable capacitance remains the same, the microphone's relatively low impedance keeps the high-frequency roll-off well above the audible audio bandwidth.

Sensitivity A microphone's sensitivity rating tells the user something about its relative efficiency in converting acoustic energy to electrical energy. Sensitivity is usually expressed in dB below a specified reference level. The two types of rating methods are: 1) the open circuit voltage rating, and 2) the maximum power rating. The majority of microphone manufacturers today use the open circuit voltage rating method. The other method, maximum power rating, is left over from the days of matching impedances for maximum power transfer.

Figure 3.34 The equivalent circuit of a microphone connected to a console input.

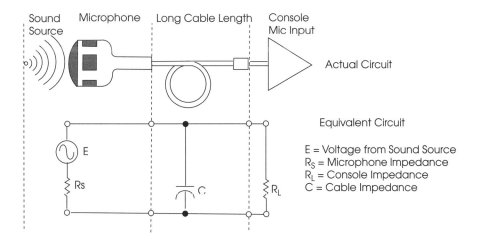

Many of today's circuits are bridging and result in the more efficient maximum current transfer.

In the open circuit rating method, either the microphone is not connected to any input, or the input impedance to which it is connected is at least twenty times higher than the impedance of the microphone itself. The reference sound pressure level can be either 1 microbar (1 dyne/cm^2 or 74dB) or 1 Pascal, abbreviated 1 Pa (10 dynes/cm^2 or 94dB). The second level of 1 Pa is the most currently used reference. The zero reference is 1 volt. In other words, if a sound pressure level of 1 Pascal resulted in a 1-volt output, the microphone's sensitivity would be 0dB referenced to 1 Pa. In practice, much lower output voltages are produced, and typical open circuit voltage ratings are on the order of 12mV to 20mV. To prevent confusion, the sensitivity specification should clearly state the reference used.

Balanced and Unbalanced Lines

A balanced line is one which uses two conductors plus a shield or drain wire, while an unbalanced line contains only one conductor, with the shield or drain wire serving as the second conductor. Figure 3-35 shows both types of lines, as well as typical balanced and unbalanced circuits. An important advantage of the balanced line is that any unwanted noise signals in the area through which the line is placed will be picked up by both conductors (one positive and one negative), and will cancel out when terminated properly. In the unbalanced line, the noise voltage will travel down the single conductor and be transmitted to the next stage of the circuit.

Providing the audio signal is of sufficiently high level (such as a +4 dBm line-level signal), the noise may not be heard as anything but a slight raising of the noise floor of the system, especially if the unbalanced line is reasonably short. But, as noted before, since microphone lines tend to be quite long, and the signal levels very low, the slightest noise or hum induced in the microphone line may become almost as loud as the microphone's output signal. For this reason, balanced lines are an absolute necessity between a microphone and its console pre-amplifier.

Polarity

Practically all professional microphones use a three-pin output connector, commonly called a Canon or XLR connector, with pin no. 1 used as the shield and the output signal appearing across pins no. 2

and no. 3. A positive pressure on the diaphragm of the microphone will usually produce a positive voltage at pin no. 2 with respect to pin no. 3. Most major microphone manufacturers adhere to this convention; however, it is by no means completely standard.

Before putting a new microphone in service, the prudent engineer will verify the relative polarity of the output pins, so that phase reversals can be avoided. This can be done by checking the documentation that came with the microphone, or by comparing the questionable microphone with a known standard. The front of the two microphones should be pointed at the same sound source and their outputs mixed together and monitored. If the combined level is less than either single microphone level, phase cancellation is

Figure 3.35 a. balanced line; b. unbalanced line; c. balanced circuit; d. unbalanced circuit.

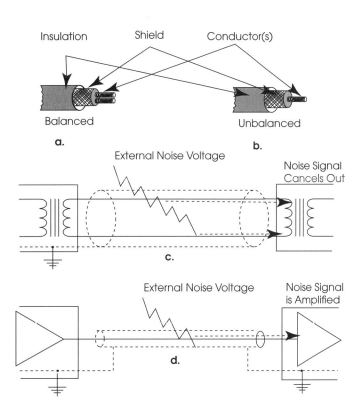

occurring. In the ideal situation, the output level of the combined microphones will be 6dB greater on the VU meters than either of them separately. If an out-of-phase microphone is found, it should be immediately rewired. When this is not possible, a short adapter cable should be permanently attached to the non-standard microphone along with a label stating the condition.

■ DISTORTION IN MICROPHONES

For all practical purposes, it is almost impossible to produce a sound pressure level indoors that will overload a professional-quality moving-coil (dynamic) microphone. Some older ribbon microphones could be overloaded, and actually damaged, by very high sound pressure levels, but more recent versions should be able to withstand most sound levels found in the studio. However, a very high microphone output level may overload the microphone pre-amplifier in the recording console. Most consoles contain a sensitivity control that allows the engineer to adjust the gain of the microphone pre-amplifier as well as a switchable pad that will insert a fixed amount of attenuation in the microphone line ahead of the pre-amplifier. In cases where such a facility is not built in, or in the case where the console adjustment is insufficient, an attenuation pad may be inserted in the microphone line. Figure 3-36a shows a commercially available attenuation pad and also gives resistance values and a diagram for 10dB and 20dB attenuation circuits.

In a condenser microphone, the voltage produced by the capacitor/diaphragm may be sufficient to overload the microphone's built-in pre-amplifier. Attenuation facilities at the console will be of no use, since the overload occurs within the microphone itself. To protect the pre-amplifier against this overload, many condenser microphones have a built-in pad that inserts 10dB or 20dB of attenuation between the diaphragm and the pre-amplifier. Figure 3-37a illustrates a microphone with a switchable pad, and Figure 3-37b shows another arrangement where a 10dB or 20dB pad may be physically inserted between the diaphragm capsule and the pre-amplifier when required.

MICROPHONE NOISES: SIBILANT, PLOSIVE, AND VIBRATION

Although a microphone's diaphragm must vibrate in order for it to produce an output voltage, some care must be taken to prevent certain unwanted vibrations from distorting the signal output. For example, a stream of air blown across a microphone would certainly create an unpleasantly distracting noise as the diaphragm is set into vibration by the wind. Although this is highly unlikely within the recording studio, a close-up voice often produces a shock wave that distorts the microphone's output. In particular, plosive words containing "p," "t," and "b" may create an objectionable pop due to the

Figure 3.36 a. A commercially available in-line microphone attenuator; b. Circuit values for a microphone line attenuator. For 10dB attenuation, R_a = 56 ohms, R_b = 150 ohms. For 20dB attenuation, R_a = 82 ohms, R_b = 39 ohms. (Caution—this type of attenuator should not be used with a phantom-powered condenser microphone since it will interfere with the voltage being supplied to the microphone.)

a.

b.

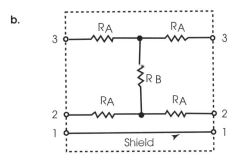

movement of air across the diaphragm. Pressure gradient microphones (cardioid and figure-8 patterns) are particularly susceptible to this, since the pressure differential between the front and back of the diaphragm is considerable.

To minimize these noises, a wind screen or pop filter may be placed over the microphone. The wind screen is made from an open-pore acoustical foam material that does not affect the microphone's sensitivity in the audible range. On the other hand, the screen does prevent puffs of wind from striking the diaphragm with full force. When a wind screen is used on a microphone with rear ports, it is important to shield these entrances as well, so that the pressure differential is reduced, rather than accentuated. Figure 3-38 illustrates several microphones protected by wind screens. In addi-

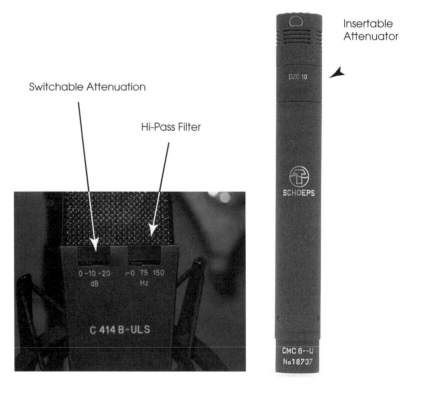

Figure 3.37 a. Condenser microphone with a built-in switchable attenuation pad, and a switchable high-pass filter (AKG C414B-ULS); b. Condenser microphone with an insertable attenuation pad (Schoeps DZC 10)(Schoeps Photo).

tion to its primary function, the wind screen also protects the diaphragm from moisture and dust particles. Especially in close-up vocal pickups, it is a good idea to use a wind screen as a matter of routine. The screen should be rinsed out frequently to remove accumulated dust and grime. Needless to say, it would be much more difficult and time-consuming to periodically clean the diaphragm itself.

Most studio structures are susceptible to at least some building vibration, particularly in heavy-traffic metropolitan areas. Even though the vibrations are inaudible to the listener in the studio, they may be transmitted via the microphone stand to the microphone, creating an audible rumble over the control room monitor speakers. Even in a rumble-free studio, certain impact noises are accompanied by considerable vibration. For example, an over-enthusiastic kick drum may set up vibrations in the floor which may travel up the microphone stand and cause the microphone diaphragm to react. Most manufacturers produce shock mounts that are designed to mechanically isolate the microphone from its stand, thereby cutting down on vibration-induced noise. Representative examples of shock mounts are seen in Figure 3-39. Cardioid microphones tend to exaggerate mechanical shock and vibration, and in hand-held applications, the noises may be quite distracting. Some microphones which are designed to be hand-held have the inner shell suspended from the outer casing to keep transmitted handling noise to a minimum.

Figure 3.38 a. An accessory wind screen placed over a microphone (Shure Beta 58)(Shure Bros. Photo); b. Two dynamic microphones with built-in wind screens (Beyer M500 and Shure Beta 57).

Figure 3.39 a. Shure shock mount (Shure Bros. Photo); b. Neumann M147 with shock mount (Neumann USA Photo); c. Neumann ultra-directional "shotgun" microphone with shock mount (Neumann USA Photo).

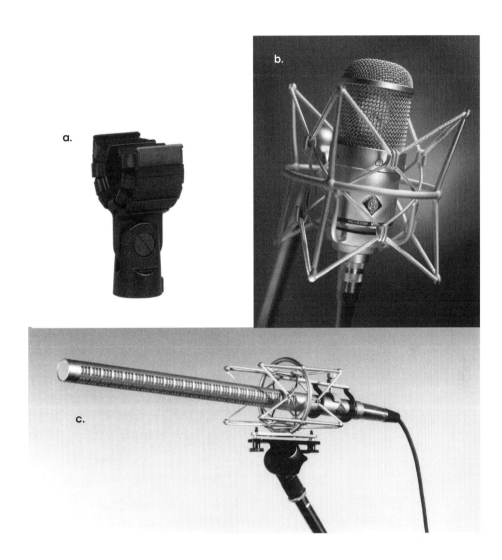

FOUR

Microphone Technique

The knowledge and understanding of microphones and microphone technique is probably the single most important area of study for the recording engineer, and although the successful utilization of microphones is largely a subjective matter, there are certain "ground rules" that should be understood before preparing for any important recording session. The engineer needs to be aware of these different philosophies of microphone placement technique, so that he or she may intelligently choose the basic approach that will yield the best results. It cannot be emphasized too strongly that the choice and positioning of microphones by the recording engineer is most often the difference between a good recording and a mediocre or bad one.

The first law of correct microphone usage has been successfully ignored almost from the day of its discovery. It is:

NEVER use more than two microphones.

Although most recording sessions would be impossible (and a great deal of this book unnecessary) if anyone took this law seriously, the engineer should understand its significance before setting up a third, fourth or twentieth microphone.

As discussed in Chapter 1, it is perhaps no accident that we have only two ears. Their positioning, and the interaction between them and the brain, allow us to form an incredibly accurate impression of the relative intensity and the location of any noise or sound within our hearing range. With the exception of those few sounds that originate quite close to one ear only (whispers, telephone conversations, and the like) we hear almost everything with both ears. Our impressions of direction usually derive from the phase shifts and time of arrival differences of a sound, as well as intensity gradients. In fact, our sense of localization for low frequencies tends to be phase dependent, while our perception of direction for high frequencies is usually determined by intensity differences.

As an example, consider the listener at a concert in a large theatre or concert hall. The orchestra is spread out across the full width and depth of the stage, with the strings nearest the front and the percussion instruments in the rear. The woodwinds and brass are arrayed in between. If there is a chorus, it may be standing behind the orchestra on risers. The larger the production, the more distance there is going to be between the front rows of violins and cellos and the last rows of the chorus. As can be seen in Figure 4-1, the sound from the chorus will reach the listener slightly later than

Figure 4.1 Sounds from the more distant orchestral sections take a little longer to reach the listener.

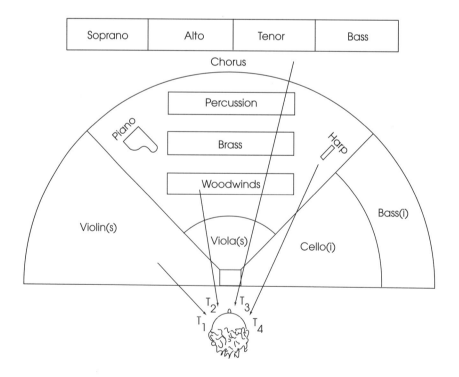

that of the front strings. As we sit and listen with our eyes closed, this time lag is so slight that it escapes conscious notice, yet our ear/brain combination processes this information and correctly concludes that the chorus is placed behind the orchestra. Even when the chorus is louder than the strings, the brain is not fooled into thinking that they have moved in front of the orchestra.

Despite our excellent perception of width and depth, we really have no control over what we hear as we sit in a concert hall. But in the recording studio, or in recording that orchestra in the hall, a microphone or series of microphones may often be placed in front of each instrumental section as well as the chorus, allowing the engineer considerable control over what he or she hears and sends to tape. The working distance between the performers and the microphones will be arranged so that each microphone favors that particular section, as in Figure 4-2. To accomplish this, microphones must be placed reasonably close to each section to better distinguish it from the rest of the ensemble. Note that the stage setup for recording is not necessarily the same as the orchestra's concert seating. But in arranging microphones in this manner, much valuable directional and spatial information can be lost for the sake of greater intensity control over the various sections of the orchestra. Unfortunately, this lost information is all but impossible to retrieve in even the most sophisticated recording systems. The engineer should clearly understand this characteristic problem with multiple microphones before concluding that it is the only way to record.

In a simple two-microphone setup, the entire ensemble is heard by both microphones. Presumably, all the subtle differences in performer-to-microphone spacings, angle and time of arrival, etc. are retained. Ideally, it will sound as if the listener were seated at the location of the microphone pair. In the multi-microphone setup, these subtleties are lost, since each performer or group of performers is recorded from closely-spaced separate microphones. The consequences of this are that time of arrival differences are minimized or completely eliminated since each microphone transmits its information simultaneously. In the control room, the various microphone outputs may be satisfactorily arranged in a left-to-right plane, but there is no completely successful way of creating the illusion of one instrument located behind another. In other words, we have created a two-dimensional orchestra, one without a sense of depth. It is rarely satisfactory to place the rear instrument microphones at a lower level than the more forward-located ones, especially if in the actual performance the music calls for the rear groups (such as the chorus) to be louder. However, some relief to

Figure 4.2 With a multiple microphone set-up, time-of-arrival differences are minimized, depriving the listener of much of the spatial information for the sake of improved control over balance.

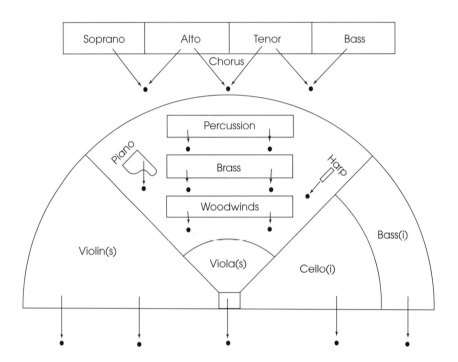

this situation can be achieved by electronically delaying the arrival of the rear-located microphone outputs by a factor approximately equal to their displacement from the front microphones. Of course, the practical day-to-day realities of the recording studio often take precedence over the ideal notions of purist techniques, particularly in the area of popular music and multi-track recording. Nevertheless, the recording engineer must understand the trade-offs between multiple microphones and stereophonic microphone techniques.

■ STEREO TECHNIQUES (BINAURAL SYSTEMS)

There are many methods in use for selecting and placing a single pair of microphones for a stereophonic pick-up. Perhaps the most obvious technique would be to space two omni-directional microphones on either side of an acoustic baffle (Figure 4-3) in an attempt to simulate the condition of actually being seated at the location of the microphones. This technique, known as binaural recording, is most realistic when wearing headphones, since the microphones seem to become extensions of the listener's ears, in effect, transporting the subject to the site of the recording. However, it can be quite a disconcerting effect when, as you move your head from left to right, the stereophonic stage moves also. Binaural recording can also be done using a model of a human head with pressure transducers placed where the ears would be located. A popular version of this is pictured in Figure 4-3b.

Figure 4.3 A binaural recording setup.

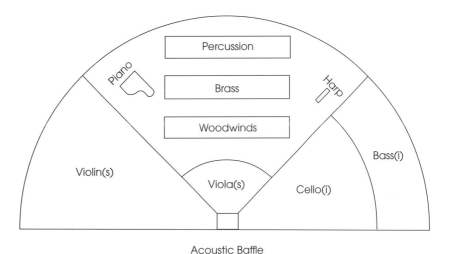

A binaural recording played over loudspeakers is even less satisfactory. To understand why, consider a sound source off at some distance to the left of the binaural pair, as seen in Figure 4-4a. Both microphones will pick up about the same intensity of sound, along with the slight phase shift or time of arrival differences that would give the listener at the site, or the one wearing headphones, the necessary directional clues. However, over the spaced loudspeakers, these subtleties will be pretty much lost, and the approximately equal sound intensity at each speaker will produce a phantom image of the sound that appears to be somewhere near the center of the loudspeaker pair. On the other hand, a sound that is close and directly on the left will be primarily picked up by the left microphone and will be principally directed to the left speaker, giving the impression that the sound is off to the extreme left, rather than close up as is actually the case (Figure 4-4b). These distortions of space and directional information rule out the effectiveness of binaural recording in professional applications.

Figure 4.3b A dummy head is often used in binaural recording to closely simulate actual listening conditions (Neumann KU-100, Neumann Photo).

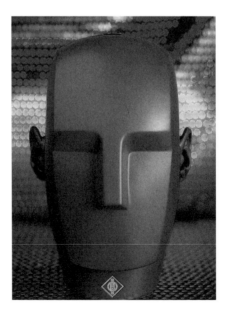

Figure 4.4 a. When loudspeakers are used, the listener may think the sound source is somewhere in the center of the room, rather than on the extreme left; b. as the sound source moves in closer, the listener hears the apparent sound source move from center to extreme left.

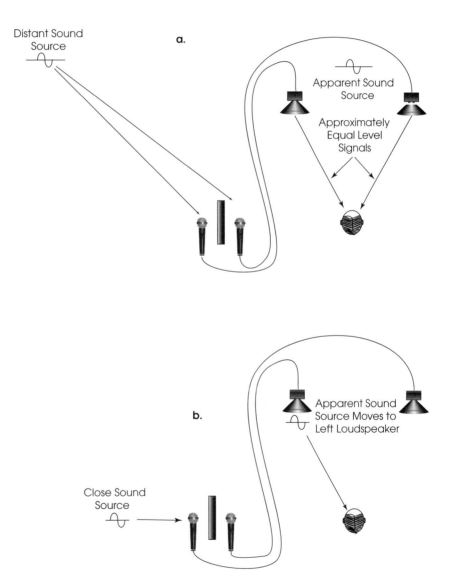

Coincident Systems

Coincident recording systems are those in which the diaphragms of the two microphones are arrayed in the same vertical plane. This means that the cues that give us our perception of direction and spacing are limited to intensity only since, as the capsules are coincident, there are no time of arrival differences. This can be accomplished by lining up two radially-oriented microphones, one on top of the other, or by using a so-called stereo microphone. The latter is actually two separate microphone systems, housed in one case, as shown in Figure 4-5. Each system usually consists of a Braunmühl-Weber capsule and has continuously variable patterns that can be selected remotely. The upper microphone capsule can be rotated through an arc of 180° or more, and the two outputs are kept electronically separate.

Figure 4.5   A condenser stereo microphone (Neumann Photo).

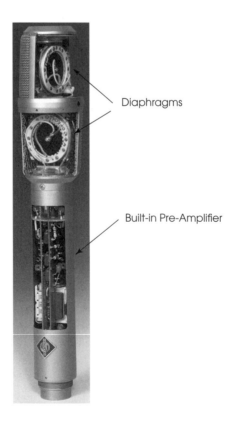

■ Blumlein System

One of the more useful systems in the category of coincident stereo recording is the so-called *Blumlein* system, named after the British experimenter Alvin Dower Blumlein. This simple but elegant system was first mentioned in a British patent by Blumlein in 1926. In operation, the stereo microphone pattern controls are both set to bi-directional or figure-8, and the capsules are oriented at a 90° angle to each other. This points one capsule toward the left side of the orchestra and the other toward the right. The null point of the left figure-8 is then facing the right side of the orchestra and the right null is facing left as shown in Figure 4-6. The patterns overlap in the center of the array at exactly the point where each figure-8 pattern has fallen 3dB in sensitivity. If these microphone outputs were to be sent to a pair of spaced loudspeakers, we would see that sounds arriving from the left would be reproduced in the left speaker and the same relationship would occur on the right. Sounds from the center would enter both capsules equally but 3dB lower in level and when reproduced by both loudspeakers would appear to be in the center and at the proper level as the signals combined. This system conveys a very accurate impression of both the width and depth of the orchestra. It is quite easy to determine which instruments are in the foreground and which are further back, although the engineer has no control over the internal balance of the ensemble. The rear lobes of the figure-8 patterns are pointing back into the hall and present an equally accurate picture of the reverberation characteristics therein. The distance between the microphone(s) and the orchestra is usually equal to about half the width of the ensemble. This ratio of direct to reverberant sound relationship conveys a sense of space and gives the listener an impression of the distance between audience and performer.

■ X-Y Systems

Sometimes, however, the ambient field in the recording location is less than perfect, or we simply wish to capture a more "present" sound. By switching the pattern selectors from figure-8 to cardioid we eliminate the back lobes and make the sound of the microphone array drier and more present in nature. This X-Y technique is often used along with flanking microphones, a technique which will be discussed later. The problem with this X-Y approach is that the overlapping cardioid patterns cross very near their maximum sensitivity point (Figure 4-7). When listened to over a pair of loudspeakers, this technique narrows the stereo image of the ensemble due to the common information in each microphone output. This problem may be alleviated by widening the angle between the capsules or by narrowing the patterns and making them closer in polar response to a figure-8. A good compromise seems to be a pair of crossed hyper-cardioid patterns arrayed at an angle of 110°. This gives more of the ambient field than the overlapping cardioid patterns, yet does considerably widen the image of the ensemble.

■ M-S System

Another system that utilizes a stereo microphone or coincident pair is the M-S system. M-S, which stands for Middle-Side, uses a cardioid microphone pointed straight ahead at the middle of the orchestra,

Figure 4.6 A stereo microphone setup for recording using the Blumlein system.

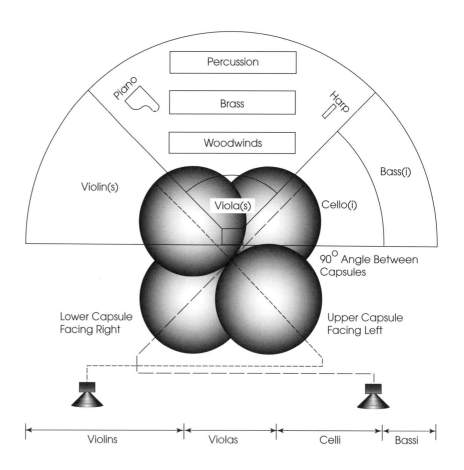

while a figure-8 microphone is pointed sideways so that one of its null points is facing the orchestra. This microphone setup is illustrated in Figure 4-8a, and it should be apparent that the cardioid microphone, which has a single output, picks up the entire ensemble, while the figure-8 microphone favors the extreme left and right sides. Center information reaches the null point of the figure-8 pattern and is pretty much canceled out. To form a suitable stereo image, the outputs of the two microphones are combined in a matrix system. To understand how this matrix works, assume that the front of the figure-8 is pointing to the left. Therefore, signals originating

Figure 4.7 Two cardioids in an X-Y configuration showing pattern overlap.

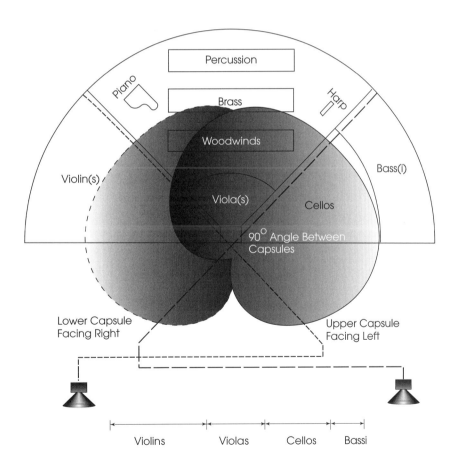

left of center will be picked up by both the forward facing cardioid and the positive front of the figure-8 microphone. If their outputs are combined, these signals will add together. On the other hand, right of center signals will be picked up by the negative rear lobe of the figure-8 pattern, and when summed with the forward facing cardioid, the signals will tend to cancel. This arrangement of cardioid and figure-8 will favor left of center information only. However, if the output of the figure-8 microphone is reversed in phase electrically, the effect will be the same as if its left-to-right orientation had been physically reversed. Now a combination with the cardioid signal will yield a right-handed signal instead. As shown in Figure 4-8b, a matrix system provides both types of combinations with separate outputs for each. By controlling the amount of the S-pattern sent to the matrix, we can effectively control the apparent width of the overall pick-up, and by adjusting the left-right gains at the output of the matrix we can steer the array in the horizontal plane of the stereo image. In this M-S system, the forward-facing microphone does not always have to be a cardioid. It can be an omni, or even another figure-8. Some combinations of different patterns and their X-Y equivalents

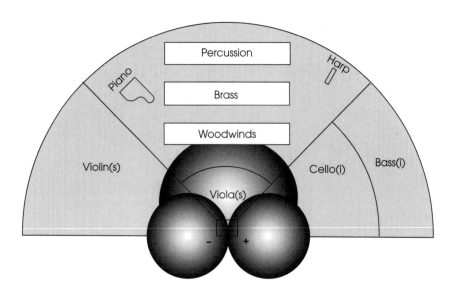

Figure 4.8a A stereo microphone has its patterns set for M-S recording. A matrix will derive the correct left and right signals.

Figure 4.8b At the top, transformers derive the equivalent left and right signals, while below an active M-S matrix using summing amplifiers provides left and right signals with a width control. The width is determined by the proportions of the M signal and the S signal sent to the summing amplifiers.

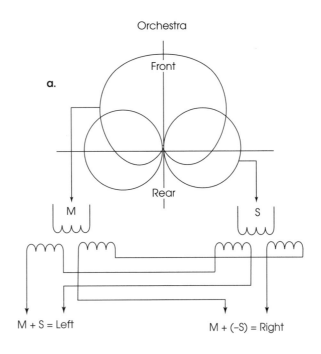

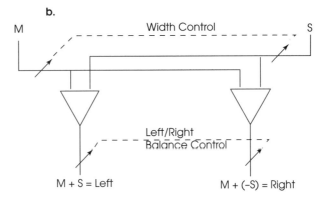

are shown in Figure 4-9a. A major advantage of this system is its ability to provide a pure monophonic signal, and for that reason it is highly favored by broadcasters. Figure 4-9b shows a dedicated M-S microphone and its control box. Figure 4-9c shows an M-S matrix that can be used with any combination of microphones.

Near-Coincident Systems

All of the preceding stereophonic microphone systems had one very important factor in common. They all relied solely on intensity differences in order to duplicate the left-right orientation of the ensemble. We have seen that the instrumental placement is extremely

Figure 4.9a M-S pattern combinations and their X-Y equivalents.

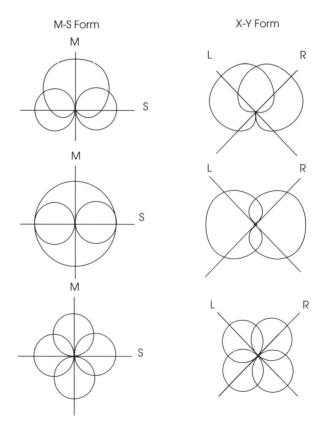

MICROPHONE TECHNIQUE   111

Figure 4.9b   M-S microphone with control box.

Figure 4.9c   M-S matrix that can be used with any combintation of microphones.

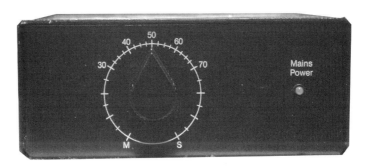

accurate and that because of the rear-oriented part of the patterns the entire stereo stage is bounded by a very even sense of ambience. In order to achieve this balance, the microphone-to-source distance has to be very precisely determined, and any inaccuracies in placement can cause a muddy, shallow sound. Yet even when properly done, many engineers feel that these coincident arrays lack a sense of warmth and perspective that we normally perceive in a room. This may be due to the absence of time cues, that seem to be so important to our ear/brain combination. Because of this, several so-called near-coincident systems have evolved. They offer many of the same advantages that are found in coincident systems in terms of instrumental placement and mono compatibility, yet may exhibit a broader sense of space and warmth with the addition of lateral time difference cues.

A system that is very popular in Europe was originated by the French National Broadcasting System whose official name is "Office de Radiodiffusion-Television Française" or ORTF. The ORTF system is formed by spacing two cardioid microphones so that their diaphragms are 17 centimeters apart and with a subtended angle of 110° between the forward-facing patterns. This spacing and angle is similar to the way our ears are positioned on our heads, yet does not resemble the earlier-discussed binaural system because of the lack of a baffle between the microphones. It does, however, simulate our listening experience. At low frequencies its directional sense comes from intensity cues only (a long wavelength produces minimal phase differences between the two microphones since they are so close together), and at high frequencies time of incidence cues are included which add a greater sense of width and spaciousness to the sound. Many engineers feel that this system rivals the Blumlein array in instrumental placement accuracy, yet without being so dependent on the sonic character of the recording space. The overall angle of pickup for the ORTF is 180°, and it seems to work very well relatively close to the ensemble, giving good "presence" without sacrificing width or depth.

Another near-coincident system has been developed by the Dutch Broadcasting System (Nederlandsche Omroep Stichting). This system, which is called the NOS array, uses two cardioid microphones at an angle of 90° with a spacing of 30 centimeters between diaphragms. This is similar to the ORTF system, but shifts the range where low-frequency phase differences become unimportant to a lower point in the frequency bandwidth. This adds even more of an apparent openness to the sound. The overall angle of coverage for the NOS system is 160°, and it therefore is very useful with smaller ensembles. Additionally, note that the frequency at which the half-wavelength of the sound is equal to the spacing between the microphone capsules varies between these two systems by about an octave. It is at this point on the frequency spectrum that the sound becomes more monophonic as the array becomes mostly dependent on intensity cues instead of time of arrival. An ORTF array reaches this point at a frequency of about 2kHz whereas the output of the NOS system remains time and intensity dependent to about 1kHz. This may be one of the reasons that the NOS array sounds slightly more open. Figure 4-10 illustrates these two near-coincident systems.

Figure 4.10 a. ORTF array; b. NOS array. Note the overall coverage angles.

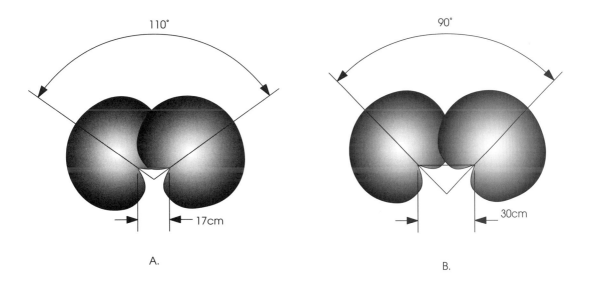

The Decca Tree A system that has been used successfully for many years by the Decca-London Company is the so-called *Decca Tree*. It is an array that uses three microphones in a triangular pattern. The original system, and the one used by many today, uses a special microphone called the M50 made by Georg Neumann GmbH. Although this tube condenser microphone is a pressure or omni-directional transducer, the capsule is mounted flush in a 40-mm plastic sphere. This gives the microphone a much more directional or cardioid pattern at higher frequencies. The low frequencies bend around the sphere and the capsule reacts like a pressure device, whereas the sphere acts as a baffle at high frequencies, creating a directional response. The transition from pressure to pressure-gradient response occurs around 2.5kHz. Additionally, the M50 microphone has a rising frequency response starting at about 3kHz. The microphone has been reissued by the manufacturer in a transistor pre-amplifier version. The microphone, with its polar plot and frequency response graph, as well as a picture of an original M50 capsule, are pictured in Figure 4-11. The circles superimposed on the drawing are there to help the reader visualize the capsule in the sphere beneath the grill casing.

The Decca Tree system uses these three microphones on a triangular bracket measuring 137 centimeters to a side. The triangle points forward, with the front microphone facing ahead and slightly down. When used with an orchestra, the other two microphones are pointed between the first and second violins and between the violas and cellos respectively. The system can be used with other microphones, although it gives the best sense of ambience and imaging if a microphone that has characteristics similar to the M50 is used. Decca has also used a two-microphone system with M50s where the separating bar is 92 centimeters long, with the microphones on the sides aimed as before. By moving the M50s closer together, a -3dB point is maintained between the capsules at the higher frequencies. The two Decca systems are pictured in figure 4-12.

Spaced Microphone Systems Notice, from the differences between the ORTF and the NOS systems, that the farther apart we move the microphones, the smaller the angle between them becomes in order to maintain the overlap at the -3dB sensitivity point, and the narrower the overall angle of coverage becomes. As we reach a distance of 4 or 5 feet another problem occurs. Let us say that we have a pair of microphones spaced apart from each other at a distance of 10 feet. When these outputs are transmitted through a pair of loudspeakers, the left-originating signals will go exclusively to the left speaker, and the opposite will be true for the right. Because of level losses due to the inverse square law discussed in Chapter 1, no left component of any significance will appear in the right speaker. Only sounds that originate exactly in the center will appear in the center, and yet this material will be lower in level because of the aforementioned inverse square law. Any sound that is slightly off-center will have a tendency to "pull" to the extreme right or left. This creates a "hole in the middle"

Figure 4.11 A Neumann TLM50 used in the Decca Tree system.

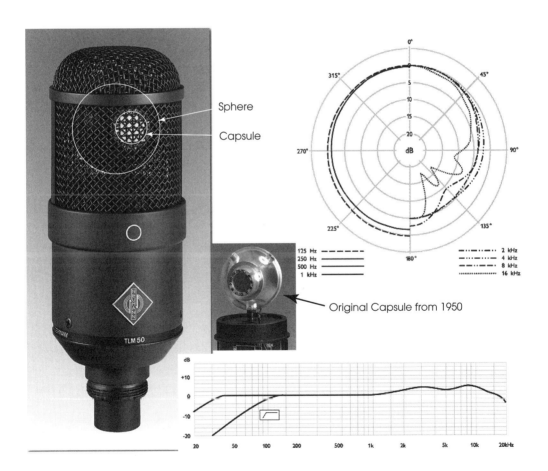

116   THE AUDIO RECORDING HANDBOOK

Figure 4.12   The Decca Tree system and the Decca Stereo array.

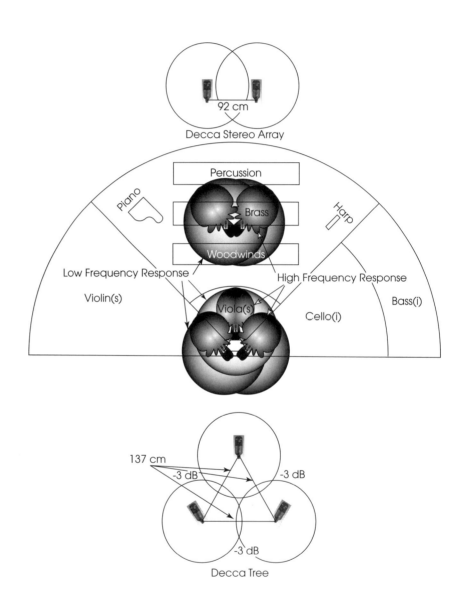

effect that can be very disconcerting. To solve this problem, a third microphone can be added with its output assigned to both left and right stereo channels. The level of this third microphone should be carefully monitored so that an even left-to-right distribution of sound is heard, with no microphone standing forward of another. With coincident or near-coincident systems, a directional polar pattern is essential for proper imaging. Obviously a pair of omni-directional microphones will create a monophonic reproduction since the outputs of both microphones will be indistinguishable from one another. However, a good choice for the spaced-apart technique would be to use omni-directional microphones. This allows the microphones to be placed closer to the ensemble without sacrificing any of the room ambience, and also does away with most of the coloration created by the off-axis response of the cardioid patterns. This technique does not yield as accurate a monophonic sum as a coincident-type system, but gives a large sense of space and good tonal accuracy which has been used successfully in many recordings.

Additional Microphones

Since the practical limitations of the recording environment may not allow the engineer to achieve a satisfactory balance with just one coincident or near-coincident pair, additional microphones may be used to supplement the principal microphones or to accentuate one or more sections of the orchestra. However, they must be used very carefully so as not to detract from the stereo perspective produced by the main array. To get accurate coverage of a very wide stage where the engineer does not wish the microphone-to-source distance to be too great, a combination of coincident and spaced-apart techniques is often useful. If a stereo microphone is giving good depth and placement but is losing the outside edges of the orchestra, a pair of flanking cardioids or omni-directional microphones may solve the problem. Placing microphones halfway between the center of the orchestra and both the extreme left and right edges effectively extends the width of the stereo microphone without overly affecting the imaging or placement. The time of arrival of each orchestral area to its associated microphone must be the same in order to prevent phase-related distortions. For omni-directional microphones to achieve the same presence as the pressure gradient central microphones, they have to be placed closer to the instruments, since they pick up a greater proportion of reverberant to direct sound (because of their circular pickup). This would change those ensemble-to-microphone time relationships just mentioned.

Therefore, it is a prudent engineer who carefully checks the phase response of the recording when mixing omni-directional patterns with directional patterns along a lateral axis. However, the flanking omni-directional microphones can add a sense of space to the recording if the performance environment is acoustically dry. Once again, it is important to monitor the total left-right distribution to ensure an even stereo perspective.

Suppose that, even after adding these flanking microphones, the woodwind melodic lines are still being covered by other instruments. At this point, we have to begin adding accent microphones. Accent microphones are microphones or pairs of microphones that are placed on individuals or individual sections of the orchestra to make them more audible in the recording. It is important to note that they are used more to "point up" the sound of these instruments than for actual level. And since the extra microphones are somewhat closer to the instruments than the main stereo array, their output will reach the listener's ear some fraction of a second earlier, focusing his attention on it. To keep these accent microphones from thus deteriorating the overall pickup, their output will probably have to be kept quite low in level. Some electronic time delay may be added to these outputs to time their arrival with the central array. Figure 4-13 shows a symphonic orchestra setup for recording with a central array, a flanking pair and several types of accent microphones.

The advantage of the various stereo techniques just described is that the sound of the orchestra as an ensemble is well preserved. There is a good representation of depth, as well as left-to-right information. Within reason, low-level instruments may be clearly heard and located within the ensemble, despite the presence of louder instruments in the same vicinity. The obvious limitation of this technique is that the engineer has only minimal control over the instrumental balance, and at a later date cannot effect major changes of individual parts without affecting the entire ensemble. In fact, with the advent of the digital recording process in the late 1970s which led to a purist's view of classical recording, this essentially "direct-to-two-channel" system caused many recording engineers to have to relearn the stereo microphone techniques discussed above, since, with the two-channel digital system, it was no longer possible to "fix it in the mix." It is, and has always been, essential for the recording or balance engineer to fully understand the musical concepts of orchestral balance and timbre, so that he can fulfill his function to the satisfaction of the producer and ensemble conductor.

SURROUND SOUND MICROPHONE TECHNIQUES

In the past there have been a number of efforts to convince the listener that stereo sound was too limited a reproduction method—that in order to fully hear the music and the environment in which it was recorded, more loudspeakers with dedicated playback channels were required. Indeed, in music history it has been shown that as the architectural style of the period changed, so did the style of music. It is certainly more difficult to hear the difference between a

Figure 4.13 An orchestral recording setup with a central array, supplemental flanking and accent microphones for the body of the orchestra and spaced directional microphones to supplement the chorus sound.

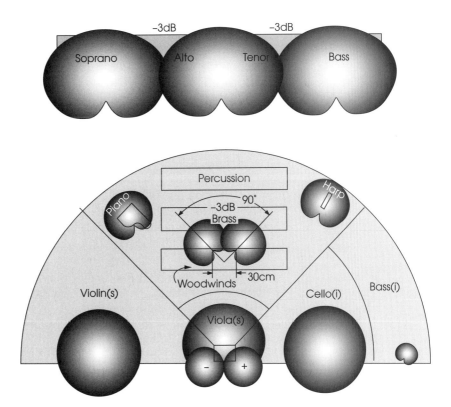

Mozart symphony and a Beethoven performance on a conventional stereophonic playback system than in a live performance in an acoustically good room. Early attempts at "surround" sound, that is, where the listener is surrounded by sound, were limited by the available playback media. Now, however, with the advent of multichannel DVD (Digital Versatile Disc) systems for theatrical playback, and DVD audio-only or SACD (Super Audio Compact Disc) playback media on the horizon, methods for recording sound for playback on more than two channels need to be developed.

Most if not all of the stereophonic recording techniques previously mentioned are quite suitable for recording information that will be played back on the front loudspeakers of a 5.1 (five dot one) surround system. Although today's surround systems include the .1 or LFE (Low Frequency Effects) channel, little if any music should be sent to that playback channel. The LFE channel was expressly designed to sonicly enhance various theatrical effects such as explosions and earthquakes, and as such should not be used for music. An exception would be, of course, the cannons in Tchaikovsky's *1812 Overture* or the musket rounds fired by the percussion section in a performance of Beethoven's *Wellington's Victory*.

The problem arises in how to determine and capture what will be reproduced by the rear channels. It would be possible indeed to put the brass and woodwinds in the rear channels with the strings in the front to simulate the feeling of actually sitting in the orchestra. Although this might be a novelty for the first hearing, the effect would soon become tiresome with repeated listening. It seems logical to assume that to reproduce what the listener hears coming from the rear in a good hall might be a reasonable approach. One of the goals of this type of recording is to place the listener in the room with the ensemble. In Chapter 6 on time-domain devices, we will discuss the makeup of the ambient or reverberant field in a natural environment and the methods that can be used to either capture or simulate these sounds. Along these lines of reasoning, placing another pair of microphones or, say, an additional M-S array further back in the hall to capture the natural ambience could be a good approach. Even with multi-tracked popular music, extra microphones are needed both in tracking and in overdubbing to bring a sense of space to the recording.

One approach is to put a pair of omni-directional microphones in the back of the hall where the ambient field predominates over the direct sound. The ambient field, made up of many reflections from various surfaces in the room, is what often gives us the sense of space in the concert hall. Capturing these sounds and directing them to the rear channels on playback can help envelop the listener in the recording space. However, if the level of the rear channels is set too high, the listener may feel too far away from the orchestra. In a situation such as this, there is no substitute for judicious listening coupled with musical training and an understanding of the music being recorded. It is extremely important that the rear sounds effectively integrate with the front sounds, and not sound like they are originating from a separate location. Later, in the section on special-purpose microphones, I will discuss the dummy head system, which has also been used effectively in surround recording systems. Note in Figure 4-14 that, although the main outputs of the rear-facing M-S array go primarily to the left rear and right rear channels respectively, a small amount is directed to the left front and right front channels. This steers sound from the rear towards the sides where it helps integrate the front channels and rear channels together. Additionally, the two omni-directional microphones that are set up in the ambient field are directed to the rear channels only.

A further consideration concerns what to do with the center channel. A full 5.1 surround system includes a center channel, normally used for dialogue in theatrical releases, as well as discrete left and right rear channels. It is an unwise move to place an instrument or sound solely in the center channel since not all surround systems have the center channel implemented in the music mode. However, in the event a soloist is featured with an ensemble, a small amount of this spot microphone can be sent to the center channel to help anchor the image. In the diagram of Figure 4-14, notice that some of the left center and right center information from the near-coincident pair is directed to the center channel, but that the majority of the center pickup is directed to the left and right.

At this time one of the problems with surround sound is the lack of a standard type for the surround loudspeakers. This means that, in some playback systems, the rear channels may be diffuse field loudspeakers, while in other systems, direct field radiators may be used. This will be discussed briefly in Chapter 5. And, although there are many techniques for recording in surround, Figure 4-14 gives a good representation of a technique to capture the way sound is heard in the hall. The prudent engineer will consider other approaches as well, including arrays that may use other coincident as well as spaced-microphone or specialty-microphone techniques.

Figure 4.14 Surround sound recording microphone setup as in Figure 4-13, but with the addition of a rear-facing M-S array and a pair of "house-spaced" omni-directional microphones.

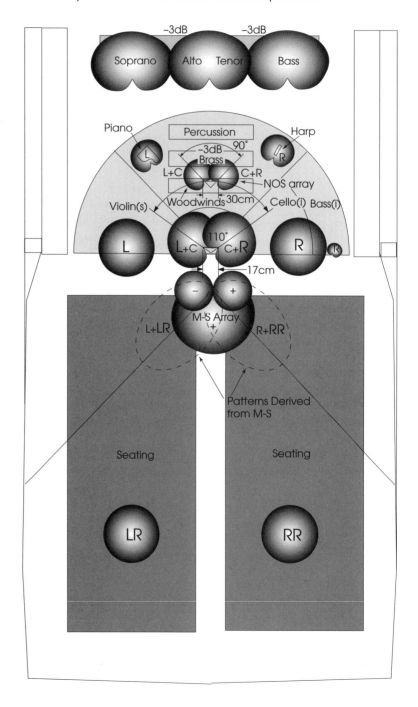

■ MULTI-MICROPHONE TECHNIQUES

Although the advantages of the various stereo microphone techniques should be clearly understood, the needs of the modern recording session may often preclude their successful application. Due to budget considerations, it may be out of the question to hire an orchestra of sufficient size to be self-balancing, even with the help of a skillful conductor. Or the musical arrangement may be such that an instrument, or instruments, will be heard out of proportion to its natural balance within the ensemble. Finally, as is found in the contemporary popular recording session, several musicians play many different parts a few at a time, and these parts may then need to be individually modified without affecting the sections recorded earlier. In many of these cases, stereo microphone techniques may not be able to do the job required, and the engineer must be prepared to mix and match these techniques with a multi-miking technique that may be more appropriate to the demands of the session.

The beginning engineer is always anxious to learn proper multi-microphone techniques, and it was not so long ago that the student of recording could find little or nothing published concerning "correct" microphone placement. Many apprentice engineers have often insisted that the experienced engineer should be able to tell him or her which microphone is best for each instrument, and where that microphone should be placed. Today, many so-called microphone "cookbooks," which purport to give the beginner the perfect microphone placement for every instrument, have appeared. However, beyond a few simple rules, microphone technique is largely a matter of personal taste, and engineers rarely agree on choice and placement of microphones. Microphone selection and placement is an art, and the engineer is the artist, creating a sonic portrait with different texture techniques and color selections from the microphone palette. When asked, "How do you mike a piano?" the experienced engineer will invariably ask, "What type of piano, what kind of space is it located in, who is playing it, and what are they playing?" The range of choices is endless, yet, to be commercially successful, there are certain parameters within which the engineer must work. Many of the rules and limitations of microphone placement and selection have been discussed earlier, and they are paraphrased here:

   1. Always take off-axis coloration and proximity effects into consideration when selecting microphones;

2. Avoid overload, either with an attenuator in the microphone line, or, in the case of a condenser microphone, by inserting a pad between the diaphragm and the microphone's own pre-amplifier, if this pre-amplifier is being overloaded;

3. Protect the microphone against wind and vibration noises;

4. Exercise caution when using acoustic baffles. Move the microphone when possible, rather than use a baffle;

5. Make sure all microphone lines are properly shielded, balanced and terminated;

6. Be sure that there are no electrical phase reversals in the signal path;

7. Don't use two microphones when one microphone will do a better job;

8. Never use equalization as a substitute for proper microphone selection and/or placement.

Beyond these rather obvious precautions, microphone usage is largely a matter of personal choice, and the remainder of this chapter should in no way be considered a "rule book" of microphone technique.

Avoiding Phase Cancellation In Chapter 1, Figure 1-13 showed how two in-phase microphones could be positioned so that one was receiving pressure maxima while the other received pressure minima from the same musical instrument, and it was noted that when their outputs were combined these signals could cancel. Unless the instrument was perfectly centered and did not move—an unlikely event—some wavelengths would reach the microphones acoustically out of phase. And with each slight movement, a different set of wavelengths, and therefore frequencies, would be affected as the relative path lengths between the instrument and each microphone fluctuated. Figure 4-15a illustrates the problem. A slightly off-center instrument creates a path length difference of three inches. The frequency with a half-wavelength of three inches (that is, a wavelength, or lambda, of six inches) is:

$$F = \frac{V}{\lambda} = \frac{1130 \text{ft} / \text{sec}}{.5 \text{ft} / \text{sec}} = 2260 \text{Hz(cps)}$$

Therefore, this frequency would be prone to cancellation whenever the path length difference approached three inches. Although it is certainly impossible to completely eliminate all interaction between microphones, most phase cancellation problems can be minimized by observing a 3-to-1 relationship. In other words, if a microphone is one foot away from an instrument, no other microphone should be within three feet of this microphone, as shown in figure 4-15b. Although small path length differences will still occur, the inverse square distance ratio is sufficiently large to minimize or eliminate any cancellations.

Another source of phase cancellation to be avoided is also caused by path length differences, but only to a single microphone. These differences occur between the direct signal reaching the microphone and a reflection of that signal from a nearby surface, a situation illustrated in Figure 4-16a. If this path difference is once again three inches, cancellation will occur at 2260Hz as mentioned above. Both reinforcements and cancellations occur, and continue to occur, at multiples of the offending frequency. The amount of signal degradation will depend on the relative loudness of each signal, with the worst case being when the signals are nearly equal in level. This causes a phenomenon called "comb filtering," so called because the resultant reinforcements and cancellations create a frequency response curve that resembles a comb. A typical situation where this can be a problem would be an instrumentalist sitting above a highly reflective surface such as a hardwood floor, or a speaker giving an address in front of a podium.

To illustrate, a trumpet player is sitting in a chair with a metal music stand in front of him, as shown in Figure 4-16b. The microphone is placed in front of him and above the stand. If the path length difference between the direct sound of the trumpet and the reflected sound from the stand is one foot, there will be cancellation at 565Hz, reinforcement at 1130Hz, cancellation at 1685Hz, etc., producing a frequency response curve similar to the one illustrated in Figure 4-16c, at the microphone output. This problem can be solved by always using a 2–to–1 relationship between the microphone and the reflecting surface. That is, if a microphone is one foot away from an instrument, the microphone should also be at least two feet away from the nearest reflecting surface. If the reflecting surface, the instrumentalist, or the microphone cannot be moved, another possible solution is to cover the offending surface with a non-reflective absorbent material.

Figure 4.15 a. When there is a slight path length difference, acoustic phase cancellations may be expected; b. To avoid phase cancellations, measure the distance between the instrument and the microphone. No other microphone should be within a radius of three times this distance.

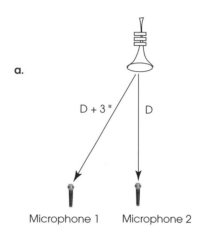

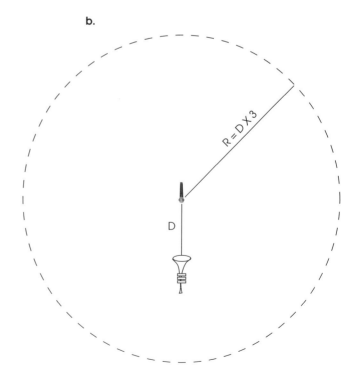

Figure 4.16 a and b. Multiple-sound pathways to a single microphone; c. Frequency response comb filtering caused by multi-path interference.

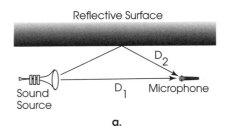

a.

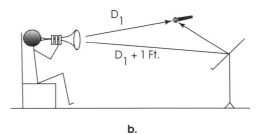

b.

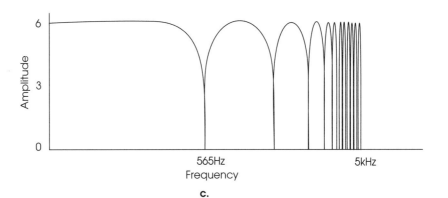

c.

Minimizing Leakage

As mentioned earlier, leakage (unwanted sounds from other instruments in the room) may be minimized by close miking, or by interposing some acoustical barriers between the wanted and the unwanted sounds. These techniques, however, must be applied with great care so that in the attempt for greater control, the overall sound quality does not deteriorate unnecessarily. Although the loss of some depth may be unavoidable when using a close miking technique, some of the other problems may be avoided through a better understanding of microphone usage and selection.

In an effort to provide greater electrical separation between instruments, the engineer's first impulse may be to use a cardioid microphone, reasoning that its comparative insensitivity to off-axis sounds may help keep leakage at a minimum. The off-axis response of the microphone itself, however, must be taken into account. Cardioid microphones, it must be remembered, are not totally deaf to rear-oriented sounds. In fact, at 1000Hz, most cardioid patterns are attenuated only 25dB to 40dB at 180°. A cardioid microphone with poor off-axis response may very well attenuate rear-originating high frequencies; that same microphone may, however, function as an omni-directional microphone at low frequencies. As an obvious disadvantage of this off-axis microphone coloration, consider this. A cardioid microphone is aimed at an acoustic guitar, and pointed away from the drum set, as seen in Figure 4-17. The higher frequencies of the cymbals and snare drum may be significantly attenuated at the microphone, yet the lower frequencies of the toms and kick drums will still be heard. The resultant drum sound, as heard by all microphones, will consequently be "bottom heavy," although the

Figure 4.17 A cardioid microphone with poor off-axis response may pick up too much low-frequency off-axis.

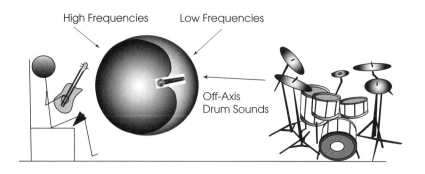

on-axis guitar itself may be entirely satisfactory. In fact, there may even be some multi-path phase distortion of the drum sound due to leakage of the low frequencies into the guitar microphone. If there are many such microphones in use, each contributing some low-frequency leakage, the cumulative effect may very well be an unpleasant muddy sound with a great sacrifice in overall clarity. In an effort to further minimize off-axis coloration, the microphone may be moved even closer to the guitar to keep the leakage to an absolute minimum. However, the poor off-axis response of some cardioid microphones is often accompanied by a pronounced proximity effect, which at very small working distances may cause an excessive bass rise in the sound of the instrument.

In many applications, these various restrictions may effectively be bypassed by using an omni-directional microphone. Free of proximity effects and most off-axis coloration, the microphone can be used very close to the sound source without excessive bass build-up, and slight changes in working distances as the musician moves will not cause frequency response fluctuations. Here, it must be remembered that the omni-directional pattern on many dual-diaphragm microphones may retain the undesirable characteristics of the cardioid microphone if, as mentioned earlier, its omni characteristic is derived from two cardioid patterns. This, however, does not rule out the use of cardioid patterns in the studio.

Many times, the sound response of a cardioid microphone is required to achieve the desired effect. Additionally, many of today's high-quality cardioid patterns are relatively free of excessive off-axis coloration and have a smooth proximity effect that may contribute substantially to the warmth of the sound. One of the benefits of the 3-to-1 rule for minimal phase cancellation between microphones is that when this rule is followed, there will usually be 9dB of electrical separation between the adjacent microphones. This may also be enough to minimize off-axis coloration from leakage in studio-grade microphones.

It should also be kept in mind that the sides of a microphone with a figure-8 pattern are generally less sensitive than the rear of a cardioid pattern. Accordingly, it is often possible to obtain good separation by pointing the dead side of the figure-8 pattern towards the unwanted sound, as long as there are no undesirable sounds facing the rear lobe of the pattern.

It is important to remember that when single microphones are used to pick up an instrument, some of the depth and three-dimensional perspective of that instrument are often lost. This phenomenon is most apparent when overdubbing a single instrument using only one microphone. It can be likened to listening to that instrument with one ear. Frequency, amplitude, timbre, and all the other characteristics of the sound may be there, but the instrument may appear flat and two-dimensional in character. It is often possible and desirable to recall the coincident and near-coincident techniques discussed earlier. Using an X-Y pair of cardioid microphones on an acoustic guitar can produce a sense of depth and space for that instrument that no single microphone can match. Obviously, this is the case only when the two outputs are kept apart and stored on separate tracks of the multi-track recorder.

In a further attempt to minimize leakage, various acoustic barriers are often set up between the microphone and the unwanted sounds. The most obvious and effective barrier is the isolation booth. This is usually a completely enclosed room (Figure 4-18) in which a musician, or small group of musicians, may be placed. Although such a room may provide the required isolation from unwanted sounds, it may also be somewhat of an artistic compromise, since the musicians thus isolated will often find it difficult to perform in ensemble with the rest of the group, regardless of the sophisticated earphone monitoring system that may be at hand.

Moveable acoustic barriers, popularly known as "goboes," a word of uncertain derivation, are frequently used in modern recording studios. The idea is to establish greater acoustical isolation between instruments by arranging the goboes as required. The goboes may be easily moved to meet the needs of the particular recording session. However, they too should be used with discretion, since an improperly used gobo can do more harm than good. Goboes are often used on the belief that they will help keep unwanted sounds from reaching the microphone. The goboes are placed between a loud sound source and a microphone that is to be shielded from this sound, so that the microphone may better hear the specific instrument placed in front of it. A typical use is shown in Figure 4-19a. There are several reasons why this gobo may create more problems than it solves. First, if the microphone is a cardioid, its side and rear entry ports must not be obstructed if the microphone is to function as intended. Assuming that the gobo is a perfect sound absorber (which it is not), it has effectively absorbed all off-axis sounds which would have otherwise reached the ports, as illustrated in Figure 4-19b. Therefore, all sounds (including the unwanted ones) reach the microphone diaphragm by the front entrance only, thereby effectively creating an omni-directional pattern (and not a very good one at that). Secondly, the gobo is not a total absorber of all frequencies. Since it is often placed quite close to the microphone, its

Figure 4.18 An isolation booth (photo by Bob Yesbek, Omega Studios, Rockville, MD).

surfaces—although absorptive—will nonetheless reflect some frequencies back towards the microphone, while others are absorbed. Such multi-path phase cancellation (similar to that mentioned earlier) can cause a deterioration of the overall sound that far outweighs whatever partial isolation has been accomplished. Lastly, the diffraction phenomenon discussed in Chapter 1 must be taken into account. As sound travels toward the gobo, low frequencies are diffracted around it, while high frequencies are either absorbed or reflected. Therefore, the microphone hears the diffracted low frequencies all too well, while the absence of high-frequency components results in an unpleasantly muddy sound pickup.

Figure 4.19   a. Some typical movable goboes (photo by Bob Yesbek, Omega Studios, Rockville, MD); b. The gobo may prevent off-axis sounds from reaching the cardioid microphone's rear entry ports.

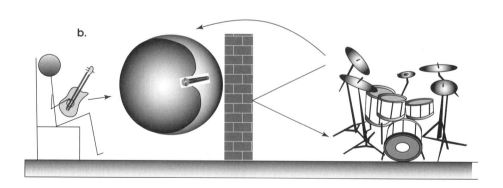

SPECIAL-PURPOSE MICROPHONES

The Dummy Head   One of the special-purpose microphones that is gaining usage as a microphone for surround-sound recording is the so-called binaural, or dummy, head. This pair of microphones is enclosed in a casing shaped like the average human head and made of a material that simulates the absorption characteristics of the average head (*sans* hat). The microphones are set at an angle that approximates the polar pattern of the human ear. As mentioned in Chapter 3, this array gives good localization and balance when listened to with headphones, but lacks good imaging, depth, and clarity when listened to on loudspeakers. The dummy head is also used in many industrial applications, such as in acoustic research and noise measurements. Some classical labels have effectively used this type of device as a primary pair for orchestral pickup, and as a sound source for the rear-channel signals in surround-sound recordings. Generally, the transducers used in this array are pressure devices and the polar response of each microphone is controlled by the dummy head pinna, or ear shape. The size and length of the ear canal can vary the polar response and frequency balance as well, just as in the human ear. The Neumann KU-100 (affectionately named "Fritz") is shown in Figure 4-20.

Contact Microphones   The contact microphone responds to the mechanical vibrations of the musical instrument to which it is attached. Therefore, it hears little or nothing of the sounds of other instruments within the studio. Specific operating principles vary from one manufacturer to another. Some actually contain a very small electret microphone, while others work like an electric guitar pickup. This type of microphone is the ultimate in close-microphone placement, as it hears only the direct sound of one instrument. Although it offers the engineer maximum control, the sound may not be as "musical" as desired. We are accustomed to hearing at least some indirect sound along with any instrument that we hear, and the contact microphone's ultra-close perspective completely eliminates this component of the total sound. On the other hand, when isolation booths are either unavailable or undesired, the contact microphone may allow the engineer to pick up a quiet instrument in the midst of a loud ensemble.

Figure 4.20 A dummy head that can be used for a stereo or surround recording system. Each ear contains a small diaphragm pressure transducer, and the head provides the proper shadowing to simulate how we hear (Neumann KU-100, Neumann Photo).

Pressure Zone or Boundary Microphones

A specialized microphone that has received considerable attention of late is the PZM, or pressure zone microphone, which is sometimes called a boundary microphone. In most cases, a small electret condenser element is located extremely close to (usually only a few hundredths of an inch) and facing a small (4-inch by 6-inch) boundary plate. The operational principal is that since the microphone is so close to a fixed and rigid plane, it can only sense changes in air pressure, as air particle velocity is zero next to a stiff boundary. Since all sound entering the microphone diaphragm is reflected from this plate into the closely-placed element, there are no multiple path length differences to create phase cancellation or random incidence effects. Therefore, off-axis coloration is non-existent, since all sound enters the diaphragm from similar angles. In other cases, the microphone is mounted flush with the boundary or surface of the casing. This prevents any path length differences between the direct and reflected sound. This is similar to the traditional footlight microphone used in theatrical performances. Most of these microphones have a polar response that is hemispherical with respect to the boundary plate. The microphone itself is often a high-impedance transducer and must be used with its own power/impedance converter. A popular version of a boundary microphone is pictured in Figure 4-22. One of the major problems with this system is that the low-frequency response is limited by the size of the boundary plate, and in order for the boundary microphone to cover the full audio spectrum, the plate must be relatively large. Mounting the PZM on a rigid wall or studio window, or laying it flat on a conference table, has produced some good results. This microphone has been successfully used as a way of amplifying a pit orchestra and sending the signal to the off-stage loudspeakers that are directed back toward the singers. This allows the singers to hear the orchestra with very little, if any, delay. Here, PZM and boundary microphone low-frequency response is achieved by mounting the microphones on the pit wall facing the orchestra.

Lavalier Microphones

Lavalier microphones are principally designed to be worn around or near the neck, and their frequency response is designed to be flat when the microphone is resting on the chest or against the body surface. They are used chiefly for sound-reinforcement applications, but have occasionally been used to record an acoustic bass.

Figure 4.22 A pressure zone microphone (Neumann GFM-132, Neumann Photo).

The older, larger type of this microphone is usually wrapped in foam rubber and inserted in the "f hole" of an upright bass, while the newer style of clip-on lavalier (pictured in Figure 4-23 and seen on your nightly news program) is fastened to the bridge of the instrument. The rising frequency response of these microphones, which is designed to bring out the consonants in the human voice, can help bring out the articulation of the bass. Most professional bass players who record on a regular basis have a contact microphone installed on their instrument. On the other hand, neither of these techniques will produce a sound that can equal a well-placed professional studio microphone.

■ MIKING MUSICAL INSTRUMENTS

As previously mentioned, there is no one correct microphone or placement spot for recording any instrument. The reason one engineer's recording sounds different from another's is that each of them subscribes to a slightly different concept of microphone usage, and it is always open to discussion as to which is the better recording. However, there are certain problem instruments which deserve mention.

Figure 4.23   A clip-on lavalier or lapel microphone.

The Electric Guitar

The electric guitar is by now such an integral part of the recording session that its use is almost taken for granted. The guitar amplifier is a fixture in most, if not all, modern studios. If we were to place a microphone on the guitar itself, as with an acoustic guitar, the results would not be very satisfactory since the sound of the electric guitar is produced by the combination of the strings, the pickup and the amplifier. A typical method for recording this instrument would be to place a microphone in front of the guitar amplifier loudspeaker. It is also possible to take the sound of the guitar "direct." This take-off point can be after the guitar but before the effects pedals, or after the pedals and before the amplifier, or even as a direct out from the pre-amplifier of the guitar amplifier itself. As always, there are advantages and disadvantages to this type of direct pickup. Since many guitar amplifiers are electronically noisy, the direct pickup gives the engineer a much cleaner signal with which to work. On the other hand, the guitar amplifier is usually considered an integral part of the electric guitar sound, and bypassing it eliminates the characteristic sound of that particular guitar amplifier's speaker. Figure 4-24 shows the signal path from an electric guitar to a console and possible direct points.

The output from the guitar itself is usually high-impedance and can vary considerably in level between different types of pickups. A transformer or "direct box" is therefore required to match the output to microphone level and impedance. For reasons cited earlier in Chapter 3, the guitar's high-impedance output cannot be fed directly through a long cable to the console microphone pre-amplifier. If the guitarist wishes to hear the amplifier in the studio, rather than from the headphone cue system, a split feed or "Y" connector can be used. Most direct boxes contain a loop-through to feed the guitar amplifier while providing a balanced low-impedance isolated output to the recording console. Direct boxes can be passive or active devices. As passive devices, they do cause some insertion loss. As active devices, they can actually improve the signal-to-noise ratio of the guitar/amplifier interface. Active direct boxes usually require phantom power or batteries. A typical direct box is pictured in Figure 4-25.

Electronic Keyboards

Many of today's bands use multiple electronic keyboards, and the transformers and direct boxes just described can be used with them as well. Many of these instruments have multiple or stereo outputs, so several boxes may be required. Several of the more sophisticated

Figure 4.24 The signal path from guitar to recording console showing possible access or "direct" take-off points.

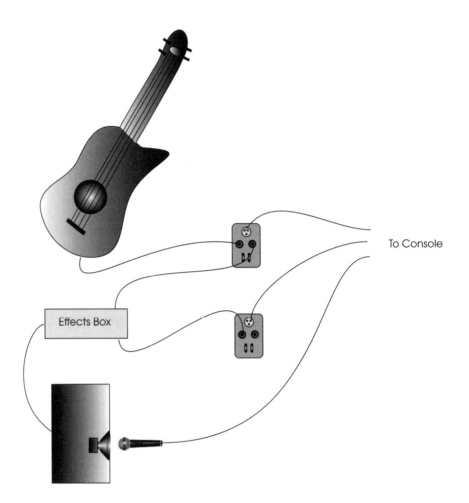

synthesizers also furnish balanced isolated microphone-level or line-level low-impedance outputs that do not require transformers or direct boxes. Many of these electronic instruments may be connected with MIDI (Musical Instrument Digital Interface, a system that will be discussed in Chapter 19); to prevent interference, it is important to keep the MIDI control lines away from the low-level high-impedance signal outputs prior to converting them at the direct box.

The Acoustic Guitar

When an acoustic guitar is used as part of a rhythm section, a single microphone is usually placed fairly close to the guitar to minimize leakage. Here, proper technique is mostly a matter of choosing the microphone that produces the most pleasing sound. Perhaps the greatest asset is the guitar player who understands the nature of multi-track recording enough to make sure that he or she remains at a reasonable constant distance from the microphone.

When the guitarist is also the vocalist, however, correct microphone placement is not quite as straightforward. Of course, one solution would be to record the guitar and the vocal on two separate passes of the multi-track tape recorder, but it is surprising how difficult this can be for a musician who usually performs both parts simultaneously. If the vocal and the guitar are miked separately, as is often done, some question arises as to how to satisfactorily balance the two microphone outputs. If one microphone is routed to the left and the other to the right, we have the unlikely situation of the musician singing and playing from opposite sides of the room. Yet if both microphones are center-placed, the monaural combination conveys little or no stereo perspective.

In many cases, a better recording could be made with a stereo microphone placed somewhere in front of the musician. Assuming the performer is capable of balancing his or her vocal and guitar

Figure 4.25 An active direct box.

performance, the stereo microphone will pick up a blend of the two and, at the same time, convey the feeling that the recording is in fact in stereo. If balance is a problem, an accent microphone may be added to the guitar and panned to the center for the extra sound required. Another possible solution is to use an X-Y or near-coincident pair of microphones for the guitar and a single microphone, sent to both channels, for the vocal.

The Drum Set There is probably more discussion about how to mike "a drum kit" than any other instrument except the piano. In fact, the drum set used for today's music can vary from the standard five-piece kit with three cymbals to a giant twelve-piece set with dual kick drums and multiple cymbals. A standard drum set includes: a bass drum, commonly called a kick drum; a snare drum; three tom-toms, two mounted on the bass drum and one floor-standing; a high-hat cymbal set, comprised of two cymbals controlled by a foot pedal; and two other cymbals, usually a "crash" and a "ride."

The engineer will often use a number of microphones for the drum setup. Since each part of the drum set has its own distinctive "sound," the engineer may balance the overall sound to suit the needs of the session. A typical setup is shown in Figure 4-26. Two overhead microphones are used as an overall pickup, with additional microphones used on the kick drum and the snare drum, as well as the two top toms, the floor tom, and the high-hat. It is not at all uncommon to find additional microphones in use for the bottom of the snare drum, for roto-toms, extra cymbals, etc.

Transient percussive peaks, especially from the snare drum and cymbals, will surely be considerably higher than the apparent levels seen on the VU meters. Whenever possible, these instruments should be measured using meters that have PPM ballistic characteristics. This type of metering, discussed in Chapter 2, will be further discussed in Chapter 17 on the modern recording console. If such metering is not available, and/or until the engineer has gained some practical experience relating the VU levels to the peaks sent to the tape recorder, particular caution should be exercised to prevent overloading any component in the signal path. It is particularly likely that a condenser microphone placed close to the snare drum will require an attenuation pad between the microphone capsule diaphragm and the microphone's pre-amplifier.

Although the choice and specific placement of individual microphones will depend—as always—on personal taste, the engineer should keep in mind the consequences of the multi-microphone setup. To better isolate the individual drum sounds for maximum control, cardioid microphones are often used. These cardioid microphones should be chosen carefully, however, since the disadvantage of proximity effect and off-axis coloration may outweigh the directional sensitivity advantage of the microphone.

When multi-microphone pickups are used, there is often little physical clearance between the microphones and the various parts of the drum set. It goes without saying that the microphones must be placed carefully where they will not accidentally be struck by the drummer as he moves from one part of the set to another. Several condenser microphone systems offer a series of extension tubes and swivels that allow considerable flexibility in microphone placement when physical clearances are a problem. Additionally, as the drummer moves, there may be a lot of generated air noise; in some cases, it is advisable to place wind screens on some of the microphones, particularly the overheads.

The overall pickup can be a source of potential phase cancellation, particularly in the case of the cymbals, which move considerably each time they are struck. The movement creates continually

Figure 4.26 A typical multi-microphone drum recording setup.

changing path lengths to the two overhead microphones; it may be a good idea to spend a little time listening to the mono mix over the monitor system to make sure there are no serious cancellation problems. Often moving one microphone slightly will clear up the problem.

On many sessions, the front-facing head of the bass drum is removed and a blanket is put inside the drum, sometimes close to, but not touching, the batter head. Although this certainly affects the tonality of the drum, it seems to produce a more percussive attack that many producers like. (A method of restoring the bass drum's tonality is discussed in Chapter 8.) The kick drum microphone may be placed inside the drum and here the proximity effect may be used to advantage.

Electronic Drums Many recordings today are made with synthesized or sampled drums. The sounds are produced by striking either tap keys or pseudo-drums containing triggers, causing prerecorded or synthesized sounds to be sent to the output of the drum machine. Some synthesizers may be programmed to play a specific rhythm by themselves for a specified period of time, and many can be MIDI controlled. The drum synthesizer or sampler may have an output for each drum or a stereo mix of the entire kit. Some give discrete outputs for each instrument, with the exception of a stereo output for the toms alone. These outputs may be high- or low-impedance and balanced or unbalanced, depending on the manufacturer. The engineer is reminded that transformers and direct boxes may be required. Most drummers who use synthesized or sampled drums still prefer the sound of acoustic cymbals, so the drum kit setup often contains a combination of electronic and acoustic instruments. It is sometimes problematic when the overhead microphones, so placed for the cymbals, also pick up the sound of the drumsticks hitting the trigger pads. Judicious use of the null point on a figure-8 microphone can get rid of these annoying sounds.

The Piano To achieve adequate overall coverage of a grand piano, a coincident or near-coincident array of microphones would have to be placed several feet away, due to the instrument's size. Since this may present problems with leakage from other instruments, the engineer will often use two or more closely-placed microphones instead to gain good coverage with maximum isolation. In such cases, some acoustic phase interaction inevitably occurs between the microphones. As before, it is important to monitor microphone outputs in mono prior to recording to make sure that there are no problems later on.

The open lid of the grand piano acts as a reflector for the sound produced inside the instrument and is an integral part of the piano sound. However, it also reflects the sounds of other instruments nearby. A microphone placed somewhere between the strings and the lid will pick up both wanted and unwanted reflections from the lid, as seen in Figure 4-27, and care must be taken to prevent these other reflections from interfering with the overall sound.

The engineer should bear in mind that most piano sound comes from the sounding board; there is thus little point in aiming the microphone(s) at the hammers, as is so often seen. In fact, one of the most resonant areas of the piano lies toward the foot, where the treble strings cross the bass strings.

Figure 4.27 A microphone placed as shown may pick up the reflections of other instruments from the piano lid.

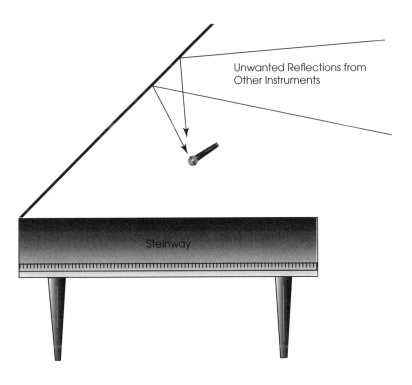

Strings, Brass and Woodwinds

The close-up perspective that so often works well on rhythm instruments (guitars, keyboards, drums, percussion and bass) is much less effective on strings, brass and woodwinds. In most cases, these instruments should sound as if they were spread over a reasonably large area. Pin-point localization of individual instruments is usually undesirable, unless required in a strictly solo sense. Keeping the microphones at some distance will create a much better illusion of listening to an ensemble, rather than to a collection of isolated point-source sounds. Normally, in the studio, these types of instruments are recorded as a group, but at a different time from the rhythm instruments, so leakage is not a problem.

FIVE

Loudspeakers

Schematically, the microphone and the loudspeaker bear a striking resemblance in operating principle if not in outward appearance. In both, the diaphragm is the center of the transducing system. In the microphone, motion of the diaphragm converts acoustic energy into electrical energy. The loudspeaker diaphragm reverses the process, converting electrical energy back into acoustic energy. A cutaway view of a typical moving-coil loudspeaker is shown in Figure 5-1. When an alternating-current signal is applied to the voice coil, the diaphragm moves back and forth, displacing the surrounding air molecules and creating a sound wave that eventually reaches the listener. Due to its structural reliability, the moving-coil system is the basis of practically all studio monitor systems. Before we can begin our analysis of the loudspeaker, some background information on loudspeaker terminology, theory and application is required.

■ LOUDSPEAKER TERMINOLOGY

Resonance   The fundamental frequency at which a tuning fork or taut string vibrates is called its *resonant frequency*. Although the device may actually be vibrating at many frequencies, the opposition of the surrounding air is at a minimum at the frequency of resonance. This frequency is a function of the device's mass and stiffness. All vibrating devices have a resonant frequency, and the loudspeaker is no exception. Depending on the mass and stiffness of the speaker cone and diaphragm/voice coil assembly, this frequency will vary. Generally, the larger the speaker diameter, the lower the resonant frequency.

Transient Response and Damping

Just as a tuning fork continues to vibrate long after it is struck, a speaker diaphragm does not instantly cease its motion the moment the applied signal is withdrawn. Accordingly, its transient response, which is its ability to follow precisely the more percussive waveforms, may suffer. This occurs especially at or near the speaker's resonant frequency. To improve transient response, some form of acoustical, mechanical, or electrical resistance may be applied to the moving system. This opposition, or damping, helps to keep the speaker movement directly related to the applied signal, and in the absence of that signal, the speaker diaphragm is brought to an almost instantaneous halt.

Figure 5.1 A cutaway view of a loudspeaker.

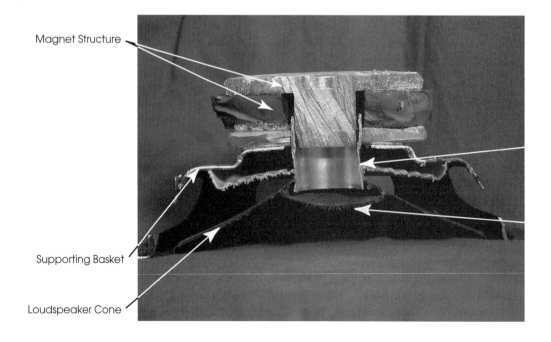

Compliance Compliance, expressed in Newtons per meter (abbreviated N/m), is the ease with which a speaker diaphragm moves. It is measured by dividing the diaphragm displacement by the applied magnetic force. A highly damped speaker will have little compliance, since the diaphragm movement per unit of applied force will be quite small. Without damping, compliance will reach a maximum at the speaker's resonant frequency. As damping is increased, compliance is reduced, and as a result, the resonant frequency of the speaker is likewise raised.

Efficiency As in other physical systems, speaker efficiency is a ratio of power output to power input. In this case, the input power is electrical, while the output power is acoustical. A speaker's efficiency is usually defined as a specific output (in dB), measured at a distance of one meter and derived from an input of one watt.

■ THE IDEAL SOUND SOURCE

In the study of acoustics, reference is often made to an ideal sound source. This is usually considered to be a point source of sound, suspended in free space. It is described as a pulsating sphere of infinitely small dimension. As the sphere releases energy, sound waves radiate away from it in all directions. This is often referred to as radiation into full space (or free space), described in mathematical terms as a radiation angle of 4π steradians. If the point source is considered to be at the center of another larger sphere, the measured sound level will be the same at all points on the sphere's circumference, as illustrated in Figure 5-2a.

Effect of Room Surfaces If our ideal sound source is now moved against a wall, as shown in Figure 5-2b, the radiation sphere will be bisected by the wall. Assuming this to be a solid, the total sound energy from the point source will now be forced to radiate into half-space only, or 2π steradians. Therefore, at any point on the hemisphere, the measured sound level will be 6dB greater than if the measurement was made in full space. (Intensity is analogous to voltage, and therefore its decibel value increases by 6dB when it is doubled). If the point source is now moved into a corner, the intersection of the two walls will cut the radiation angle in half again, and the intensity rises another 6dB, as shown in Figure 5-2c. Finally, if the point source is placed at floor or ceiling level, the intersection of the three surfaces will again halve the radiation angle, and the intensity will rise once more by 6dB, as in Figure 5-2d.

Loudspeaker Radiation (Polar) Patterns   Polar patterns may be drawn for each of the theoretically ideal conditions just described. Although similar in principle to the microphone polar pattern, the loudspeaker pattern is an indication of the intensity radiated from the sound source into the surrounding space. The patterns are shown in Figure 5-3, and the room surfaces that have influenced the radiation are indicated in the illustration. Since both vertical (walls) and horizontal (ceiling or floor) surfaces may affect the radiation angle, patterns are often given in both the horizontal and vertical planes, as seen in the figure.

■ THE PRACTICAL LOUDSPEAKER

As with other theoretically ideal systems, the point source of sound does not exist in reality. Practical loudspeakers may not, by any stretch of the imagination, be considered to be spherical radiators. For example, if we draw a speaker diaphragm schematically, as in Figure 5-4, we can anticipate that it may radiate into the angle shown, rather than into full space. For one thing, the speaker housing gets in its own way, just as the body of a flashlight prevents its light from radiating backwards.

Diffraction   A speaker's radiation angle varies considerably with frequency, for once again the diffraction phenomenon described in Chapter 1 must be considered. The size of the speaker itself becomes an obstacle to small wavelengths (high frequencies), tending to focus them into a narrow radiation angle. On the other hand, long wavelengths (low frequencies) are readily diffracted around the speaker assembly. Therefore, the radiation pattern of a practical loudspeaker suspended

Figure 5.2 a. An ideal sound source radiates into full space (4π steradians); b. when the point source is placed against a wall, the total energy radiates into the half-space (2π steradians); c. the point source is at the intersection of two surfaces and the total energy radiates into quarter space (π steradians); d. the point source is placed against the intersection of three surfaces, so the total energy radiates into one-eighth of the space ($\pi/2$ steradians).

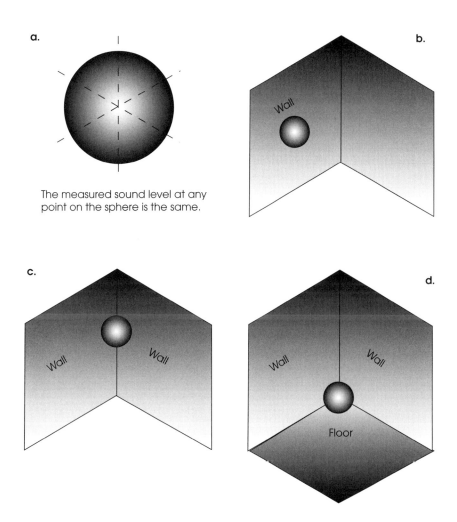

The measured sound level at any point on the sphere is the same.

in free space may be as shown in Figure 5-5. Note that sounds are focused into a progressively narrower beam of sound as the frequencies rise. Provided the listener is directly in front of the speaker, the frequency response is flat, as seen in the illustration. But, as with some microphones, off-axis response may be quite distorted, or colored.

Effect of Room Surfaces In evaluating a microphone, the environment in which it is placed may often be ignored, as ultra-close miking reduces the contribution of the room's acoustics to a negligible factor. On the other hand, the effect of the listening room on the loudspeaker must never be ignored. From our brief study of the ideal sound source, it is clear that speaker placement within the room will greatly affect what the listener hears.

Figure 5.3 Polar patterns for speaker placements in different positions.

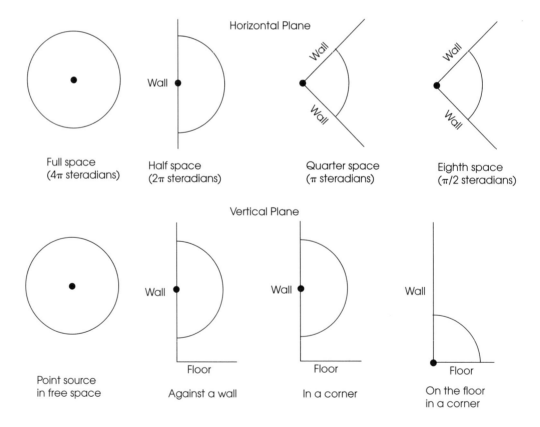

Figure 5.4 The radiation angle of a practical loudspeaker.

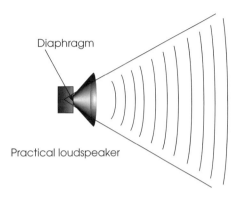

Figure 5.5 Radiation patterns of a practical loudspeaker, showing the effects of diffraction.

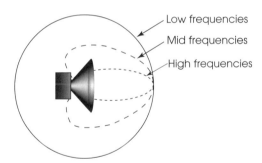

For example, if the loudspeaker represented in Figure 5-6 is placed against a wall, this surface will reflect whatever rear energy there is back towards the front. Since this rear energy is predominantly longer wavelengths (due to diffraction) the apparent on-axis low-frequency response will rise, while the high-frequency response remains unaffected. And if the speaker is moved into a corner, or to the floor or ceiling, the low-frequency response will rise even more. When the speaker is placed at the intersection of three room surfaces, the low-frequency response will be boosted to a maximum. Figure 5-7 illustrates the effect that the room exerts on the speaker for each of these conditions.

SPEAKER ENCLOSURE SYSTEMS

The Direct Radiator   Just as the walls, ceiling and floor influence the loudspeaker's performance, so does the cabinet in which it may be enclosed. Speaker enclosures may be classified as direct or indirect radiators. As its name implies, the direct radiator system is designed so that the speaker radiates directly into the listening room. This result can be achieved using a variety of enclosure designs.

The Infinite Baffle   The infinite baffle, shown in Figure 5-6, is the simplest form of direct radiator, consisting simply of a hole cut in a wall. Assuming that the other room surfaces are sufficiently distant, the speaker's full space radiation pattern is minimally affected, since the speaker is free to radiate much as it would in free space. Of course, the energy that is dissipated in the rear is wasted, from the point of view of the listening room. Furthermore, use of the infinite baffle makes the impractical assumption that a large space behind the speaker wall is readily available.

Figure 5.6   The infinite baffle. The speaker's full-space radiation pattern is minimally affected by the wall.

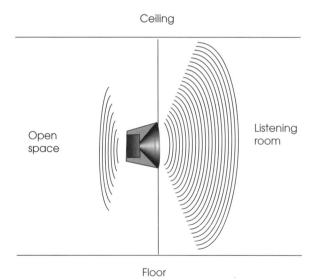

Figure 5.7 Each additional room surface reflects more of the refracted low-frequency energy into the room. The more directional high frequencies may not be affected by the room surfaces.

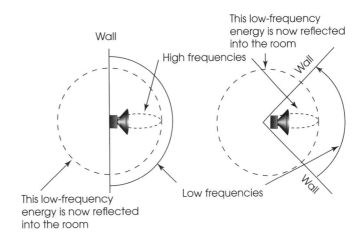

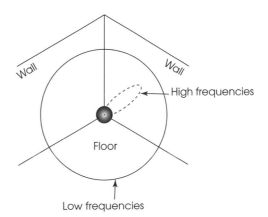

■ **The Open Baffle**   The open baffle system is likewise impractical, but should be discussed at some length since it forms the basis for the practical speaker enclosure. To understand its purpose, consider an unmounted loudspeaker, as shown in Figure 5-8. As the diaphragm moves forward, the air molecules directly in front of it are compressed, while at the same time there is a rarefaction immediately behind the diaphragm. As the diaphragm moves back and forth, sound waves should be radiated away from the speaker, as indeed is the case with high-frequency sounds. However, when the diaphragm is moving relatively slowly (low frequencies), the sound pressure wave on one side of the diaphragm may simply cancel the wave on the other, as seen in the figure.

To prevent this type of low-frequency cancellation, the speaker may be mounted in an open baffle, as shown in Figure 5-9a. Now, sound waves from the rear must travel around the baffle before they are combined with those from the front. The longer the baffle length, the lower will be the frequency at which cancellations begin. The baffle length should equal one quarter the wavelength of the lowest frequency that is desired to be reproduced without cancellation. As seen in Figure 5-9a, a frequency whose wavelength equals four times the baffle length ($\lambda = 4D$) will now arrive at the front of the speaker in phase with the forward wave, producing a maximum reinforcement.

Figure 5.8 An unmounted speaker. As the diaphragm moves forward, a long-wavelength forward pressure wave, P_1, may be canceled by combining with the rear wave, P_2.

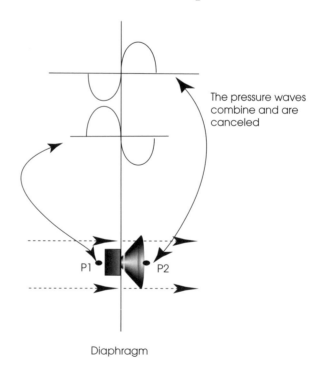

Longer wavelength signals will be progressively attenuated, due to both phase cancellations and the speaker's own inherent low-frequency limit. Frequencies whose wavelengths are somewhat shorter than 4D will also tend to be attenuated somewhat, although as the wavelength decreases, the rear wave has less and less of an effect due to the diminishing radiation angle described earlier.

The frequency response (or rather, wavelength response) of the open baffle system will appear as shown in Figure 5-9b. Note that there is a peak at the frequency where $\lambda = 4D$, for it is at this point that the rear and front waves add to produce a maximum sound pressure level, or resonance peak. Another disadvantage of the

Figure 5.9 Frequency response of an open-baffle system.

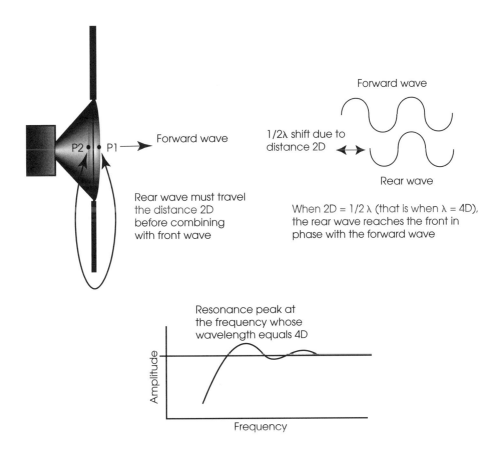

open baffle is the dimension required for good low-frequency response. A little arithmetic will reveal that the quarter wavelength of 50Hz is about 5.5 feet. Therefore, an open baffle tuned to this resonant frequency must be 11 feet in diameter, that is, 5.5 feet in radius from the speaker cut-out; hardly a practical dimension!

■ The Folded Baffle

The folded baffle seen in Figure 5-10 offers some relief in size over the open baffle system. By folding the baffle as shown, it is possible to achieve a smaller enclosure without raising the resonant frequency described earlier. However, when the sides take on any appreciable dimension, the speaker cabinet acquires the acoustic properties of an open column of air. That is, it produces a resonance at the frequency whose wavelength equals four times the length of the column created by the baffle sides. Therefore, if the baffle sides are 1 foot long, the enclosed air column will resonate at 275Hz, while the baffle itself will continue to resonate at $\lambda = 4D$, as described earlier.

■ The Sealed Enclosure, or Acoustic Suspension System

The sealed enclosure, shown in Figure 5-11, eliminates the acoustic resonance peaks of the open and folded baffle systems. This elimination occurs because it is impossible for rear waves to reach the front of the speaker. In addition, the enclosed volume of air acts as an acoustic resistance (damping) against the rear of the speaker cone, thus lowering the system's compliance. As the compliance is reduced, the speaker's resonant frequency is raised as was noted earlier. Depending on the design parameters, the resonant frequency of the speaker/enclosure system may be an octave or more above the resonant frequency of the speaker itself in free space.

Since in the sealed enclosure the speaker is, in effect, resting on the enclosed air column, it is often referred to as an acoustic suspension system. A representative example is shown in Figure 5-12. The sealed enclosure may also be called an infinite baffle, since, like the true infinite baffle described earlier, the system prevents rear waves from reaching the front of the speaker. However, the true infinite baffle has an unlimited air space behind it, while the sealed enclosure does not.

Figure 5.10 The folded baffle. Acoustically, the enclosure resembles an open pipe and may resonate at λ = 4L.

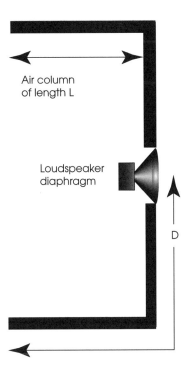

The Vented (Bass Reflex) Enclosure

One limitation of the acoustic suspension system is that the rear waves are trapped within the enclosure. Consequently, the speaker's efficiency rating is reduced since this portion of the total output power is lost. The vented enclosure, shown in Figure 5-13a, allows the rear waves to reach the front, thus improving the system's efficiency.

The Tuned Port

The volume of air within an enclosure has a certain compliance. That is, it acts as an acoustic capacitance, affecting the system's resonant frequency as was discussed earlier. If a port is cut into the enclosure, the opening allows sound waves to escape. However, the port has a certain acoustic inertness, analogous to inductance in an electrical circuit. The system becomes, in effect, a capacitance (the enclosure) in parallel with an inductance (the port). The dimensions of both may be designed to create an acoustic resonance at a specified frequency. It has been found that if this frequency is the same as the speaker's own resonant frequency, the system resonance will drop in amplitude. In fact, it will exhibit two resonance points, one above and the other below the original speaker resonant frequency. As a result, the usable low-frequency response of the system is extended downward. However, due to open baffle-type cancellation effects, the low-frequency attenuation below resonance will probably fall off at a sharper rate than in the sealed enclosure system.

In the open baffle systems described earlier, baffle dimensions had to be quite large to keep phase cancellations at as low a frequency as possible. In the vented system, the acoustic phase shift of the enclosure and port accomplishes the same thing, and low frequencies (above resonance) emerge from the port in phase with the front of the speaker, as is illustrated in Figure 5-13b. An example of a vented enclosure system is seen in Figure 5-14. This type of enclosure is also often referred to as a phase-inversion enclosure.

Figure 5.11 The sealed enclosure, or acoustic suspension, system.

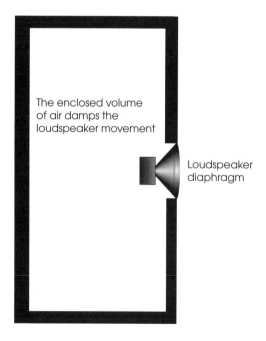

Figure 5.12 An acoustic suspension system (BM6, Dynaudio Acoustics photo).

Figure 5.13 a. The open port allows rear waves to reach the front of the system; b. rear waves reach the front.

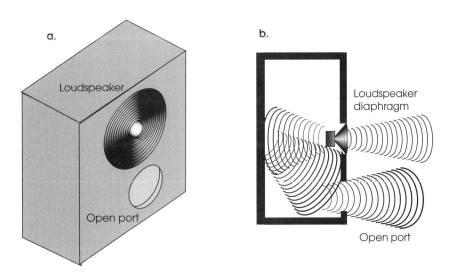

■ The Vented Enclosure With Passive Radiator

As sound waves travel through the open port just described, a certain amount of acoustic friction acts against the air particles. The friction is greatest at the edge of the opening, diminishing toward the center. To make the movement of the air particles more uniform across the entire opening, a passive radiator may be placed in the port. Sometimes called a slave or drone cone, the passive radiator may be nothing more than a non-powered speaker assembly. The air pressure against the back of the passive radiator forces it to move in and out. In the front, a sound wave is radiated that is more uniform than one from a simple open port. Figure 5-15 shows a vented enclosure system using a passive radiator.

■ Efficiency of the Direct Radiator

A direct radiator is a speaker that is coupled directly to the air in front of it. Although at each stage in the development from infinite baffle to vented enclosure with passive radiator, the efficiency of the acoustic coupling of diaphragm-to-air increased, the direct radiator still remains a relatively inefficient system. The poor efficiency is a function of the high mechanical impedance of the speaker diaphragm attempting to transfer energy to the low acoustic impedance of the surrounding air.

Figure 5.14 A vented enclosure system. In this speaker system, several small ports are used to prevent structural weakening of the front baffle (Dynaudio Acoustics photo).

Figure 5.15 A vented enclosure with a passive radiator.

In an electrical circuit, when maximum power transfer is required, an impedance-matching transformer may be used between any two stages where there is a significant impedance mismatch. For example, without such a transformer, a high-impedance microphone will function with minimum efficiency when connected to a low-impedance microphone pre-amplifier. In a direct radiator speaker system, the diaphragm-to-air coupling is pretty much a "brute force" method. A high electrical input power applied to the speaker terminals produces a low acoustical output power in front of the diaphragm. Even a so-called high-efficiency direct radiator system may have an efficiency of not much more than 5 percent.

The Indirect Radiator (Horn-Loaded Systems)

In an indirect radiator, the speaker diaphragm is coupled to the surrounding air by a device that functions as an acoustic transformer. The horn system shown in Figure 5-16a is a well-known example of this type of indirect radiator. Note that the speaker diaphragm with a horn in front of it is no longer able to radiate into a large air mass. The throat of the horn is considerably smaller than the diaphragm diameter, and like the action of water being forced into a narrower diameter pipe, there is a resultant high-pressure area developed at the beginning of the throat. This high acoustic impedance is closer to the mechanical impedance of the speaker, resulting in a more efficient transfer of power.

As the sound wave travels the length of the horn, the horn's gradually flaring shape disperses the sound pressure over a progressively larger cross-sectional area. Eventually, the sound wave reaches the open air, by which time its pressure has been reduced substantially. The original high-pressure/high-impedance sound wave has become a low-pressure/low-impedance wave front. The

Figure 5.16a The horn acts as an acoustic impedance-matching transformer.

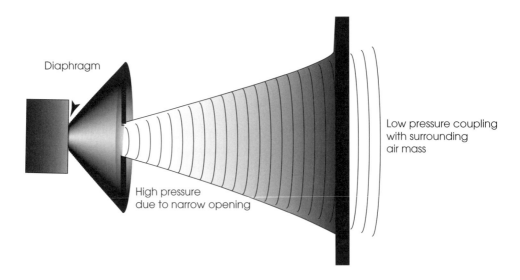

transformer action of the horn results in a more efficient coupling with the surrounding air. As a result, the sound level will be considerably higher than with a direct radiator system. This type of indirect radiator is often referred to as a horn-loaded system. Figure 5-16b shows a drawing of a horn of this type. Note the narrow diameter throat and large flare of the horn.

■ **Radiation Characteristics of Horn-Loaded Systems**

Although the horn-loaded indirect radiator system has a higher efficiency than a direct radiator, its radiation pattern tends to narrow appreciably at high frequencies, as the neck and inner surfaces of the horn focus the high frequencies into a narrow beam, as shown in Figure 5-17.

Recent developments in horn-loaded indirect radiators have produced a new horn called the BiRadial horn. This horn provides constant coverage with respect to frequency from a crossover point of about 1000Hz to beyond 16kHz, and has a 100 degree by 100 degree coverage pattern. This radiation pattern is more closely related to that of a direct radiator mounted in a proper enclosure, and as such can provide a two-way horn-loaded system with relatively wide coverage angles in all planes. A horn of this type is shown in Figure 5-18.

■ **Folded Horns**

A straight horn designed for low-frequency operation would be quite long, and physically impractical. However, the horn may be folded as shown in Figure 5-19 to fit within a physically manageable area. The long-wavelength efficiency of such a design is excellent, since the diffraction phenomenon allows low frequencies to readily pass through this acoustic labyrinth. However, high-frequency output will be very inefficient, and a separate high-frequency system will be required. The speaker system shown in Figure 5-20 is a dual folded-horn bass loudspeaker system.

■ **The Compression Driver**

The compression driver is a special-purpose transducer designed to be used in mid- and high-frequency horn systems. The unit shown in Figure 5-21 has a 4-inch diameter aluminum voice coil/diaphragm and a 2-inch diameter throat. The relatively small diaphragm has a high resonant frequency and is generally more efficient in horn-loaded applications than a massive speaker cone assembly. The comparatively narrow throat diameter is designed for maximum efficiency coupling with the horn assembly.

Figure 5.16b A representative horn design.

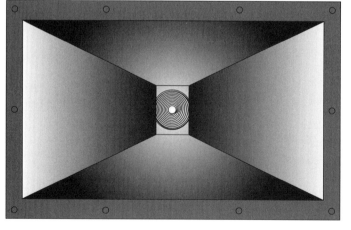

Figure 5.17 The tapered horn tends to narrow the radiation angle of high frequencies.

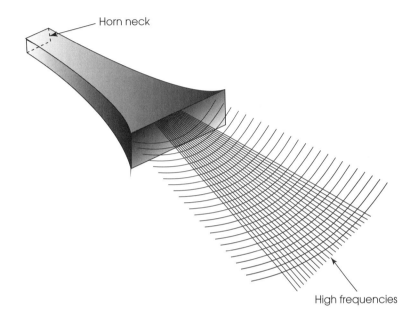

Figure 5.18 A BiRadial horn (JBL 4430; JBL photo).

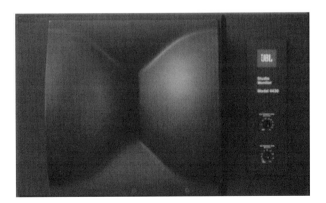

Figure 5.19 Cutaway view of a folded horn system.

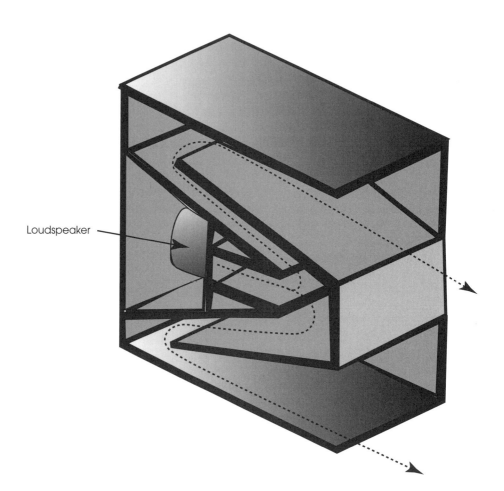

Figure 5.20 A speaker system using the folded horn principle.

Figure 5.21 Cutaway view of a compression driver (JBL 2440; JBL photo).

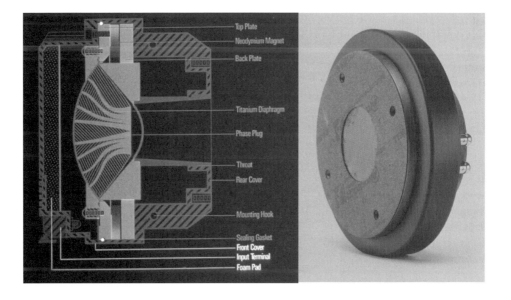

■ MULTI-SPEAKER SYSTEMS

Like most musical instruments, a given loudspeaker may be most efficient over a portion of the audio bandwidth. For example, a family of stringed instruments ranging in size from the physically small violin to the oversized double bass is required to cover the entire musical spectrum. Likewise, a small high-frequency speaker may be incapable of reproducing low frequencies, and vice versa. Accordingly, many studio speaker systems employ two or more speakers, each optimized for a specific segment of the audio bandwidth. In its simplest form, a wide-range system may comprise a high-frequency tweeter and a low-frequency woofer. (The terms describe accurately, if irreverently, the sound output of the two speakers.) These systems may or may not combine direct and indirect radiators.

Crossover Networks

In most cases, a crossover network is used with a multi-speaker system. In a two-way system, as shown in Figure 5-22, the crossover uses high-pass and low-pass filters to route the signal to the proper radiator, where in a three-way system, a band-pass filter is added for the mid-range. The network passes only the appropriate frequency band to each speaker, while suppressing frequencies outside the band. Thus, no speaker receives unusable power from a signal outside its own pass-band. In addition, the tweeter, and mid-range if applicable, is protected against overload from high-level low-frequency signal components.

Crossover Phase Distortion

In Figure 5-22, note that there is some overlap in the frequency passbands, and that the output from both speakers is down 3dB at the crossover frequency. Therefore, both speakers deliver the same amount of power at the crossover frequency, when in fact this is most often not the case. Each filter introduces a phase shift as it attenuates frequencies outside its bandwidth. In one typical situation, each filter shifts the output signal by 90 degrees at the crossover frequency. However, the phase shifts are in opposite directions, giving a net phase shift of 180 degrees between speakers at the crossover frequency. Therefore, there will be a severe acoustic phase cancellation in this region unless one of the speakers is electrically reversed in phase. This reversal puts the speakers

Figure 5.22 A two-way crossover network.

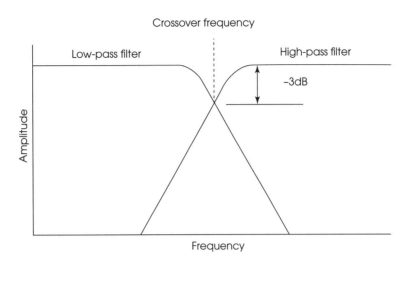

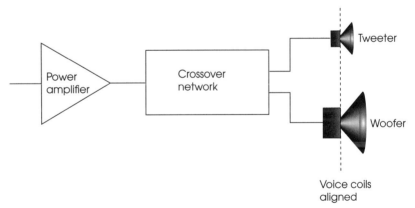

back in phase at the crossover frequency and out-of-phase within their pass bands. However, since the speakers are only reproducing the same signal in the vicinity of crossover, there will be little cancellation over the rest of the audio bandwidth. Of course, if the phase shift at the crossover frequency is not exactly +/- 90 degrees, the wiring reversal will be ineffective.

Acoustic Phase Shift

The acoustic center of a loudspeaker is the point at which sound waves appear to originate. Especially in horn-loaded systems, the acoustic center may be somewhat removed from the actual center line of the diaphragm at rest. When more than one speaker is to be used in an enclosure, it is important that their acoustic centers all lie in the same vertical plane, to prevent acoustic phase cancellations in the vicinity of the crossover frequency. These cancellations cause poor imaging and increase listener fatigue, and are often called "time smear." If the electrical phase shift at the crossover frequency cannot be corrected by the wiring reversal described above, one of the speakers may be moved slightly so that its acoustical center is shifted by the distance required to cancel out the crossover network's electrical phase shift. Another method called Time-Align™ electrically adjusts the crossover to compensate for phase cancellations caused by speaker displacement. Studies by Blauert and Laws have determined the minimum amount of acceptable time delay discrepancy, and most manufacturers today strive to keep their systems below this limit.

The Coaxial Speaker

Many loudspeaker systems are of the two- or three-way variety and, by eliminating time smear, achieve excellent results. However, as mentioned earlier in this chapter, the ideal sound source is a single point in space radiating equally in all directions. Obviously, no loudspeaker system can truly be a single point radiator, but by mounting two radiators together on the same speaker frame, this ideal is approached. This loudspeaker, called a coaxial loudspeaker, uses two concentric radiators—one mounted inside the other—to cover the audio spectrum. The system may use two direct radiators—one for the treble range and one for the bass—or may contain a direct radiator for the low frequencies and an indirect radiator or horn-loaded system for the highs. The system can be mounted in any of the previously described enclosures, and may or may not use a passive radiator depending on the individual manufacturer's design. A typical coaxial loudspeaker is shown in Figure 5-23.

Multi-Speaker and Coaxial System Characteristics

Excellent results can be obtained from both multi-driver speaker systems and from coaxial speaker systems. However, both systems have advantages and drawbacks. In coaxial systems, both high and low frequencies radiate from the same physical area and have a

Figure 5.23   A coaxial loudspeaker system with an indirect radiator for the highs (Tannoy Coaxial; Tannoy/APK drawing).

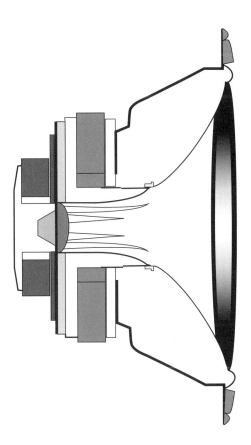

tendency to interfere with each other. This is called intermodulation distortion. On the other hand, the coaxial system maintains the best imaging since the entire audio spectrum is radiating from the same point. Even though a multi-driver system is phase aligned, the distance-to-listener relationship between the speakers shifts with the vertical movement of the listener. Another consideration is the overall dispersion angle of the system. Because most coaxial speakers use a small compression driver and horn for the upper frequencies, the overall coverage angle of the system is relatively narrow. This is particularly true in the vertical angle, so placement of most coaxial systems must be very precise.

Bi-amplification In a bi-amplification system, the tweeter and the woofer are driven by separate amplifiers, with the crossover network placed before the amplification, as shown in Figure 5-24. When a single amplifier drives a passive crossover network, the network itself must be capable of sustaining the full power output of the amplifier. In addition, harmonic distortion components of the low-frequency output may be delivered to, and reproduced in, the high-frequency output section.

On the other hand, an active crossover network placed ahead of the amplification completely isolates low- and high-frequency components. Distortion in one side will not be transmitted to the other. And, since woofers characteristically require more power than tweeters, a lower-power amplifier may be used in the high-frequency leg of the system. In fact, a well-designed bi-amplified system will require less total power than the same speaker system driven by a single amplifier.

Just as bi-amplification has increased the efficiency and accuracy of two-way systems, tri-amplification is beneficial to three-way monitor systems. However, bi- and tri-amplification are not possible with coaxial radiators, but additional amplification could be used to drive a supplementary woofer.

Powered Loudspeakers The next logical step after bi-amplification is to place the amplifiers in the loudspeaker enclosure itself. These are often referred to as powered or active loudspeakers. The advantages are many. First, the amplifier for each speaker can be optimized specifically for that driver. Additionally, the length of cable between the amplifier and loudspeaker is very short, thereby permitting maximum power transfer and minimum loss of signal. The console +4dBm-balanced monitor outputs are simply fed directly to the loudspeaker. The loudspeaker system pictured in Figure 5-25 utilizes a 9.5-inch woofer with a 4-inch voice coil, and an approximately 1-inch tweeter with a 1.7kHz electronic crossover, and is capable of a frequency range from 40Hz to 21kHz. The unit contains amplifiers with over 200 RMS watts of power and is capable of 120dB at a distance of three feet. The total package only weighs 41 pounds, and is therefore very portable. Often a unit such as this will be paired with a matching powered subwoofer, as shown in Figure 5-26, to provide a large full-range (20Hz to 21kHz) loudspeaker system. As more control rooms feature 5.1-channel and 7.1-channel surround systems, these smaller active monitors and subwoofers are becoming more widely used.

Figure 5.24   A bi-amplification system. The crossover network is placed ahead of the amplification.

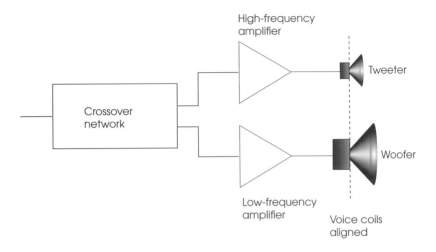

Figure 5.25   A powered studio monitor system (BM15A, Dynaudio Acoustics photo).

Figure 5.26 A powered studio subwoofer system (BX30, Dynaudio Acoustics photo).

■ THE ROOM/SPEAKER INTERFACE

As we have already seen, room surfaces have a considerable influence on loudspeaker performance. This influence, however, extends far beyond the effects on bass response that have already been discussed. Earlier, it was seen (Figure 5-7) that nearby surfaces reflect low-frequency energy back into the room. As the sound energy radiates away from the speaker, it eventually reaches the more distant surfaces. These in turn reflect some portion of the sound wave back into the room. The listener hears not only the direct sound coming from the speaker, but also the multiplicity of reflections from all the room surfaces. In fact, in most cases, the amplitude of the reflec-

Figure 5.27 The listener hears mostly reflected signals, although the direct sound does arrive first.

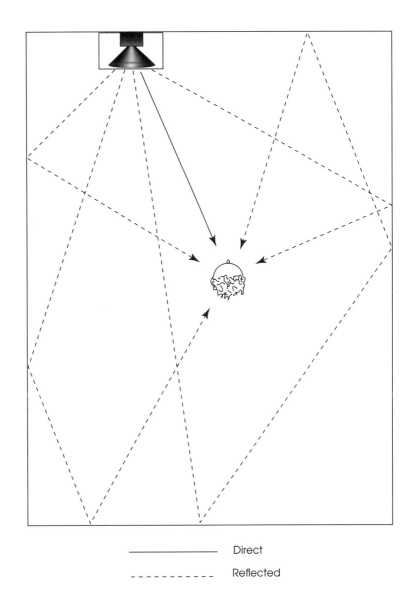

tions near the listener's ear is greater than that of the direct sound, as shown in Figure 5-27. Therefore, the character of these reflections will play a very important part in the listener's impression of what he or she hears.

Room surfaces are neither perfect reflectors nor perfect absorbers. The frequencies that are efficiently reflected—or absorbed—are influenced by surface textures and density, the air space behind the surfaces, rigidity, and so on. A change in any one of these parameters, in fact, may very well exert an influence on the overall sound that is greater than would be heard if the speaker system itself was replaced with a different type. Depending on the complex interaction between the speaker and the room, the overall sound quality may be described as bright, dull, live, dead, boomy, shrill, or by any of a seemingly endless number of similar adjectives. In other words, the room becomes part of the monitor system, and its effect must be taken into consideration when evaluating any speaker.

Standing Waves

When a sound wave strikes a wall and reflects back into the room, there may be areas in the room where the direct and reflected waves interact to form a stationary waveform, or standing wave. The phenomenon may be quite apparent when listening to a single frequency tone from an audio signal generator. To explain the standing wave, consider a sound wave reflected back on itself, as shown in Figure 5-28. The sound source, S, eventually strikes the wall, W. The listener, L, standing somewhere on the line S-W, may hear either a strong signal or none at all, depending on the wavelength of the signal and the listener's distance from the wall. When the round-trip distance from L to W is a whole number multiple of the wavelength, the reflected sound wave reinforces the direct wave, and the listener hears a strong signal. But when the distance brings the wave back shifted by a half wavelength, the waves cancel and the listener hears nothing.

The first dead spot will be 1/4 wavelength from the wall; there will also be other dead spots at half-wavelength intervals (3/4λ, 1 1/4λ, 1 3/4λ, etc.) as the listener moves away from the wall. In between these dead spots will be points at which the waves always reinforce each other. Since these live and dead areas remain stationary, the condition is known as a stationary, or standing, wave.

Standing waves make it difficult to impossible to intelligently evaluate what one hears over the monitor system. For example, the producer, although sitting next to the engineer, may hear an entirely different balance due to the different cancellation and reinforcement patterns at each location. An effective means of minimizing standing waves is to construct non-parallel wall surfaces in the listening room, as shown in Figure 5-29. In the illustration, sound waves striking the walls are reflected back into the room at an angle, thus helping to minimize the build-up of any standing waves.

Wall Treatment As noted earlier, the major portion of the sound energy within the room is made up of reflected signals. If there is an excessive amount of reflection, clarity will be sacrificed, and it may be difficult to pinpoint the actual location of each sound source. On the other hand, when wall treatment reduces the reflection to a minimum, the room becomes acoustically lifeless. Many well-designed control rooms have adjacent wall surfaces that are alternately reflective and absorptive. The reflective walls give the room a reasonable amount of liveness, while the absorptive surfaces prevent multiple reflections from interfering with clarity.

Room Resonance Modes Just as each speaker system enclosure has its characteristic resonance point, the listening room itself will have certain resonances, or room modes. These occur at frequencies whose wavelengths are twice one of the room's dimensions. Thus, a room that is 15 feet in length will have a main resonance frequency of $F = V/\lambda = V/2L = 1100/30 = 36.67$Hz. If the room is ten feet wide, it will have a secondary resonance at $1100/20 = 55$Hz. Not unlike the other speaker enclosures discussed earlier, the room enclosure will attenuate frequencies whose wavelengths exceed 2L. In other words, an extended low-frequency response is a function of both the design of the loudspeaker and the dimensions of the room. Regardless of the speaker system in use, it is unlikely that a small room will have a satisfactory bass response.

Room Dimension Ratios The length, width and height of a listening room should be unequal to avoid creating the same resonance frequency in two dimensions, which would accentuate it even more. In addition to the fundamental resonance frequencies of the room's dimensions, there may also be resonances at the harmonics of those frequencies. Therefore, length-to-width-to-height ratios should be chosen to avoid common harmonic resonances in two or more dimensions. Over the years, several "ideal" ratios have been proposed by acoustical engineers. An often-quoted formula is the "golden section," (1:1.62:2.62) first recommended by the ancient Greeks, and still much favored today. Needless to say, there is very much more to room acoustics than choosing the right ratio. The point is to try to avoid those dimension ratios that will obviously cause room mode problems.

Figure 5.28 The creation of standing waves. As the wall reflects sound energy back into the room, there are points where the direct and reflected waves always reinforce each other, and other points where they always cancel.

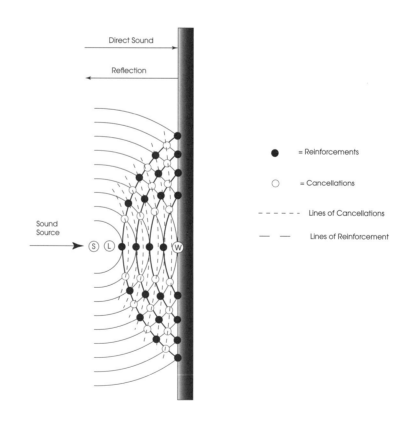

Figure 5.29 Non-parallel surfaces help minimize standing waves.

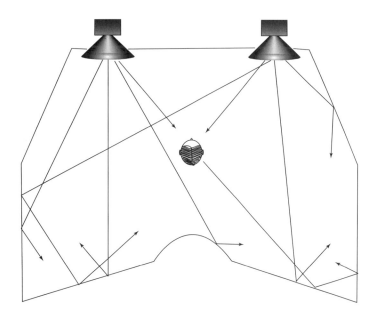

System Power Requirements

It is a relatively easy task to determine the amount of power required to reach a specified monitoring level in a given room. As mentioned earlier, most loudspeaker systems state an efficiency rating based on an input of one electrical watt. From this information, using the inverse square law and the power formula from Chapter 1, the amount of amplification power for a given system can be calculated.

For example, a loudspeaker system is rated at a sensitivity of 99dB for an input of 1 watt at a distance of one meter. The loudspeaker specifications may abbreviate this as 99dB/1w/1m. If we want to listen, say, at a distance of four meters at a maximum sound level of 107dB, we can calculate the required amplifier power required as follows:

Original specification	99dB	1 watt	1 meter
Double the Distance (Inverse Square Law)	93dB	1 watt	2 meters

Double the Distance	87dB	1 watt	4 meters
10 times the power (10log 10/1 = 10dB)	97dB	10 watts	4 meters
10 times the power	107dB	100 watts	4 meters

This, of course, does not take into consideration the fact that peak levels may be as much as 9dB above the required listening level, in which case additional power will be needed. Actually, an additional 3dB of level is usually obtained since most monitor loudspeaker systems are used in stereo pairs.

| 2 times the power (10log 2/1 = 3dB) | 110dB | 100 watts (each system) | 4 meters |

Even so, for our system to accurately reproduce peak levels of 116dB, 400 watts (two more doublings) of power per loudspeaker system should be available.

If a system containing a compression driver and horn for the high frequencies and a large woofer for the low frequencies is bi-amplified, two smaller power amplifiers may be used. As mentioned earlier, a bi-amplified system with an active crossover may require less total power than the same system with a passive crossover driven by a single amplifier. Let us say that we are using a woofer with a sensitivity of 102dB/1w/1m and a compression driver/horn combination rated at 112dB/1w/1m. For the same listening level as above (116dB at 4 meters with both speakers driven), the power requirement for the lows is 200 watts per channel, while only 20 watts per channel is required for the highs.

Although the described systems are capable of these extremely high levels, it is not advisable to listen at such levels under normal conditions. The above examples show the maximum output possible for the listed systems including a headroom factor. Monitor levels in the 85dB range are closer to the norm, which allows a headroom of at least 15dB above the average level, making these systems capable of handling any peak information that may come along. In fact, the recommended monitoring levels in use today are 85dB for large rooms and 79dB for small ones. This specification applies to both stereo and surround monitor systems.

Room Equalization

Room equalization is the practice of tailoring the frequency response of the signal delivered to the speaker to correct for certain frequency anomalies created by the room. For example, if the room modes create a substantial resonance peak at 100Hz, a filter in the speaker line may be tuned to attenuate the power output at this frequency, thus flattening out the system response. Peaks throughout the audio spectrum may likewise be attenuated, and to a lesser extent, slight dips in response may sometimes be brought up. Most room equalizers are of the graphic type and have frequency centers spaced one-third octave apart. This type of equalizer will be explored in greater depth in Chapter 7 on frequency domain processing.

As may be expected, the room equalization process is ineffective against standing wave cancellations, and in general should not be used in an attempt to cover up serious deficiencies in room design. Recall that analog equalizers may create as many problems as they solve due to phase shift error. Today, it is much better to use some of the excellent available physical acoustic modification tools such as diffusers, reflectors and absorbers, rather than to make use of electronic frequency adjustment with its associated phase problems.

System Power Requirements

With the advent of surround sound for home theaters, audio has moved into a new phase. Due to the popularity of these systems, wide distribution of multi-channel audio media is imminent, and while setting up a good stereo monitoring system in a good environment was once relatively easy, additional problems occur when the engineer is faced with surround.

At this time, there are two potential standards for surround monitoring: 5.1 and 7.1. The most common is the 5.1 which uses five full-range loudspeakers designated as the left, right, center, left surround and right surround. The .1 (dot one) channel is a dedicated subwoofer channel used mainly for special effects in movie soundtracks. There are very specific requirements for loudspeaker placement with these systems. With the center channel directly in front, the left and right loudspeakers should be 30 degrees off center for a spread angle of 60 degrees. The left surround and right surround should be behind the mixing position between 110 degrees to 120 degrees from the center axis. Figure 5-30 shows a typical control room setup. The 7.1 system adds two additional loudspeakers between the L&R and the LS&RS, or between the L+C and R+C.

One consideration is that some engineers make recordings designed to be used with direct radiators as the surround loudspeakers, while others favor the use of diffuse sound loudspeakers. The diffuse variety can be either bi-polar (figure-8) or dipolar with two positive lobes. There are even some tri-poles currently available. These three types of diffuse surround channels are pictured in Figure 5-31. Diffuse field loudspeakers are often placed at a 90-degree angle to the listener's axis instead of the 110-degree location of the direct radiator. This places the null of the bi- or tri-pole array on line with the listener's ear, thus providing as little direct sound as possible in that direction.

■ SUMMARY

Loudspeaker performance is a complex function of speaker design, enclosure style, and the room in which the system is placed. In effect, the room becomes part of the speaker system and has a profound effect on what the listener hears.

While microphones may be (and are) moved or replaced with ease, the speaker and its location tend to become somewhat of a permanent fixture in the control room. The engineer who insists that only his favorite speaker tells what's "really on the tape" is no doubt blissfully unaware of all the variables at work in his control room that help him form his impression of the perfect system. One need only hear the same program over the same speaker in another room to realize that the speaker itself is but a small part of the total monitoring system. Some typical monitoring loudspeaker systems, both large and small, are shown in Figure 5-32.

Figure 5.30 A 5.1 surround monitoring system (LFE not shown).

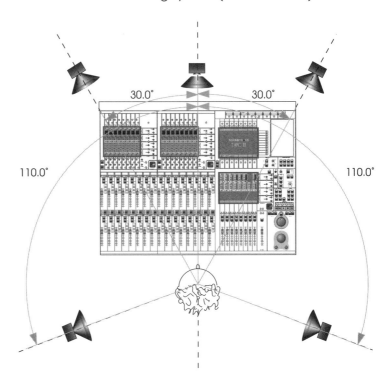

Figure 5.31 Bi-pole, di-pole, and tri-pole loudspeakers.

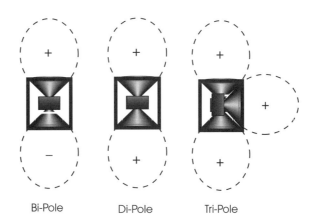

Figure 5.32 Several typical monitor loudspeaker systems: a. Nearfield powered studio monitors (Tannoy photo); b. powered midfield studio monitors with powered subwoofer; c. JBL 4430 two-way system and 4435 two-way with dual woofers; d. Tannoy DMT15 coaxial systems.

SECTION III

Sound Processing Devices

■ INTRODUCTION

The path between the microphone and the loudspeaker is rarely an electrically straight line. Along the way, the signal may pass through one or more processing devices designed to make the sound more, or perhaps less, realistic.

Time domain processing such as artificial echo and reverberation may be applied to simulate the concert hall, or to create a new and otherwise unobtainable effect. Frequency domain manipulation may subtly correct, or deliberately distort the natural sound and response of an instrument or group of instruments. Amplitude domain processing includes compression, limiting, and expansion, which not only allow us to shape the dynamic range of a signal, but give us the basis for most noise reduction systems.

The beginning of Chapter 6 deals with time relationships and the creation of reverberation systems, while the last section talks about time related effects such as phasing, flanging, chorusing and harmonizing. Chapter 7 discusses frequency processing and looks at equalizers and filters. Chapter 8 shows how we can change the dynamic range of a signal, either to fit the storage medium or as a corrective measure, while Chapter 9 takes this dynamic processing into the realm of noise reduction.

In addition to the more traditional signal processing equipment found in the modern recording studio, many multi-use signal processing devices have appeared. They may combine amplitude and frequency domain processing or time and frequency domain processing, or even allow signal manipulation in time, frequency and amplitude domains singly or in combination. Since they do not fall into any one particular category, but have operating parameters similar to the more traditional devices, a special section discussing this type of equipment is included at the end of Chapter 8.

SIX

Time Domain Processing: Echoes and Reverberation

■ ROOM ACOUSTICS

With the exception of an anechoic chamber, all monitoring environments have some effect on what the listener hears. The perceived sound of a musical instrument will vary from one room to another, and an acoustically dead studio will never be confused for a concert hall with its superb natural acoustics.

When the listener is at a reasonable distance from a musical instrument, he hears something quite different from the direct output of the instrument. Although some sound certainly reaches the ear by way of a direct path from the instrument, other signals are also present as the sound radiates from its source to the various room surfaces of walls, ceiling and floor. Each of these surfaces absorbs some portion of the sound and reflects other portions back into the room. Consequently, the listener hears an incredibly complex mixture of direct and reflected information with properties that vary according to where he or she is seated. If very close to the sound source, the listener will hear mostly the direct signal. On the other hand, the listener at the rear of a large concert hall may hear a signal that is made up of almost totally reflected sounds.

The character of the reflected sound is influenced by the construction of the room surfaces and the room's volume. Some materials are more reflective than others, and all have some influence on the frequency response and dynamic range of the reflected signal. Carpeting, for example, tends to absorb higher frequencies, while glass may efficiently reflect these same frequencies. In the older concert halls one often finds paneling, thick plaster surfaces and parquet wood floors, all of which reflect sound and contribute to that illusive concert hall realism. On the other hand, the surfaces of a well-designed popular music studio may reflect very little sound back into the room. Because so much recording is done with close microphone placement, the microphone "hears" a signal that is, for all practical purposes, only the direct sound of the instrument. Figure 6-1 shows three examples of room acoustic environments.

Whether the primarily direct pickup is due to close miking, a dead room, or a combination of both, the resultant sound is usually described as "dry," or "dead," or by some other subjective term that suggests the absence of reflected information. As a corrective measure, some sort of signal processing is often required to simulate a more natural sound. Or, for a special effect, an apparently greater than normal amount of reflected information may be required. The engineer of today has many tools at his disposal to either simulate the ambience of a concert hall, or to create a unique effect that would otherwise be unattainable naturally.

Before describing these methods, a few definitions and a brief discussion of room acoustics are required:

Echo: one, or at most a few, repetitions of a sound.

Reverberation: any repetitions of a sound, becoming more closely spaced (denser) and diminishing in amplitude with time.

Delay: the initial time interval between a direct sound and its first echoes.

Decay: the time it takes for the sound and its echoes and reverberation to die away.

The Late Sound (Reverberant) Field

The total energy present within any listening environment is a mixture of three components: 1) the original or direct sound, 2) the early reflections (echoes), and 3) the later, more diffuse reflections or "reverberation." These three components are illustrated in Figure 6-2. Although there is of course only one direct path from the sound source to the listener, there will be several paths for the early reflections, and many more for the later reflections within the typical music listening room. A graphical representation showing the distribution of a typical sound field, related to time with respect to amplitude, is shown in Figure 6-3.

Figure 6.1 a. Griswold Hall, Peabody Conservatory of Music, Baltimore, MD. The hard plaster walls, wooden floors and large volume make the room quite reverberant. There is an area above the acoustically transparent ceiling that is nearly as large as the lower part of the hall; b. Friedberg Concert Hall, Peabody Conservatory of Music; a medium-large hall designed for small orchestras and chamber music recitals; c. Peabody Recording Studio 220; A small studio designed for minimal reverberation.

First, the direct sound (1), is heard. After a very short interval (T1), a repetition is heard as the direct sound is reflected from a nearby surface and reaches the listener's ear. Still later (T2) another repetition is heard (probably from a different direction), and then (T3), perhaps another, as the sound reflects from other nearby surfaces. As time passes, more and more echoes and repetitions of echoes occur as the sound waves spread through the listening environment, striking and being reflected from the myriad surfaces within the room. Now, at (T4, T5, T6, ...), the reflections have become so closely spaced that they fall almost on top of one another and the listener is no longer aware of their individual identities. These multiple repetitions of echoes are known as the late-sound

Figure 6.2 Echo and reverberation within a room. The listener hears a combination of: 1. Direct sound; 2. Early reflections (echoes); and 3. Later reflections (reverberation).

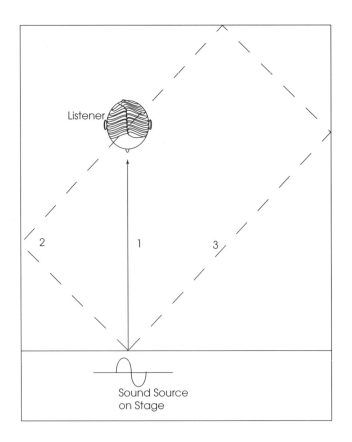

field or reverberant field. Reverberation time is defined as the period of time it takes for the original sound pressure level to drop to one-millionth of its original intensity. If we were to convert that figure to a decibel amount using the formula for acoustic power (Chapter 1),

$$\text{NdB} = 10\log \frac{P}{P_{ref}} = 10\log \frac{1}{.000001} = 10\log 1{,}000{,}000 = 10\log 10^6 =$$

$$10(6) = 60\text{dB},$$

the equivalent drop in level would be 60dB. Therefore, it can be said that reverberation time is the time it takes the reverberation to diminish 60dB from its original level, hence the popular representation T60.

Figure 6.3 Graphical representation of a sound field, showing direct sound, echoes, and reverberation.

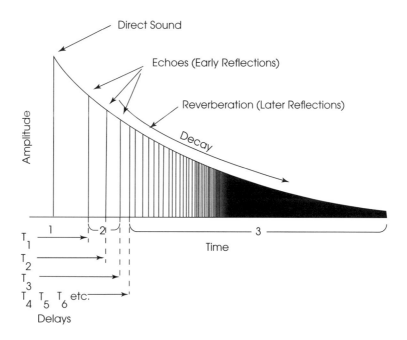

As a matter of fact, even the very earliest reflections (T1, T2, T3) that make up the early sound field are seldom consciously noted as such. Their presence is instead sensed rather than distinctly heard. Our ear/brain combination takes all these factors into account and tells us the kind of space we appear to be in. Actually, if a listener were blindfolded and led into an unknown room, he or she could, simply by hearing an impulse sound in the room such as a snare drum, pretty closely determine the size of the room, the distance from the snare drum, and the probable composition of the surfaces in the room. Our brain interprets the time between the direct sound and the first perceived reflections to tell us the size of the room. The proportion of direct sound to reverberant sound defines the apparent distance we are from the sound source. Finally, the length of the reverberant field gives us an idea of the apparent reflectivity of the surfaces. Needless to say, all these factors are interrelated. We can summarize by saying that:

1.) The time interval between the direct sound and the early reflections give us the apparent size of the room.

2.) The apparent distance from the source of the sound is determined by the proportion of the direct sound to the reverberant sound.

3.) The length of the decay tells us how reflective or absorbent the surfaces of the room tend to be, and in conjunction with the audio bandwidth of the late sound field implies the composition of those surfaces.

To simulate this natural condition, the engineer is called upon to artificially produce one or more of these variables.

■ SIMULATING REVERBERATION IN THE STUDIO

Electro-Mechanical Systems

The engineer of today is extremely fortunate to have at his or her disposal many sophisticated reverberation computers that, using complicated pre-programmed algorithms, can simulate almost any natural space, as well as a few that cannot be found naturally. However, to have a complete understanding of all the variable parameters found in these units, the beginning engineer should be aware of the precursors of these highly developed systems. In

TIME DOMAIN PROCESSING: ECHOES AND REVERBERATION

many cases, these devices that simulate a part of the total reverberant field may be put to good use to create other effects.

■ **Simulating Early Reflections (Echoes)**

To produce a discrete echo, some sort of signal duplicating and delaying process is required. Formerly, this delay was accomplished with an auxiliary tape recorder. In addition to the normal recording process, the signal to be treated would also be sent to this extra machine. Depending on the tape speed and the distance between the record and playback heads, the output from the tape would be delayed by some fraction of a second. Figure 6-4 illustrates the tape delay process, and lists the delays available when the heads are two inches apart, which is typical of many professional analog tape recorders. If it is possible to continuously vary the speed of the machine, additional delays become available. This system is certainly usable, but could be rather inconvenient since the "echo tape" needs to be rewound frequently and not everyone has an "extra" tape recorder to dedicate to this task.

With the introduction of digital technology to audio, the digital delay line has taken over the task of the tape delay system. The obvious advantage of the digital delay line is that the delay is provided electronically, with no moving parts in the system. The input signal is filtered, and then sampled similar to the process used for digital recording, which will be discussed in Chapter 16. This digitized signal is then delayed by putting it through a shift registrar containing over a thousand sections. The signal is shifted through these sections by clock pulses and may be retrieved at any point in time, resulting in a delay of that amount. Many of the newer types of digital delay lines have substituted RAM (Random Access Memory)

Figure 6.4 Artificial echo, produced by a tape delay system.

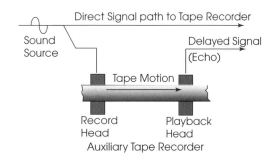

Tape Speed	Resulting Delay
3 3/4 ips	528 ms
7 1/2 ips	264 ms
15 ips	132 ms
30 ips	66 ms
60 ips	33 ms

for the shift register technique, and this along with VLSI (Very Large Scale Integrated) DSP circuits, has considerably lowered the cost of longer delays. As in a digital audio tape recorder, the frequency response of the system is limited by the sampling rate, and the dynamic range or signal-to-noise ratio is fixed by the number of bits in the system (approximately 6 dB per bit). Maximum delay time is limited by the number of shift registers or the amount of RAM available. Figure 6-5 shows a modern digital processor that includes a delay-only function, as well as other effects that will be discussed below.

Delays are continuously variable over a very wide range, and can be found from as low as one or two milliseconds to over one-half second. Many units have additional memory available as an option; this can extend the delay to as much as two or three seconds. Also, many delay lines have two or more outputs, so that the input signal may produce two or more echoes, each one delayed for a different amount of time.

There were some analog delay lines in use at one time, but the higher quality and lowered cost of the digital delay line has led to their disappearance from the professional studio.

■ Simulating Later Reflections

Later reflections, collectively known as reverberation, were described earlier as a series of closely spaced reflections diminishing in amplitude with time. In fact, they are so closely spaced that it is impossible to perceive them as discrete echoes. Their cumulative effect creates an impression of room ambience instead. The time intervals between the multiple reflections are entirely random. To create such an irregular pattern using multiple taps on a delay line is not a very satisfactory solution, since the delay outputs are all derived from the same regularly paced clock. Early experiments, using a feedback loop to provide delays of delays, did not achieve the random patterns that are necessary to produce convincing reverberation. A more practical solution is to set some kind of elastic

Figure 6.5 A versatile digital effects generator containing a high-quality delay line (Lexicon PCM82; Lexicon Photo).

medium in motion with an applied audio signal. This simulates the random multiple-echo pattern of natural reverberation without actually producing discrete echoes. We can then pick up the back-and-forth vibrations of the medium using a special type of microphone attached to the vibrating medium. When the applied signal is removed, the vibrations gradually diminish, simulating the natural decay of room reverberation.

Several types of vibrating media have been used, and are still being used, with excellent results. Some of these devices have been so successful that we identify their sound with certain types of music. Perhaps the most widely used device of this type is the artificial reverberation plate. This unit consists of a steel plate protectively suspended within a wooden cabinet, as seen in Figure 6-6a. The element is driven by an audio signal and sets the plate in motion. As the plate flexes back and forth, the motion is sensed by two contact pickups, which are also attached to the plate. The audio signal produced by these two pickups simulates the reverberant field of a large space. When the applied signal ceases, inertia keeps the plate in motion for several seconds. Reverberation time of the suspended plate is approximately 7 seconds; this decay time can be reduced as desired by using a mechanical damping system. It is important to locate the plate in an acoustically quiet environment, as any vibration induced in the plate by an unwanted source will end up in the reverberation outputs. Reverberation plates come in various sizes. Some have also used materials other than steel for the plate successfully. However, the density of the reverberation is directly dependent on the size and material of the plate itself and reducing this size can degrade the character of the reverberation. Although the plate reverberation device produces a remarkably good simulation of the late sound field, it is a rather large device and no longer available as a new unit today. During the 1970s, the plate technology included the gold foil plate. Operation was similar to the steel plate, but instead of a damping mechanism to control the length of the decay, a mechanical shutter system was used. Figures 6-6b and c show a plate of this type.

Another type of mechanical vibration system uses a single or a series of coiled springs. Although specific construction details vary from one manufacturer to another, this type of system uses a long coiled spring suspended between a driver and a pickup element. The applied signal sets the spring in motion, and this mechanical motion is sensed and converted to an electrical signal. The simplest spring units are easily recognized by a characteristic metallic sound quality, which is unconvincing as a simulation of natural reverberation.

Figure 6.6 a. A steel reverberation plate. The cabinet is eight feet long and four feet high (EMT 140 ST); b. A gold-foil reverberation plate—front view. A thin gold-foil sheet replaces the original steel plate; c. A gold-foil reverberation plate—rear view. The vertical "venetian blinds" are a mechanical damper used to vary the reverberation time (EMT 240).

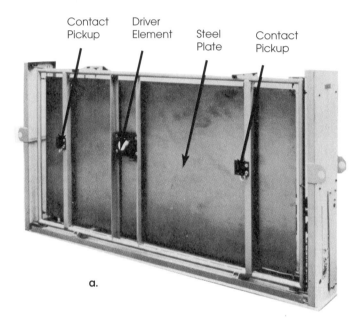

a.

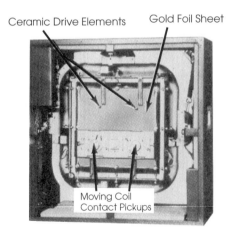

b.

c.

However, as a special effect on an electric guitar, for instance, the "springy" sound may create a unique sound texture not otherwise attainable. These inexpensive spring units are found built into a wide variety of guitar amplifiers and analog electric keyboard instruments. To capture this sound, the engineer must remember to insert the direct box or transformer at the output of the amplifier rather than at the output of the guitar, as previously mentioned in Chapter 4.

The most elegant reverberation device is the acoustic reverberation chamber, which is actually a highly reverberant room in which all surfaces have been treated for maximum reflectivity. A loudspeaker placed within the room transmits the signal to be processed, and two microphones placed some distance away pick up the multiplicity of echoes produced as the signal is reflected again and again within the room. Such a room is pictured in Figure 6-7. Although the room is unsatisfactory for normal listening, the overly reverberant sound, combined with the direct unprocessed or "dry" signal, creates the illusion that the recording was made in a very large room with normal reverberant characteristics. The acoustic chamber may well be the very best way to simulate a natural ambience, but it is quickly disappearing from use in professional recording studios. As digital reverberation units become better and less expensive, it is hard to justify the luxury of a room dedicated exclusively to creating reverberation.

■ Reverberation and Stereo Sound

By its nature, natural reverberation is diffuse, and the listener is barely conscious of a recognizable location from which the reverberation is coming because the reverberation surrounds the listener from all sides. This means that any reverberation device with only one output is limited in its effectiveness to simulate the natural condition. A single output will be just that: a point source of sound, reverberant in quality but not diffuse. Consequently, the definitive stereo reverberation system should have at least two outputs, both derived from the same mono or stereo input. Note that the steel plate system described earlier had two pickup devices located at different distances from the driver element. If the two outputs were compared, one at a time, they would sound practically identical since they are products of the same input. But if one output is routed to the right and the other to the left, the subtle variations in phase will create an overall ambient field that closely resembles the natural condition. It must be clearly understood that any single source of sound panned to the center is not the same as a stereo signal, in which left and right may be practically identical, yet different, because of random phase shifts. Although the center-panned mono signal comes out of both speakers, it still lacks the stereo "dimension."

A totally flexible echo-reverberation system will provide separate control of delay time, echoes, reverberation, and decay. Figure 6-8 illustrates the basic routing paths of a well-designed system that uses a delay system to create a discrete early reflection, with its second output being used to drive the reverberation unit. This system creates the effect of a center-placed direct sound, followed by a left discreet reflection, followed by a dense stereo reverberant field. The illustration is a simplified drawing of a fairly involved system.

Figure 6.7 Although the room is not large, its highly reflective surfaces simulate the natural reverberation characteristics of a concert hall, and the convex surfaces help create a more diffuse sound.

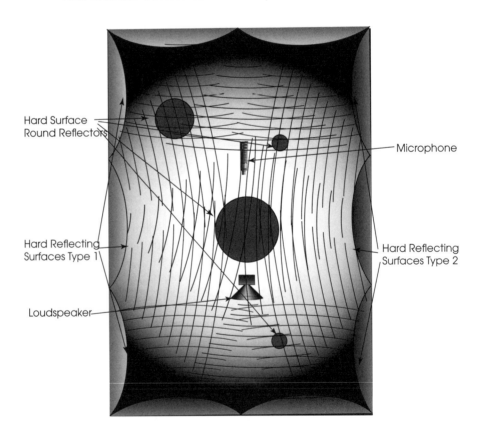

In practice, the various signal paths feeding the loudspeakers would first be combined so that there would be only one feed to each side of the system. Once the desired mixture of direct sound, echo, and reverberation was established, this complex program would then be fed to a two-track tape recorder to produce a master tape.

Reverberation Computers With the advent of modern digital circuits and microprocessors, it has become possible to integrate a complete stereo reverberation system into one package. These systems combine digital signal delay technology with sophisticated computer algorithms to simulate the reaction of a sound in a defined architectural space. They give the user control over the early reflections, the reverberant or late sound field, as well as the ability to control the perspective of the listener with regard to the source of the sound. The output of these systems is remarkably smooth and natural, and they have developed to a point where many people cannot discern the differences between the product of one of these devices and a natural concert environment when the created reverberation is properly mixed back in with the direct sound.

Figure 6.8 A complete stereo echo/reverberation system.

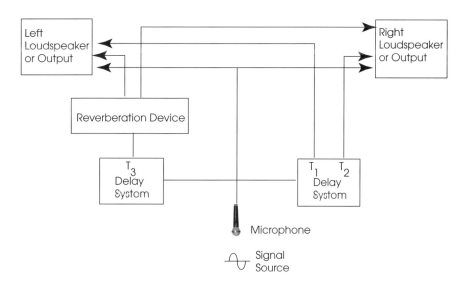

The software programs are located on DSP chips, and most offer the flexibility of changing or updating the ROMs (Read Only Memories) as new and better programs are introduced. A typical reverberation computer may contain: several concert hall programs, with varying sizes and simulations of early reflections and air absorption; several "plate" programs, which mimic the response and sound of the more popular steel and gold foil plate reverberation devices; acoustic chamber programs to simulate the sound of a highly reflective diffuse room; room programs to simulate a vocal overdub or announcer's booth; and special effects programs such as chorusing, infinite echo, gated reverberation, reverse reverberation, sound-on-sound effects, and many more. All of these predetermined programs can then be manipulated locally from either the control panel or a small easily positioned remote control device. The possible changes within the program usually include: the decay time of high- and low-frequency bands; time between early reflections and the onset of reverberation; high-frequency roll-off to simulate damping; depth perspective; density of the reverberant field; and the frequency envelope of the overall output. After each of these parameters has been changed to the engineer's liking, the modified programs can then be stored in non-volatile memories for later retrieval.

Many of the available computer reverberation systems are discrete stereo devices with two inputs and as many as four outputs, while others contain monophonic inputs and stereo outputs similar to the more familiar plates and springs. Figure 6-9 shows a current reverberation computer and its associated control head.

■ SPECIAL EFFECTS

Doubling Digital delay lines were discussed above regarding their use as a producer of echoes to simulate the early sound field in a reverberation system. As mentioned, it is possible to produce a delay so short that the ear cannot recognize it as such, even if its level equals that of the direct sound. Generally, as the time interval between the two signals is increased from 20 to 40 milliseconds, there comes a point within that range at which the echo separates from the direct signal and becomes clearly audible. The actual time value at which this happens depends on the nature of the signal in question. The echo of a drum figure might be clearly heard at 20 milliseconds, while a legato string line may require twice that delay or more before the echo is heard.

These very short delays may often be used to simulate the effects of a larger violin section, for example. In a string ensemble, the playing is never precisely synchronized. If it were, the group would sound like one very loud violin, and it is the very imprecision of attack and release within the ensemble that gives the listener an aural cue as to the size of the section. (Of course, if this effect is overdone, the listener will object to the sloppy playing.) By selecting a delay that is too short to create an audible echo, the engineer can increase the apparent size of the violin section. The almost instantaneous repetition of the signal simulates that very slight impression which is characteristic of a large group. The result may be particularly effective if the delayed signal is placed somewhat away from the direct signal to create the illusion that the ensemble is spread out over a wider area. A judicious amount of reverberation added to the phantom violins will also help, especially if the reverberation is generated by the real violins.

Figure 6.9 A reverberation computer with remote control head. This system is capable of many effects in addition to multiple types of reverberation. It also provides six channels of reverberation for surround work (Lexicon 960L; Lexicon Photo).

Phasing (Flanging)

Many attempts have been made to describe in print the unique effect known as phasing, or flanging. This effect is created by canceling some frequencies within the audio bandwidth, while others are reinforced. Moreover, the frequencies at which cancellation and reinforcement take place are continuously shifting up and down the audio range. For the person who has not heard the effect first hand, descriptive phrases may not contribute much to the understanding of just what phasing and flanging really sound like.

In the studio, the effect was first produced when a signal was fed to two tape recorders simultaneously, and whose outputs were then combined. The speed of one of the recorders was varied just a little bit by applying a slight pressure to the flange of the supply reel (hence the term "flanging"). As the machine's speed varied, so did the tape's transit time between the record and playback heads. Compared to the other tape recorder, the very slight time-delay differential produced a series of phase-shift cancellations and reinforcements when the outputs of the two machines were mixed together. As time delay varied, the cancellations and reinforcements moved up and down the audio bandwidth, producing the effect that is now known as flanging.

In an effort to produce a suitable flanging effect without the use of an auxiliary tape recorder, electronic phasing systems were developed. In these systems, cancellations were produced by combing the outputs of a series of phase shift networks (analog) with the direct signal. The nulls could be moved up and down the audio bandwidth by varying the resistance value in each phase shift network, the number of cancellations and reinforcements being dependent on the number of phase shift networks built into the circuit. However, there were no constant harmonic relationships between the nulls, as in flanging. Although electronic phasing units do produce an interesting effect, they lack the depth of sound of the flanging effect.

Digital technology has made possible a variable delay line that more closely resembles tape recorder flanging. Here, the tape recorder's time delay differential is simulated by a delay system that is continuously variable between about 200 microseconds and 15 milliseconds. The speed of the time delay variable is controlled by a VCO (voltage-controlled oscillator), and when combined with the direct signal, produces a series of nulls at regular harmonic intervals over the entire audio bandwidth. As the delay time increases to maximum, the spacing between adjacent nulls decreases. Figure 6-10 is a series of frequency response curves, showing the effects of phasing and flanging.

Figure 6.10 a. Phasing. The output of an 8-section phase network is combined with the direct signal. Varying a resistance produces the phasing effect as the cancellations move up and down the audio bandwidth; b. Flanging. The output of a delay line is combined with the direct signal. Varying the delay time produces the flanging effect (Graphs courtesy of Eventide Clock Works).

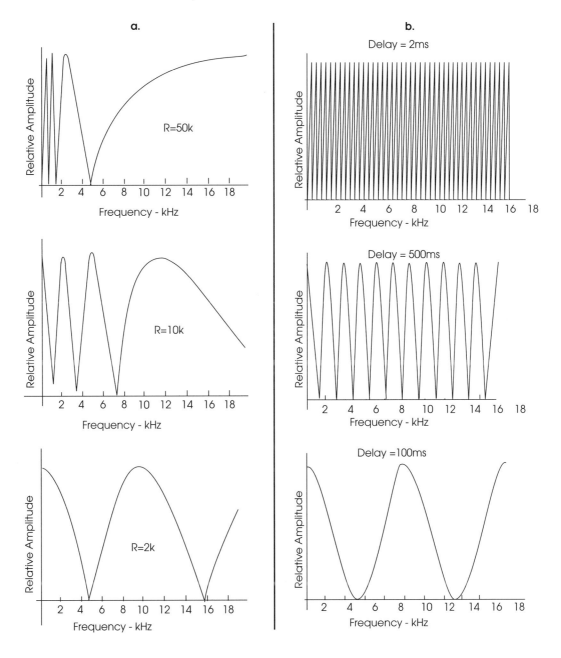

Chorusing

Chorusing is an outgrowth of the techniques used for doubling and phasing. Instead of creating just a single delay and adding it to the direct sound, a number of delayed signals are derived from the input signal and then electronically phased. This creates a random series of small tonal and rhythmic errors that are reminiscent of a large ensemble. Many electronic organs make good use of this effect to simulate a pipe organ by chorusing a single oscillator to simulate a number of different ranks playing together. Chorusing is not particularly effective on singles or duos, but can add a very interesting texture to larger groups.

The Harmonizer

A harmonizer is a digital device that generates one or two signals that are variable in pitch above and/or below the input signal. In a normal analog tape playback process, raising or lowering the speed of the tape machine will not only affect the pitch of the music, but will also raise or lower the tempo accordingly. However, if we manually edit out some of the duration of each and every note over a given period of time, and play that tape back at the original speed, the tempo will appear to increase without affecting pitch. Conversely, if we use this laborious process to edit in extra note values, this will lengthen the duration of each note. Then, if tape speed is increased, pitch will have risen without changing tempo. Obviously, this razor-blade editing and speed manipulation process can be accomplished much more efficiently. In the harmonizer, this is done by converting the input signal to a series of digital samples and by varying the clock speed while duplicating or deleting parts of notes as needed within the digital domain. A harmonizer is shown in Figure 6-11 and will typically provide an output that has a range of two octaves above or below the original input signal. It is also possible to delay the output by a varying amount and/or change the phase relationship between the signals, so that the harmonized signal can be phased or flanged as well.

An extension of this process also allows tempo and pace compression/expansion of the spoken word. For instance, keeping the voice pitch the same and varying the delivery speed will allow the engineer to fit that 32-second commercial into that 30-second time slot. Modern stand-alone systems of this type, as well as many DAWs, allow this to be done over a wide frequency range. The length of storage time, the amount of time compression and the frequency bandwidth are all a function of DSP processor speed and the amount of available RAM.

Figure 6.11 A sophisticated modern harmonizer (Eventide H969; Eventide Photo).

Harmonizer-type devices today are capable of many effects, and it is indeed hard to categorize them as one type of processor. In many units, along with pitch shifting, there are compression/expansion effects, reverberation effects, equalization, as well as phasing and flanging. One of the more notable units available can pitch shift four separate voices while adding stereo reverberation to the mix at its output. It is also capable of very subtle eight-voice pitch shifting that can be used to fatten a vocal chorus sound or enhance a string section. The harmonizer is a very versatile tool indeed.

Sampling   A further extension of digital delay technology allows the engineer to record a sample of a sound and store it in RAM or on disc for later use. Since the stored sample is in the digital domain, it can be reproduced at any desired speed or pitch. Many digital keyboards use samples of real musical instruments for many of their sounds. The original sound is stored in a ROM memory or EPROM and called upon as desired. Once again, the length and quality of the storage and playback of the sample are directly related to the amount of RAM available. Libraries of samples are available from many sources, usually on CD-ROM, and can be uploaded into the sampler or keyboard as needed.

Often the recording engineer will be called on to record samples for use in various projects. The prudent engineer will use all the proper techniques for microphone placement and selection when recording these samples even though they will later be digitally manipulated and processed. A perhaps unnecessary reminder to engineers is: the better the source material, the better the final processed product. Coincident and near-coincident microphone techniques work particularly well for collecting samples, and the reader is referred to Chapter 4 for a review of these procedures.

It is of course possible to sample sounds from other recordings but it must be noted here that copying or sampling someone else's work for inclusion in a different production without their permission is a violation of International Copyright Law.

SEVEN

Frequency Domain Processing: Equalizers

The term *equalization* may be somewhat misleading, since—like alignment—it seems to imply some sort of adjustment process necessary to bring an audio signal within published specifications. Some equalization is, of course, done for this reason. For example, frequency adjustments are made to overcome the limitations of the recording medium. These adjustments, when correctly made, should have no apparent effect on the program as heard by the listener. On the other hand, equalization may be a form of signal processing, with adjustments made to noticeably modify the frequency response of the signal being treated. When this is done, there is no intention to conform to a standard. Changes are made according to the taste of the listening engineer or producer, and the signal is equalized to suit the standards of the moment. These standards may again change at the very next moment, as a result of listener subjectiveness.

The well-equipped studio will have a variety of program equalizers on hand. Some are built into the recording console and permanently assigned to specific signal paths. Others may be installed in an auxiliary equipment rack, to be used when and where required. Both types are shown in Figures 7-1 and 7-2. The equalizer may provide one or more variable controls with which the engineer can modify various portions of the audio bandwidth. Typically, these controls are distributed among low, middle and high frequencies.

Figure 7.1 An equalizer designed for in-console installation. This particular model is controllable with a touch screen (Soundtracs illustration).

BANDWIDTH AND "Q"

In Figure 7-3 the three boost settings at a 1,000Hz center frequency illustrate narrow, medium and wide bandwidths. Note that when the same narrow bandwidth curve is drawn at 10,000Hz, the bandwidth is arithmetically much greater. At 100Hz, it is, of course, much smaller. Therefore, a bandwidth expressed in hertz is only meaningful when the center frequency is also known. To eliminate this ambiguity, the

Figure 7.2 A self-contained equalizer designed to be mounted in a standard equipment rack. This model offers filtering as well as four bands of equalization per channel (Urei 546).

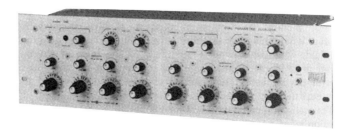

Figure 7.3 Typical mid-frequency equalization bandwidth curves.

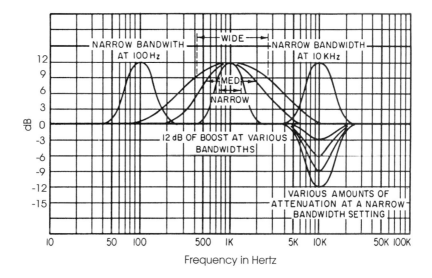

bandwidth is often expressed in terms of Q, a ratio that is equal to the center frequency, fc (in hertz), divided by the bandwidth (in hertz), or Q = fc/bw. From the graph, it should be seen that the Q of all three narrow bandwidths remains the same, regardless of the center frequency chosen. Some equalizers have a variable Q control, allowing the engineer to vary the bandwidth as required. Typical values of Q are .5 to 3, with 3 representing a very narrow band of frequencies and .5 designating a wide bandwidth.

■ TYPES OF EQUALIZERS

Parametric On a parametric equalizer, the frequency selection control is always continuously variable over a wide band of frequencies. In addition to having selectable frequencies and variable amplitude, the parametric equalizer offers the flexibility of variable Q. This allows the engineer to select the width of the bell-shaped curve for boosting or attenuating. For convenience, this width is usually defined in terms of octaves, but may be designated by some manufacturers as Q. Q values commonly range from a wide band of four octaves down to a narrow band of half an octave. These may be continuously variable or switch selectable between the two-bandwidth extremes. Therefore, a parametric equalizer allows the engineer to select the amount of boost or cut, the center frequency of the equalization curve, and the Q, or bandwidth, of the curve. A typical outboard parametric equalizer is shown in Figure 7-2. Note that as well as having two channels of four-band full parametric equalization, each band can be switched on and off individually. In addition, a bandpass filter (discussed later in this chapter) is furnished for each channel.

Graphic The graphic equalizer is so called because the physical positioning of the controls gives a graphic display of the resulting frequency response, as shown in Figure 7-4. Here, each slide control boosts or attenuates the response around the frequency assigned to it. The shape of the response for each switch peaks like the mid-frequency boost/attenuate settings shown in Figure 7-3. Individual bandwidths

Figure 7.4 A dual-channel graphic equalizer with one-third octave intervals. This unit also contains high-pass filtering (dbx photo).

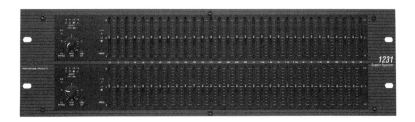

are not adjustable, but should be sufficiently small to limit the effect of each control to a narrow bandwidth around its frequency. Frequencies on professional graphic equalizers are based on the ISO (International Standards Organization) standard frequencies. The distances between the frequency bands can be from as small as one-sixth octave to as wide as one octave, with octave and third-octave spacing the most common. Some graphic equalizers combine one-third octave spacing in the lower range with one-sixth octave spacing in the higher frequencies for maximum control.

A full-range graphic equalizer is usually found as an auxiliary piece of equipment, since its size and cost preclude it from being used in each separate signal path. However, smaller graphic equalizers are now available in a format that makes them physically compatible with other inboard equalizers, and they can therefore be built into the recording console input channel. Figure 7-5 shows a typical example.

The Notch Filter   The notch filter is a specialized form of frequency attenuator, generally used to tune out a very narrow band of frequencies. A typical application is shown in Figure 7-6, where a notch filter has been set at 60Hz to remove A.C. hum from a program. The very narrow bandwidth prevents the severe attenuation at 60Hz from unduly affecting the rest of the audio bandwidth. The notch filter usually has two bands available so that the octave harmonic above the troublesome frequency can also be attenuated if necessary.

Figure 7.5 Drawing of an inboard graphic equalizer. The unit illustrated has ten bands with octave spacings (Sony figure).

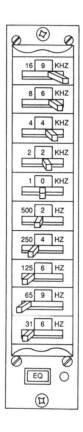

The Band-Pass Filter

A band-pass filter is an equalizer with both low- and high-frequency attenuation. The band of frequencies between the two cut-off points is "passed," and therefore not affected by the equalizer. Although any such arrangement of low- and high-frequency filtering constitutes a band-pass filter, the term is usually reserved for applications where the "pass band" is fairly narrow.

The band-pass filter is a frequency attenuating device, while the mid-frequency equalizer boosts the frequencies in its pass band. For comparison, the two are drawn in Figure 7-7. Note that the mid-frequency equalization rises to a maximum at its center frequency and then quickly falls off again to a flat response. In contrast, the band-pass filter remains flat over its pass band and falls off at either end. Therefore, the mid-frequency equalizer passes all frequencies, while the band-pass filter does not.

Figure 7.6 A notch filter, set for 35dB attenuation at 60 Hz.

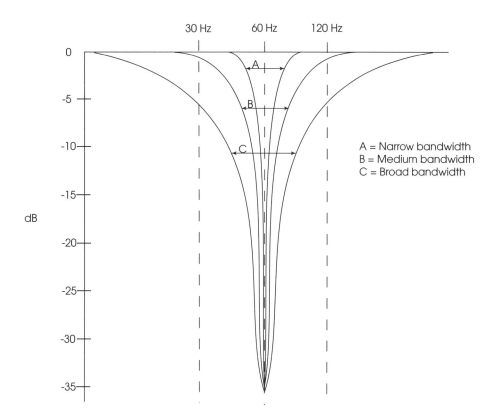

Active and Passive Equalizers

An active equalizer is actually an amplifier, designed so that certain frequencies are amplified while others are not, according to the equalizer settings chosen. The basic active equalizer may be a unity-gain amplifier, with additional gain supplied only to those frequencies that are to be boosted.

A passive equalizer, on the other hand, does not amplify any frequencies, because it contains no active elements (transistors or op-amps). In a passive equalizer, a boost at 1,000Hz, for example, is actually accomplished by attenuating the rest of the audio band using only passive elements (resistors, capacitors and inductors). This is usually followed by a fixed-gain amplifier that heightens the level of the entire bandwidth, effectively raising the selected frequency above the others.

Figure 7.7 Comparison of a band-pass filter and a mid-range boost equalizer.

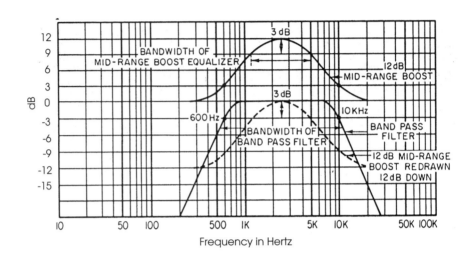

■ APPLICATIONS

Low-Frequency Equalization

Figure 7-8 shows the effect of two types of low-frequency equalization. The downward-sloping solid lines illustrate typical high-pass filter curves, while the upward-sloping solid lines are shelving curves that show low-frequency boost. The dashed line also illustrates a shelving curve, but with low-frequency attenuation.

The high-pass filter is usually identified by its cut-off frequency, that is, the frequency at which output level has fallen by 3dB. Beyond this point, the level falls off at a steady rate, or slope, expressed in decibels per octave (dB/8va). Generally, a high-pass filter offers a choice of several cut-off frequencies. The dB/8va slope is usually given in the published specifications and is not adjustable by the user.

Note that the shelving curves in Figure 7-8 show a response that rises (or falls), and that the slope eventually flattens out, or "shelves," at some low frequency. This type of equalization is identified by its turnover frequency, that is, the frequency at which the

Figure 7.8 Typical low-frequency equalization curves.

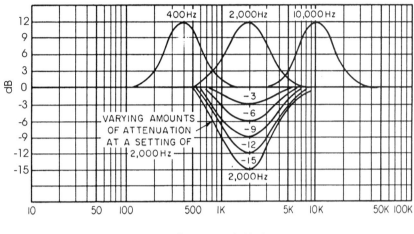

slope begins to turn over or flatten out to a shelf. This frequency generally lies 3dB below the maximum amount of boost or attenuation. Thus, if the shelving equalizer has been set for a maximum boost of +12dB, and the turnover frequency selected is 100Hz, then the level at 100Hz will be +9dB (3dB below +12dB).

Mid-Frequency Equalization

Figure 7-9 shows some typical mid-frequency equalization settings. Note that the response reaches a maximum boost or attenuation at the frequency selected, and then returns to zero as the frequency is raised or lowered beyond this point. This type of equalization is often referred to as a "haystack" due to the characteristic shape of the response curve. Its correct name, however, is "peaking equalization."

High-Frequency Equalization

As with low frequencies, equalization at high frequencies may be either shelving or low-pass, such as a high-frequency filter. Typical curves for both are drawn in Figure 7-10. As before, the shelving curve is identified by the turnover frequency, or the frequency at which the response is 3dB from maximum. The cut-off frequency of

Figure 7.9 Typical mid-frequency equalization curves.

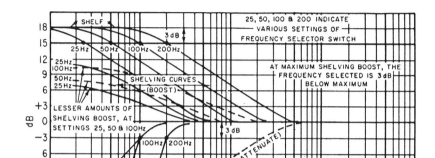

Figure 7.10 Typical high-frequency equalization curves.

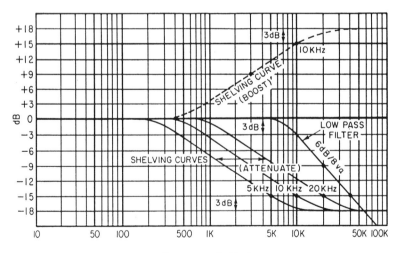

the low-pass filter is the point at which the response has dropped by 3dB, beyond which it falls off at a fixed dB/8va slope.

Inboard console equalizers may have some overlap in the available settings of the low-, middle- and high-frequency equalization. Figure 7-11 is a response curve for an equalizer with a combination of low-frequency cut-off, low-frequency shelving boost, some mid-frequency attenuation, and a high-frequency shelving boost. The individual effect of each equalizer section is shown, together with a dashed curve representing the resultant composite response of the four sections. Built-in console equalizers are often found with four bands: high, upper middle, lower middle, and low. Many of them allow the engineer to switch the high and low sections to either peaking or shelving characteristics. The frequency selector control on most recording console equalizers is a rotary-stepped position switch, or may be a detented vertical or horizontal sliding switch. Each position selects a different turnover or cut-off frequency, and depending on the equalizer's versatility, there may be between two and six or more frequencies available per section. Some console equalizers also have a continuously variable frequency select, with each band of equalization covering a specific but overlapping range.

Figure 7.11 A composite equalization curve. the dashed line illustrates the overall equalization that results from the addition of the low-, mid-, and high-frequency curves shown on the graph.

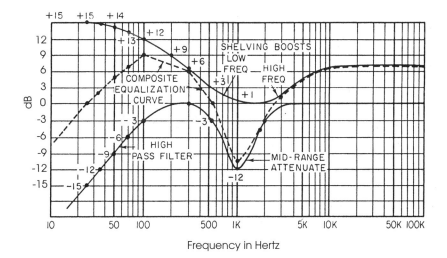

This allows the engineer to tune the equalizer to any frequency within the equalizer's range. Or, while recording, the frequency selector may be swept back and forth as a special effect.

■ EFFECT OF EQUALIZATION ON DYNAMIC RANGE

In applying any sort of equalization to a high-level signal, it is important to bear in mind that since a frequency boost raises the level of certain parts of the audio band, it places these frequencies just that much closer to the maximum permissible level. If the overall signal is already at a maximum, equalization could cause noticeable distortion of amplifiers or tape. Consequently, the overall level may have to be brought down to keep the equalized band of frequencies within safe limits. Although the overall dynamic range remains the same, the maximum permissible level is now enjoyed only by those frequencies that were equalized (or boosted). This means that the dynamic range may seem to be somewhat less than before the equalization was applied. On the other hand, some attenuation of a troublesome band of frequencies may permit the overall level to be brought up, resulting in an apparently (and actually) louder program.

■ EQUALIZER PHASE SHIFT

Since the equalizer achieves its frequency boost or attenuation through the use of reactive components (inductors and capacitors), a certain amount of phase shift accompanies any equalization added to a signal path. Figure 7-12 shows typical phase-shift characteristics for shelving and cut-off filters. Figure 7-13 illustrates the effect of a mid-frequency boost equalizer. It is for this very reason that minimal equalization should be used whenever possible. And as mentioned in Chapter 4 on Microphone Technique, "equalization is not a substitute for poor microphone selection or placement."

Figure 7.12 Typical phase shift characteristics for shelving equalization and cut-off filters.

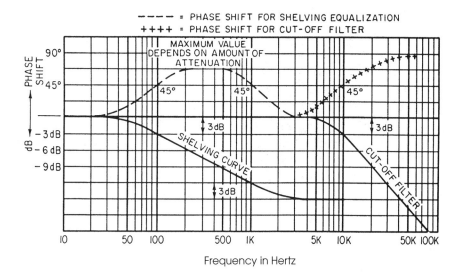

Figure 7.13 Typical phase shift characteristics for mid-range boost equalization.

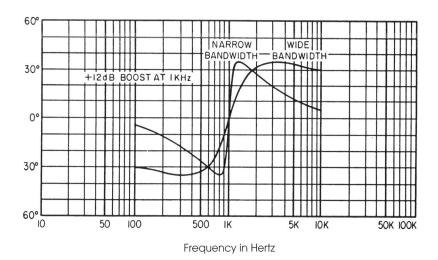

EIGHT

Amplitude Domain Processing

At the present state of the art, there are still some practical restrictions on the dynamic range available to the engineer using analog magnetic tape as a storage medium. Digital storage methods also have their limitations. The sound level of a symphony orchestra with a chorus can reach 110dB, and enthusiastic rock groups may be even louder. By comparison, the dynamic range available on the analog magnetic medium may be about 72dB, or less if a safety margin for peaks is allowed. Even with the 98dB signal-to-noise ratio of 16-bit professional digital recording, an exuberant performance can exceed the dynamic boundaries of the storage medium. Although 24-bit digital recordings offer even greater dynamic range, restrictions are still placed on the recording by both the analog and digital storage media (to be discussed at great length in Chapters 10 and 13, respectively). Whatever system is used, the dynamic range of the recorded program must be compressed so that it does not exceed the limitations of the recording storage medium. Although a 24-bit digital recording can theoretically yield an almost 146dB signal-to-noise ratio, few if any home playback systems could handle that kind of range. Even professional monitoring systems seldom have more than a 130dB dynamic range.

While recording, the engineer may raise the recorded level of a low-level program, and decrease that level during louder passages. Such "gain riding" is an important part of many successful recordings, yet it will not solve all dynamic range problems. In many cases, the engineer cannot anticipate and control every high-level peak that comes his way. Occasional high-level signals from a few instruments may not be troublesome in themselves, but when from time to time they occur simultaneously, the cumulative signal will exceed the maximum permissible level. In many cases, particularly with popular music, it is unlikely that the engineer can anticipate these conditions in time to take corrective steps before they occur, because two consecutive versions of a song may not be the same. During the classical recording session, the engineer's familiarity with the music and his or her ability to read the score (a musical road map) can assist in the anticipation of these loud sections. This, along with pre-recording rehearsal, can often alleviate the troublesome levels.

■ COMPRESSORS AND LIMITERS

As a complement to the method just discussed, the engineer may need to use signal-processing devices known as compressors and limiters to meet the restrictions of the recording medium. The terms "compression" and "limiting" have been in the audio vocabulary for years, yet there is often some confusion over their definitions. This arises from the fact that both the compressor and the limiter are devices that restrict the dynamic range of a signal, and the difference between them is one of degree, with the limiter having the most effect. To simply define:

Compressor: An amplifier whose gain decreases as its input level is increased.

Limiter: A compressor whose output level remains constant regardless of its input level.

Both definitions are valid only after the signal being processed reaches a certain level. We therefore have one more definition:

Threshold: The level above which the compressor or limiter begins functioning.

These terms are further defined in Figure 8-1, which is a graph of input versus output for an idealized combination compressor/limiter. Notice that as the input level increases from -10dB to 0dB, the output level likewise increases from -10 dB to 0 dB. Here, the device is functioning as a simple unity-gain amplifier, with no effect on the signal level. However, once the signal level exceeds the compression threshold of 0dB, the output level is found by following

Figure 8.1 A graph of compressor and limiter functions. T_C = compression threshold; T_L = limiting threshold.

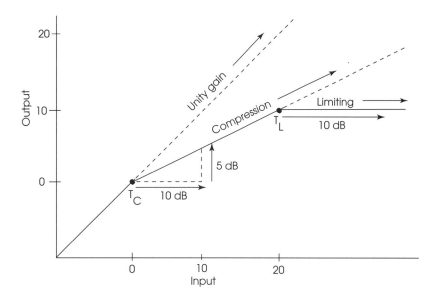

the compression curve. A further increase of 10dB in input will yield only 5dB more output level. In other words, the device now has a compression ratio of 10 to 5, or 2:1. Since this 2:1 compression ratio took effect only after the signal level exceeded 0dB, we call 0dB the compression threshold, the point at which the compressor begins functioning.

Once the input level reaches +20dB, there is no further increase in output level. Hence, the device is now operating as a limiter, with a limiting threshold of +20dB. As drawn in Figure 8-1, the limiter's compression ratio is (inf):1. In actual practice, compression ratios of 10:1, 12:1 or greater are usually considered as limiting. Notice that once the +20dB limiting threshold has been reached, output level remains at +10dB, despite further increases in input level. Therefore, the limiter threshold does not necessarily indicate the maximum allowable output level of the device, but rather, it indicates the input level at which the limiter begins working.

In Figure 8-2, note that the same compression ratio of 2:1 will have different effects on the overall dynamic range, depending on the point at which the compression begins. The position of the compression threshold will also influence the point at which limiting must begin if a certain maximum output level is not to be exceeded. For example, when output levels are to be kept below +10dB, the lower the compression threshold, the higher the limiting threshold may be.

In Figure 8-3, the faceplate of a combination compressor/limiter is shown. Here, the compression threshold may be varied from -40dB to +30dB, allowing considerable control over the point at which compression begins. In the case of a compressor that does not have such a control, the same effect may be realized by inserting an amplifier/attenuator combination before the compressor, and a complimentary attenuator/amplifier, if necessary, after it. For example, consider a compressor with a fixed threshold of -10dB. In Figure 8-4, input signal (A), with a dynamic range of 30dB, is compressed to an output signal, (a), with a dynamic range of 20dB and a maximum level of 0dB. However, if the input signal is amplified 10dB before

Figure 8.2 To prevent output levels in excess of +10dB, the limiter threshold setting will vary, depending on where the compression threshold was set.

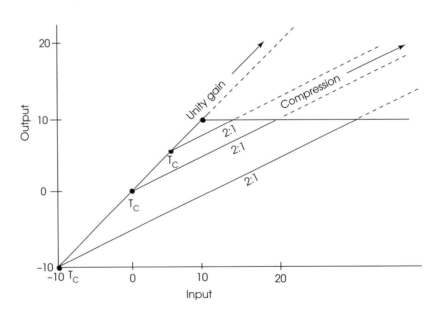

Figure 8.3 A dual-channel compressor/limiter with separate threshold adjustments for compression and limiting (dbx 160XL, dbx photo).

Figure 8.4 In a compressor with a fixed threshold, the output dynamic range may be varied by boosting (or attenuating) the input signal level before compression.

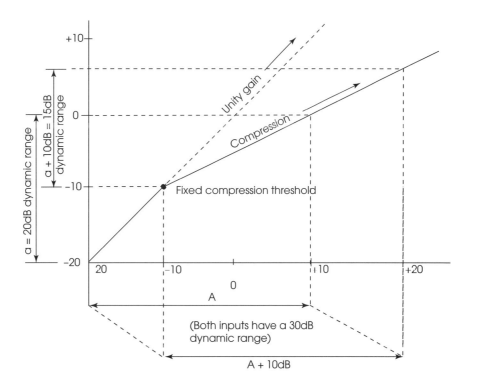

reaching the compressor as shown in the figure by (A + 10dB), the output signal will have a dynamic range of only 15dB and its maximum output level will be +5dB. This increased output level may now be attenuated by 5dB if it is necessary to keep the maximum output level at 0dB as before.

When there is some amount of gain before compression, Figure 8-4 can be redrawn as shown in Figure 8-5. Here, the dB gain before threshold raises the transfer characteristic above the unity gain line as shown. Beyond the threshold, the system gain steadily decreases as the compression curve approaches the unity gain line. The point at which they intersect is known as the rotation point. At this point, the gain of the compressor is 1:1, or unity. Beyond the rotation point, the gain continues to decrease. It should be noted that as a consequence of the gain before threshold, the compressor raises low-level (below the rotation point) signals, and lowers high-level (above the rotation point) signals. Thus, low-level signals may be brought up above the residual noise level while at the same time, high-level signals are prevented from saturating the recording medium.

Figure 8.5 Graph of a compressor, with gain before threshold. Input levels below the rotation point are raised. Input levels above the rotation are lowered.

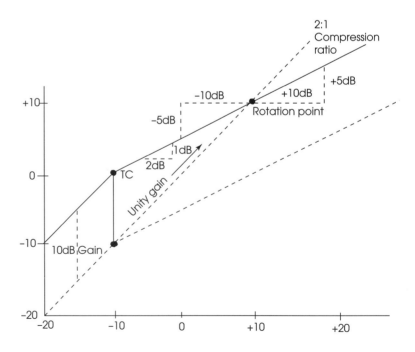

Many compressors offer the engineer a variety of compression ratios from which to choose. Figure 8-6a illustrates four different compression ratios, all beginning at the same threshold point. Assuming that the output level must be kept below +10dB, the greater the compression ratio, the wider the dynamic range of the input signal may be. On the other hand, the four ratios of 8-6a might each begin at a different threshold, as seen in Figure 8-6b. In this illustration, each threshold/ratio combination passes through the same rotation point, and each has a different effect on the total program. The relatively gradual 1.5:1 ratio affects the program from -20dB, while the more severe 4:1 ratio does not begin until the input signal level reaches -3dB.

It is an easy matter to graphically determine the compression ratio and threshold required to prevent a wide dynamic range program from exceeding a specified output dynamic range. However, it should be realized that the action of the compressor may become audibly obtrusive, especially at higher compression ratios. To understand why, remember that the compressor is a variable-gain device. As defined earlier, the compressor is an amplifier whose gain decreases as the input level increases; the higher the compression ratio, the greater the gain change.

The gain or attenuation of an amplifier may be expressed either as a positive number (0.25, 1, 3.5, etc.) or as a dB value (-10dB, 0dB, +10dB, etc.). The positive number value is a ratio of output-to-input voltage. Thus, if input voltage is 0.5 volts, and output is 0.2 volts, gain is 0.2 divided by 0.5, or simply 0.4. If input and output are equal, gain is of course 1, or unity. However, it is usually more convenient to express output in terms of dB above or below the input level. Thus, a unity gain amplifier has a gain of 1, which would be expressed as a decibel gain of 0dB. A compressor will have a negative dB gain, usually called a gain reduction. Figure 8-7 graphs the gain reduction from several compression ratios. It will be seen, for instance, that as the compression ratio is increased, a constant high-level signal of +10dB, will cause more gain reduction. When the high level is removed, the amount of gain reduction decreases as the compressor returns to unity gain. If the gain reduction fluctuates rapidly, it may be quite audible as the background noise goes up and down in time with the compressor action, causing a breathing-like sound. On the other hand, if the compressor takes a relatively long time to restore itself after a high-level signal has caused gain reduction, then other low-level signals following the high-level signal will also be reduced in gain.

Figure 8.6 a. Four compression ratios, all beginning at the same threshold; b. Four compression ratios, each passing through the same rotation point. Note that each ratio begins at a different threshold.

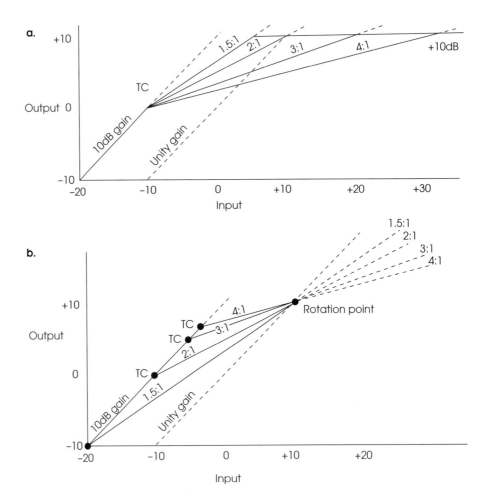

AMPLITUDE DOMAIN PROCESSING 229

Figure 8.7 a. The solid lines show the amount of gain reduction at various compression ratios. The corresponding compression ratios are drawn as dashed lines.

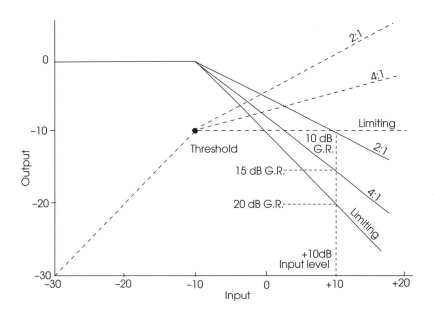

	2:1 Ratio		4:1 Ratio		Limiting	
Input	Output	Gain Reduction	Output	Gain Reduction	Output	Gain Reduction
-30	-30	0	-30	0	-30	0
-25	-25	0	-25	0	-25	0
-20	-20	0	-20	0	-20	0
-15	-15	0	-15	0	-15	0
-10	-10	0	-10	0	-10	0
-5	-7.5	2.5	-8.75	3.75	-10	5
0	-5	5	-7.5	7.5	-10	10
+5	-2.5	7.5	-6.25	11.75	-10	15
+10	0	10	-5	15	-10	20
+15	+2.5	12.5	-3.75	18.75	-10	25
+20	+5	15	-2.5	22.5	-10	30

Compressor Release Time

The amount of time it takes for a compressor to return to its normal gain-before-threshold is known as its release, or recovery, time. Some compressors have an operator-adjustable release time. Figure 8-8 shows the effect of various release times on a series of beats. In the Figure 8-8a, the first beat is above the compressor's threshold, while the next four beats are all below the threshold and will not cause the compressor to react. However, if the release time is relatively long, these beats may be affected by the initial gain reduction caused by the first beat, as shown in Figure 8-8b. Here, the compressor gradually returns to unity gain. At the time of the second beat, there is still 6dB of gain reduction; at the third beat 4dB, and so on. As a result of this gradual release time, the dynamic range of the series is changed considerably. In Figure 8-8c, the release time has been considerably shortened, and only the second beat is affected by the gain reduction. In this case, the rapid return to unity gain may cause an audible breathing sound as the system gain rapidly increases. There is no inherently correct release time setting as it really depends on the type of signal being processed. More often than not, release time is varied by the engineer to produce the least objectionable effect.

Compressor Attack Time

The attack time of a compressor is the time it takes for it to react to a signal above the threshold level. On some compressors it, too, may be adjustable. In many cases, a long attack time will allow the first part of a sustained note to pass through the compressor unaffected. If the note persists in duration, the compressor will attenuate it after a fraction of a second. Subjectively, the note may sound more percussive due to this type of compression. For example, a longer attack time may help to accentuate a compressed bass line that is otherwise difficult to distinguish, especially in a busy instrumental arrangement.

Often, compression may be applied to the overall program rather than to a particular instrument. Known as program limiting (or compression), this practice will prevent the cumulative levels of the various instruments from getting too high or too low. This type of gain reduction must be approached with care, since the adverse effects of the compression action are heard on the entire program, and excessive compression will probably be audibly distracting.

Program limiting is often used to raise the apparent loudness of a signal. In addition, the type of detection circuitry used in a compressor or limiter can vary. Many compressors use a detection circuit

Figure 8.8 a. A series of beats, only the first of which is above the compressor's threshold; b. Progressively less gain reduction as compressor recovers from the initial beat; c. A relatively fast release time prevents the initial gain reduction from having a long-time effect.

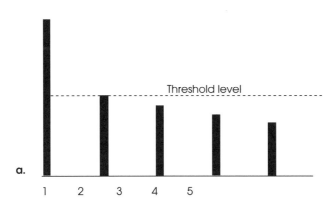

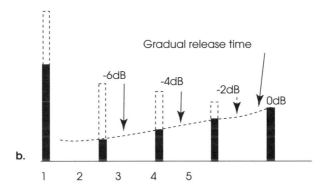

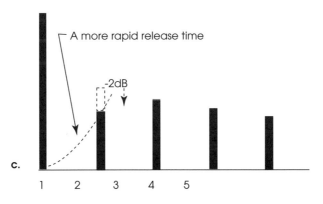

that responds when the RMS (Root Mean Square) value of the signal is above the threshold point, while others respond to the peak value of the signal. The RMS detector reacts in a manner similar to the way our ear responds to changes in air pressure (as discussed in Chapter 2), and has a reaction or attack time of about .3 (three-tenths) of a second. On the other hand, a compressor/limiter that is designed to respond to peak amplitudes has a detector with a response time of around .000025 seconds (25 microseconds) and will cause gain reduction to occur immediately for nearly any signal. This can accentuate the modulation effect known as breathing mentioned earlier.

Many broadcast facilities use these two types of compressor/limiters in tandem. The RMS section is used to compress the signal and raise the gain to as close to 100% modulation of the transmitter as possible. This is followed by a peak-detection device that prevents any sudden highly transient signals from exceeding the legal power limit licensed to the station. Thus, an average- (but boosted-) level program with no peaks will seem louder than a low-level program with occasional high peaks. Figure 8-9 illustrates this. Program B will sound much louder than program A, despite A's higher peaks, since B's average level is obviously much higher. In an attempt to produce louder-sounding compact cassettes, compact discs and FM broadcasts, such loudness boosting is often overdone, much to the detriment of the finished product.

When a stereo tape, disc or broadcast is program-limited, the gain-regulating sections of the left and right compressors must be electronically interlocked, so that compression in one side causes an equal amount of gain reduction in the other. This keeps the overall left-to-right stereo program in balance. Consider a stereo program in which the right channel occasionally needs some compression. During that compression, a center-placed soloist would appear to drift to the left whenever the gain of the right channel is reduced by the compressor. To prevent such center-channel drift, the stereo interlock function reduces the gain of both channels whenever either one goes into compression. This stabilizes the information that is common to both channels and keeps the soloist from shifting right or left during processing.

Compressor/Limiter Special Effects

As mentioned earlier (Chapter 2), the energy distribution of the human voice is such that sibilants are apt to be significantly louder than other voice sounds. The high-frequency/high-energy content of many "s" sounds may overload an amplifier even when the apparent listening level does not appear to be unduly loud. Some compressors feature a de-essor function that helps to keep these sibilants under control. There are also devices specifically designed to reduce these sibilants that do not function as broadband compressors. These de-essors incorporate a high-frequency equalization boost in the compressor's gain-reduction control circuitry so that frequencies in the sibilant range will cause more compression action than do other frequencies. Therefore, during those fractions of a second when a distortion-prone sibilant sound passes through the compressor, gain is reduced by a more than usual amount. On most compressors, gain reduction lowers the level of the entire signal.

On more sophisticated devices, however, the audio bandwidth may be split into several sections with each section being treated independently. The signals are then recombined after processing. In this way, the sibilant band may be isolated and compressed more than the rest of the audio program. When the signals are recombined, the sibilant

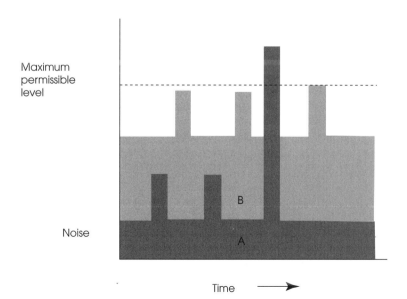

Figure 8.9 Program B will sound a lot louder than program A despite the latter's peaks, since its average level is higher.

sounds are reduced in level without affecting the rest of the program.

Some compressors have a second input, which accesses the detector input only. This "side-chain" or "key" input allows the engineer to use a separate signal to trigger the gain reduction of the program material passing through the compressor. This type of device is often used as a voice-over compressor or "ducker," and permits an announcer's voice to take precedence over a musical background. The musical program is routed through the compressor, and the voice signal (or a multiple of it) is routed to the key input. The voice signal is what actuates the compressor, causing gain reduction in the musical background, but not in the voice path itself. Voice and music are then combined into a mono or stereo program as the case may warrant. As the announcer speaks, the background level of the music is automatically lowered. As a further refinement, the threshold of the compressor can be adjusted separately from the key input threshold. This provides voice-actuated gain reduction only when the musical program is above a certain level. Therefore, if the music is already sufficiently low in level so as not to obscure the announcer's voice, no further gain reduction is provided.

■ EXPANDERS

Like the compressor, the expander affects the dynamic range of a program. But, as its name suggests, the expander widens or expands the dynamic range, rather than restricting it. Two definitions describe its operation:

Expander: An amplifier whose gain decreases as its input level is decreased.

Threshold: The level below which the expanding action takes place. (Note that the expander functions below the threshold—just the opposite of a compressor.)

Figure 8-10 is an input/output graph for an expander with a 1:2 expansion ratio. Note that when the input level lies below the expander threshold, the output level changes 2dB for every 1dB change in input level. As drawn, an input signal with a 20dB dynamic range produces an output with a 30dB range. Although maximum output level equals maximum input level, the dynamic range has

Figure 8.10 Graph of an expander with a 1:2 ratio.

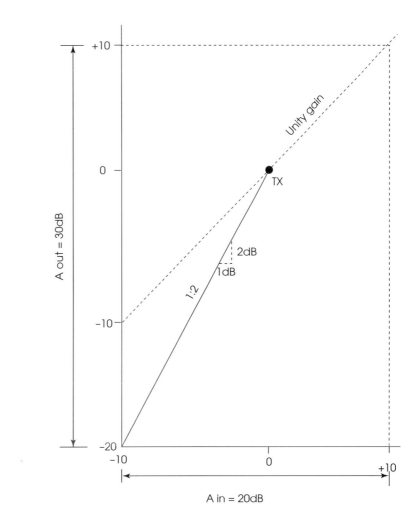

been expanded since minimum output level is now 10dB less than minimum input level. In other words, the dynamic range has been expanded downward.

Expansion, when used prior to recording to increase the dynamic range of a program, must be approached with caution since the low-level components of the signal are now even closer to the residual noise floor or minimum level of the recording medium. On the other hand, if the overall level is raised after expansion, the high-level components of the signal are brought closer to the medium's maximum permissible level. Because it is so often necessary to compress the dynamic range of a program to meet the restrictions of the recording storage medium, this application of expansion is rarely used.

Expanders as Noise Gates

In many typical recording situations, there is often a certain amount of undesirable low-level sound present in the studio. Air conditioning noise, chair squeaks, and the sounds of nearby instruments are some of the many noises that a microphone may hear. These sounds may be masked adequately when the sound level from the instrument in front of the microphone is above zero. However, when the instrument is silent, this background noise can become audible. An expander may be used to reduce this noise by selecting a large expansion ratio, with a threshold level at a point above the noise level, but just below the lowest output level of the instrument in front of the microphone. As seen in Figure 8-11, low-level noise signals are practically eliminated as they are expanded below audibility. This function of the expander is often called a "noise gate," since the effect is to cut off the noise without affecting the musical program. In fact, many expanders manufactured today are simply called noise gates, and have as many as six gates on one mainframe. The effect is often referred to simply as "gating." Of course, the line between the noise and the music is not often clearly defined, and it is all too easy to misadjust the noise gate and cut off the very quietest musical passages along with the noise. As before, this type of expansion must be approached with care.

An expansion threshold that is set too high will cut off the low-level end of a program along with the noise. Although this certainly is not desirable in most cases, it may be effective on some percussive sounds (drums, hand claps, etc.). As an example, a snare drum produces a series of high-level transients, each of which should quickly fade away. If each drum attack sustains too long, the overall sound may lack the subjective tightness that is usually sought in many popular music recording sessions. An expander threshold set at a relatively high level, well into the drum sound itself, will bring the level down quickly after each attack. This effect is shown in Figure 8-12. The same technique may be applied to most percussive instruments. It is not uncommon when using electronic drums to hear extraneous sounds at the end of a note, particularly when that sound was created by sampling. In a live performance, this objectionable sound may be lost in the noise floor of the sound system or

Figure 8.11 The expander as a noise gate. Noise signals at a level below the expander threshold are significantly attenuated.

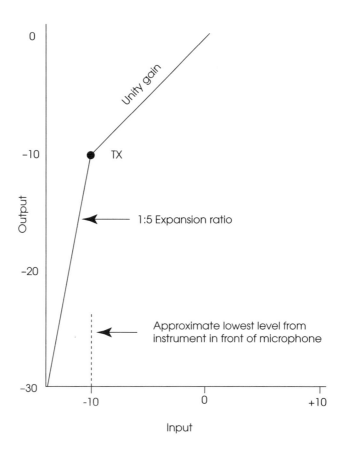

in the audience noise. But in the recording studio, when using a direct box to access the sound of the electronic drums, the extraneous noise at the end of the sample will be clearly heard. An expander or noise gate can be used to attenuate, or remove, these low-level sounds between each attack.

During the attack itself, the expander cannot remove noise since it reacts to the overall program level and has no way of distinguishing wanted from unwanted sounds. For this reason, the noise gate function of an expander may only be marginally effective on sustained instruments such as organ and strings. When neither instrument is playing, the expander may attenuate the noise, but as before, the noise will reappear whenever the instrument begins playing. Due to the sustained nature of the instrument, noise may still be heard through the music. Consequently, its sudden disappearance and reappearance in time with the music may be more of a distraction than a help. No matter how effective a noise gate can be, it is not a substitute for good musicianship and should not be expected to transform mediocre players into skilled artists.

Most expanders have controls that govern ratio, threshold, attack time, release time, and range. Ratio and threshold have been discussed; attack and release time adjustments react the same as the similar controls on the compressor, to determine how quickly the expander reacts to changes in level and how quickly the gain returns to unity. As with the compressor, very short release times are more noticeable than longer values. Since it may be desirable to restrict the amount of low-level attenuation provided by the

Figure 8.12 a. A series of transient attacks with long decay times; b. A high-level expansion threshold shortens the apparent decay time of each transient.

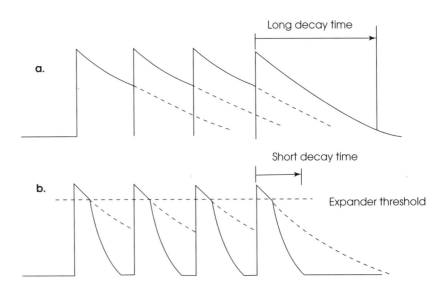

AMPLITUDE DOMAIN PROCESSING 239

expander, a variable range control may be provided. Its effect is shown is figure 8-13, where low-level signals are attenuated by only 5dB or 10dB. The range control may be continuously variable from 0dB to about -80dB. It may also be desirable to control the duration of time attenuation between attack and release; some units provide a variable sustain control for this purpose.

Figure 8.13 Graph of an expander, showing 5dB and 10dB attenuation ranges.

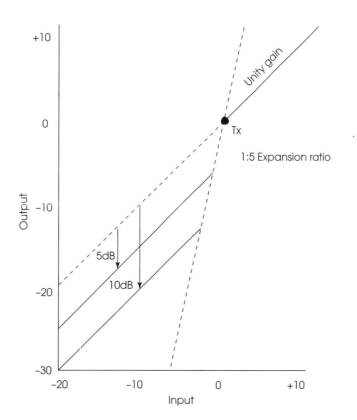

Expanders and Special Effects

So far, the expander has been described as a self-keying device—meaning that the expanding action is controlled by the program passing through the expander, as shown in Figure 8-14a. Some expanders may instead be controlled, or keyed, by an external signal, as in Figure 8-14b (similar to the ducking application mentioned in the discussion on compressors). Here, it is the level of the external keying signal, rather than the signal being expanded, that regulates the amount of expansion. If the keying signal is at a constant level above the threshold, the expander will simply stay on, functioning as a unity gain amplifier, regardless of the dynamic range of the signal passing through it. It should be clearly understood that the keying signal is not heard at the expander's output. It simply provides a control voltage to regulate the gain of the expander and thereby the level of the signal passing through it. When the keying signal is removed, the audio signal passing through the expander will fade out. The length of the fade will depend on the sustain and release time of the expander. An automatic fade-out may be accomplished by setting the release time between 3 and 6 seconds. When the keying signal is suddenly removed, the program through the expander will gradually fade away over that period of time. A typical unit of this type is shown in Figure 8-14c.

Unique effects can be achieved with an expander by keying a sustained musical instrument with a control voltage derived from a percussive instrument. With a maximum range setting and a relatively short release time, the sustained instrument, for instance, an organ setting on a synthesizer, is routed through the expander. The expander is then keyed on and off, perhaps by a drum track. The expander switches on and off with every drum transient attack, and imposes this percussive envelope on the sound of the organ, as shown in Figure 8-15.

In another application, a bass drum or tom-tom (and some other percussion instruments) may be "tuned" by feeding a low-frequency audio signal through the expander. The intention is to hear the low-frequency tone each time the expander is keyed on. The expander is keyed by the signal of the bass drum, for example, thus passing the low-frequency tone each time the drum is struck. The signal generator may be tuned as desired, and the output of the expander is mixed with the regular drum sound to impart a definite tonality to the sound.

AMPLITUDE DOMAIN PROCESSING 241

Figure 8.14 A popular KEPEX (**KE**yable **P**rogram **EX**pander) noise gate (dbx 904, dbx photo).

a.
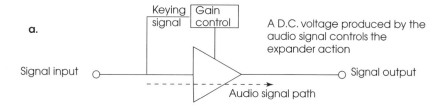
A D.C. voltage produced by the audio signal controls the expander action

b.
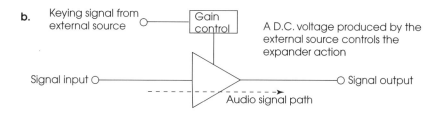
A D.C. voltage produced by the external source controls the expander action

c.

MULTI-FUNCTION PROCESSORS

As mentioned in Chapter 6, digital signal processors have been developed which combine many different functions in one processor. It is not uncommon to find devices capable of multiple types of time-based effects, such as reverberation, gated reverberation, chorusing, harmonizing, time compression, and pitch shifting, all in one package. There are other multi-use devices that operate in the amplitude domain. These so-called dynamics processors are capable of

Figure 8.15 An expander used to modify the sound of one instrument with the envelope of another.

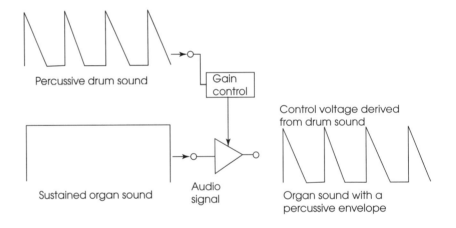

Figure 8.16 A digital multi-effects processor (dbx DDP, dbx photo).

operating as a compressor, a limiter, a de-essor, and often include auto-pan and auto-fade functions, along with expanders and noise gates. Variable filtering can also be found in many of these units. Today, a number of multi-function processors can meet most signal processing needs instead of needing a stack of compressors, a rack of gates, and several groups of reverberation and delay devices.

Newer generation processors let the engineer combine several functions in any logical order. For instance, an engineer may choose to gate and then de-ess an overly sibilant vocal prior to compressing it. It is certainly more productive to accomplish all of these functions in one unit rather than chaining together three separate units, each with its own additive noise level. Dynamics processors of this type usually have the traditional side chain or key inputs as well; these inputs can be filtered, gated or compressed as necessary. A multi-processor of this type is shown in Figure 8-16.

Further, digital consoles often have these effects built into the individual input/output channels. These effects can be accomplished in the digital domain without worrying about additional noise from voltage-controlled amplifiers or gain followers. It is not uncommon to have as many compressor/expander/gate units as there are inputs on the console. A typical software-based dynamics processor is shown in Figure 8-17.

Figure 8.17 A virtual dynamics module with soft-assign controls and touch-screen control. Contains compression, expansion/gating, keying and filterable keying, as well as the ability to chain devices together in several permutations.

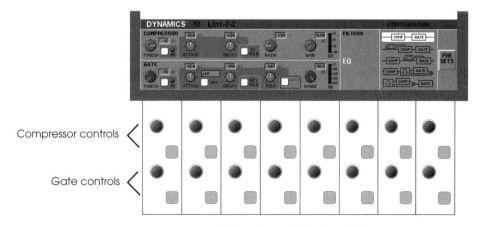

As complicated and comprehensive as these multi-effect processors can be, their use is still governed by the same operating parameters as the simpler single-use equipment. A thorough understanding of each component device discussed in Chapters 6, 7 and 8 is still essential for the proper use of the more sophisticated combination analog or digital multi-use systems.

NINE

Analog Noise Reduction

Noise, like music, is sound. However, unlike music, it is unwanted sound. For our purposes, we will ignore the fact that one man's music may be another man's noise, and concentrate on those noises that everyone agrees do not belong in the recording studio.

With care, some of these noises may be minimized or eliminated completely. Coughs, chair squeaks, and dropped mutes may be removed by editing or doing a retake. There are not yet available, however, any devices that will remove the sound of a chair squeak from a violin solo, or a falling music stand from a trombone solo. And if the percussionist strikes the microphone instead of the instrument in front of him, the noise will be painfully obvious. Other noises are a function of the recording medium itself. For example, it is well known that every component in the signal path, from microphone to amplifier, introduces a little noise into the system. This noise may be in the form of a hum resulting from faulty shielding, or a hiss from an amplifier output. However, in a well-designed and properly functioning studio, the cumulative effect of noise in the signal path should remain negligible, and the greatest noise source will probably be the magnetic tape medium itself. Even if a totally noiseless program could be recorded, on playback, the analog tape will contain modulation noise and asperity noises that were not there before.

These noises are a function of the magnetic recording process, and even digital recording systems have some by-products that may be found objectionable. Major advances in analog tape formulation and manufacture have lowered this noise to an impressive minimum, yet it is still there and will probably never be totally removed. Just as a loud thunderclap will momentarily block out the sound of falling rain, loud programs will effectively mask low-level noise and the listener will be entirely unaware of its presence. At lower levels, however, the program material may not be loud enough to mask the noise and the listener will hear it along with the music. Although asperity noise is often masked by the signal causing it, the increase in modulation noise may be apparent over the rest of the audio bandwidth. In fact, even during a rather loud passage of low-frequency music, the program may not completely mask the noise present in the higher frequencies.

The advent of digital tape recording in professional recording studios has not negated the need for analog noise reduction systems. Many compact discs are produced without going through an analog storage medium, and thus do not require noise reduction. However, many film and television media, as well as the smaller recording studios, still rely heavily on analog noise reduction principles. Recent advances in noise reduction technology have nearly equaled the digital medium in performance.

■ WHITE NOISE

Since noise is sound, it follows that it should have some frequency or frequencies associated with it. Indeed, certain kinds of noise are identified with certain frequencies. For example, a hum induced in a poorly-shielded cable is readily recognized as a 60-Hz tone. Rumble, either in a studio building or on an analog turntable, will also have measurable frequency.

But what of the ever-present tape hiss? The term "hiss" itself suggests that this particular noise is high pitched. There are other quite similar sounds, or rather noises, like those made by an amplifier or found between stations on the FM radio dial. All three, however, are a form of white noise. The term is analogous to white light, which is light that contains energy at all wavelengths, such as sunlight. When passed through a prism, white light is dispersed into hundreds of equal-intensity hues, each representing the light energy of a particular wavelength. Likewise, white noise contains equal sound energy at each frequency within the audio bandwidth. But despite the equal energy present at each and every audio frequency, white noise is thought of as being mainly hiss, or high-pitched in sound. This is because each successively higher octave contains twice as many discrete frequencies as the octave just below it. For example, the octave that begins at 1000 Hz (1 KHz to 2 KHz) contains 1,000 discrete frequencies, which is twice the number of frequencies for the octave that ends at 1000 Hz (500 Hz to 1 KHz), which has 500 discrete frequencies. Therefore, any octave contains twice the energy of the octave immediately below it, and this seems louder. Higher octaves mask the sound energy of lower ones, and the ear perceives a sound aptly described as hiss. There is, however, definitely low-frequency energy or noise present with tape hiss.

With white noise, the doubling of energy per octave signifies a power gain of 3 dB/8va, or, if monitored on a spectrum analyzer, a voltage gain of 6 dB/8va. The spectrum analyzer (discussed in Chapter 2) is a device that meters the energy level of a number of frequency bands simultaneously. Although generated white noise is often used in acoustical measurements, it is more meaningful to use a wide-band noise source that maintains a constant energy level per octave. This may be realized by inserting a 6 dB/8va filter at the output of the white noise generator, thus canceling its inherent energy rise. This filtered white noise is known as pink noise, also named for its relationship to light of the same color. White and pink noise energy spectra are illustrated in Figure 9-1.

■ RESIDUAL NOISE LEVEL

The residual noise level of a device is its output level in the absence of an applied signal at the input. A sensitive meter might read a very small voltage, which represents the amount of white noise present in the system. Recording consoles, processing equipment, and tape recorders will usually include within their specifications some reference to residual noise level. For example, a tape recorder may have a noise level of -90dB. Obviously, this means that the noise voltage is 90dB below some reference level. When comparing specifications, it is important to make sure that the reference level being used is the same for both devices, as different references can produce vastly different readings. It should be understood that the noise level of a tape recorder represents the output voltage level when the machine is on, but not running. If a roll of new unused tape is played, the measured noise level will be somewhat higher. When the machine is placed in record mode, still with no input signal applied, the noise level will increase again.

■ SIGNAL-TO-NOISE RATIO

Noise measurements are often discussed in terms of a signal-to-noise ratio (abbreviated S/N). Since a ratio is either a fraction or a comparison of two quantities, it might seem that a signal-to-noise ratio should include two values, one for the signal and another for the noise. For example, if the level of our reference signal is +4 dB, and the noise level is measured at -65dB, we might expect to find the signal to noise ratio written as +4dB/ -65dB. This fractional format suggests that some form of division is possible, but since decibel values cannot be directly divided, they must first be converted into

Figure 9.1 a. White noise: equal sound energy at each frequency; b. Pink Noise: Equal sound energy within each octave.

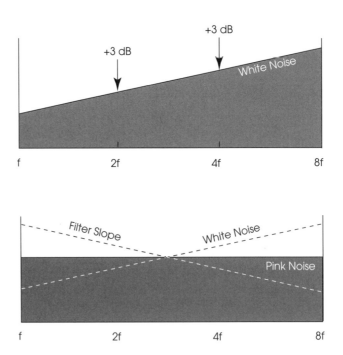

voltages (or powers if more convenient). For this example, the equivalent voltages are +4dB = 1.226 volts and -65dB = 0.435 millivolts. Therefore:

$$S/N = \frac{+4dB}{-65dB} = \frac{1.226 \text{ volts}}{.435 \times 10^{-3} \text{ volts}} = 2.818 \times 10^3$$

Now, although 2.818×10^3 is a mathematically correct signal-to-noise ratio, it is not of much practical use to the recording engineer who is concerned with the decibel difference between signal and noise. The decibel quantity equal to $20 \log 2.818 \times 10^3$ is a more useful figure.

$$NdB = 20 \log S/N = 20 \log(2.818 \times 10^3) = 69dB$$

Because of the nature of the logarithm, it will be noted that 69dB is the simple arithmetic difference between +4dB and -65dB, and it tells us exactly what we wish to know: how far the noise level lies below the reference signal. Although this decibel value is really twenty times the logarithm of the signal-to-noise ratio, in practice it has come to be popularly known simply as the signal-to-noise ratio. As with other popular misnomers, little harm is done provided that the engineer is not confused by the practice.

■ NOISE REDUCTION SYSTEMS

Good engineering practice demands that noise levels be kept as low as possible. Particularly in multi-track recording, the engineer must be concerned with the build-up in noise level as the number of tracks increases. It is a common procedure to record each individual track at a high level, regardless of the eventual track-to-track balance, to obtain as high a signal-to-noise ratio as possible on each track. When, during mix-down, the track is mixed at a lower level, the noise level is also reduced. This helps keep total noise within reason. Even so, the noise level of the total program (when mixed down to the stereo master tape) will increase by 6dB for every doubling of tracks, similar to the rising energy level of white noise discussed earlier.

Although signal processing devices are usually thought of as a means of altering the signal to produce a desired effect, there are several devices whose sole purpose is to reduce the noise level without audibly affecting the quality of the signal itself. Basically, these noise reduction devices can be classified as static or dynamic, and as complimentary or non-complimentary.

The terms static and dynamic refer to the way in which a noise reduction device (or any other device, for that matter) reacts to the signal passing through it. An equalizer is a static device. Its settings do not change once they have been set by the engineer. On the other hand, a compressor reacts to the program material, and is therefore considered a dynamic device.

In a complimentary system, some type of processing is done before recording, with equal and opposite (complimentary) processing done on playback. In a tape recorder, pre- and post-emphasis might be considered a complimentary equalization process. In a non-complimentary system, processing is done only once, either before or after recording. The signal processing devices discussed in previous chapters are all non-complimentary devices.

Static Systems

In its simplest form, noise reduction may consist of a filter adjusted to attenuate the frequency band in which the noise is found. This filter would be considered a static, non-complimentary system. Such filters may be marginally effective, for with the exception of a few noise sources such as 60-cycle hum, most noise is wide band, i.e., it exists over all or most of the audio spectrum. Filtering also effects the signal frequencies as well as the noise. In the case of a very low-frequency cut-off filter, used to minimize structurally transmitted rumble, the effect on the program may be negligible. In most other cases, however, the filter becomes musically objectionable long before the noise is removed.

In a static, complimentary noise reduction system, high-frequency boost is applied before recording in an effort to keep these frequencies well above the noise floor. Then during playback, a complimentary roll-off restores the high frequencies to normal, while lowering the tape noise level. This form of noise reduction may also be minimally effective. High-frequency pre-emphasis puts these signals just that much closer to the saturation point of the tape. Therefore, if the program has a significant amount of high-frequency energy to start with, the overall record level will probably have to be lowered, negating any noise reduction advantage. For most typical music programs, the pre- and post-emphasis built into the tape recorder itself furnishes as much of this type of noise reduction as is practical.

Both of the above examples of noise reduction are called static, since their noise reduction properties are fixed. Here, the nature of the program material in no way affects the action of the filter or of the pre/post-emphasis.

Dynamic Systems

The expander may be considered a dynamic, non-complimentary noise reduction device. The expansion threshold may be set just above the residual noise level. Therefore, as the signal falls below threshold, system gain is reduced, sharply attenuating both signal and noise. Another way of looking at the same effect is when a gradually increasing low-level input signal is quickly boosted above the noise floor. Higher level inputs (above threshold) follow the unity gain slope and use the masking effect to cover residual noise. Expanders are also frequently used as noise gates to attenuate a microphone or tape recorder output in the absence of a program signal. However, since the device is reacting to overall program level, there is always the risk that the very low-level signals will be lost if they fall below the threshold of the device. Therefore, as a noise reduction device, the expander must be used with caution. Its effectiveness varies greatly according to the nature of the program being treated. The preceding chapter deals in detail with the expander's operating characteristics.

A compressor may also be considered in the same category of noise reduction devices. By compressing the dynamic range of the program, the overall gain can then be brought up, thereby raising the lower-level segments above the noise while preventing peaks from saturating the tape.

Both expansion and compression generally leave something to be desired in the area of noise reduction. The signal processing effects of either device alone become noticeably audible long before any significant noise reduction has been accomplished. Therefore, while compressors and expanders are valuable studio tools, they are rarely considered for noise reduction.

The Compressor/ Expander (Compander)

As just described, neither the compressor nor the expander is an effective noise reducer. However, a complimentary combination of the two devices can achieve a degree of noise reduction not anticipated by observing the characteristics of either unit used alone.

The compressor/expander combination is called a "compander," and it is the basic principle behind the operation of most commercially-available studio noise reduction systems. Analog noise reduction techniques have not become standardized, nor are the various systems currently in use compatible. This means that a tape that has been recorded using one system cannot be played back properly on another. Even systems manufactured by the same company, but designed for different applications, are not compatible and may not be intermixed.

In theory, the compressor/expander (compander) noise reduction system is reasonably straight forward, though in practice, some highly sophisticated circuitry is required. To help understand the basic principles, consider a music program with a dynamic range of 100dB, and a recording medium whose maximum dynamic range is 75dB. A simple compressor with a 100:75 (4:3) compression ratio will keep the high-level program peaks at or below 75dB. However, as pointed out in Chapter 8, such compression will also attenuate low-level signals, placing them even further below the residual noise floor. As a noise reducer, this system does not look too promising. After recording, we may recover our 100dB dynamic range with a complimentary expander set to a 3:4 ratio. But the expansion will also bring up the noise level along with the program material. A more likely approach would be to select a convenient compressor rotation point so that input levels above this point would be reduced in level, while levels below the rotation point would be raised, thereby bringing them up above the noise floor.

Whenever a dynamic, complimentary noise reduction system (compander) is used, it is particularly important to ensure that the tape recorder's electronics are properly aligned. In the playback mode, the noise reduction system's expander restores the recorded signal to its proper level and frequency response, while noise level is brought down. For optimum operation, the tape recorder itself must be well aligned so that the expansion does not react to errors of gain or frequency response within the tape recorder. If the noise reduction system's output does not match the original input, the deviation is known as tracking error.

Because noise reduction is so important within the multi-track recording system, and in light of the differing design concepts and incompatibilities among devices, the various systems will now be described below in some detail.

■ dbx Noise Reduction System

The dbx noise reduction system uses a compander similar to the one just described. Certain refinements have been made, however, to minimize the audible effects of compression and expansion. In the basic compander system, the noise level continually varies according to the degree of expansion in the playback circuit. Although the human ear may eventually disregard a steady level of noise, a fluctuating one-even at low level-becomes a distraction. As mentioned earlier in the chapter, hiss may be quite audible when the program is primarily in the lower frequency range. Therefore, although the basic compander may allow us a greater dynamic range, its noise reduction capabilities require improvement.

In the dbx Type I noise reduction system, that improvement is realized through a 12dB high-frequency pre- and post-emphasis network. High frequencies are boosted prior to recording (Figure 9-2) and then attenuated on playback. Since tape noise occurs after the pre-emphasis, the post-emphasis reduces the noise while restoring the original frequency balance. If the tape's saturation and self-erasing characteristics (discussed in Chapter 10) could be ignored,

Figure 9.2 The 12db high-frequency pre-emphasis in the dbx system's record circuit.

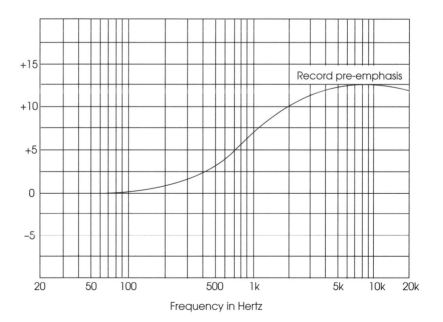

this high-frequency pre- and post-emphasis might be the extent of the noise reduction system. High frequencies would be boosted above the hiss before recording and then returned to normal on playback, with the resultant post-emphasis dropping the hiss level below audibility. However, as mentioned earlier in the discussion on static complimentary systems, the dbx's considerable high-end boost will certainly drive the tape into saturation in the presence of any sort of high-frequency program material. To prevent this, the level-sensing circuitry in the compressor (record side) contains an additional high-frequency boost. Consequently, when the program contains high-frequency components, the compressor overreacts and in effect, works against the program pre-emphasis by bringing the entire level down, thereby keeping the boosted high frequencies from overloading the tape.

Figure 9-3 is a block diagram of the dbx Type I record/playback system, and Figure 9-4 is a graph of the compressor/expander functions. In Figure 9-4A, a program with a 100dB dynamic range is

Figure 9.3 Block diagram of the dbx noise reduction system: a. High-frequency equalization boost is applied to input signals; b. Input signal is compressed; c. Significant high-frequency components cause greater amount of expansion; d. Input signal is recorded; e. Output signal is reproduced; f. Output signal is expanded; g. Significant frequency components cause greater amount of expansion; h. High-frequency equalization attenuation restores original flat frequency response of program.

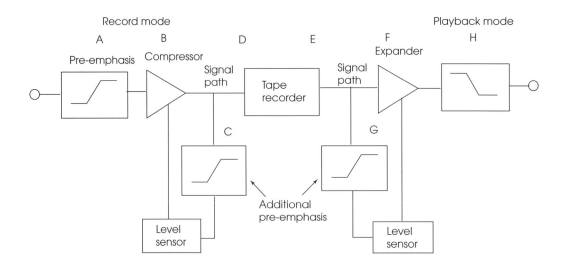

applied to the dbx system's input. In the record mode, the system functions as a compressor with a 2:1 compression ratio. The program's -80dB to +20dB range is compressed to -40dB to +10dB. As a result, low-level program material is raised above the noise level of the tape, while high-level peaks are kept below the saturation point. When the compressed program is played back, the dbx functions as an expander, as shown in Figure 9-4B. The dynamic range of the signal is restored to 100dB, and the -60dB noise floor is reproduced at -120dB, which is well below the threshold of audibility.

Another benefit of the dbx Type I noise reduction system is in the area of headroom improvement. Note that the compressor reduces the input levels that lie above the rotation point. Therefore, higher level output signals from the console may be handled by the tape recorder. Although the magnetic capacity of the tape is not

Figure 9.4 a. In the record mode, the input signal is compressed by a 2:1 ratio. The adjustable rotation point has been set for a 0 VU (=+4dBm); b. In the playback mode, the signal recorded on the tape is expanded at a 1:2 ratio. The rotation point remains at 0 VU.

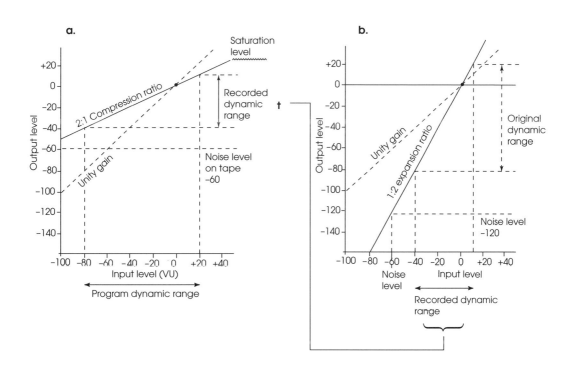

improved, the dbx expander restores the compressed high-level signals to their original value, thus improving the tapes' apparent headroom capabilities.

Figure 9-5 is a partial listing of the dbx Type I noise reduction system's specifications. Note that the specifications claim 10dB of noise reduction on signals with dominant energy below 500Hz. This type of signal takes full advantage of the system's pre- and post-emphasis circuit, since most of the musical energy lies below the high-frequency equalization boost that was shown in Figure 9-2. The specifications also claim varying amounts of additional noise reduction depending on the level of the input signal. This may be seen in Figure 9-6, which shows that the dbx's compressor raises the input levels cited in the specifications by 10dB, 20dB and 30dB, respectively. Therefore, the complimentary playback expander will lower these levels the same amount, simultaneously reducing the noise level by 10dB, 20dB and 30dB.

Another type of dbx noise reduction system, the dbx Type II, has been used in the consumer marketplace on compact cassette recorders, dbx-encoded FM radio broadcasts and dbx-encoded phonograph records. The dbx Type II system was developed as a lower cost alternative to the Type I professional system, and relies on similar noise reduction principles. The main difference is the absence of the additional pre-emphasis circuit that affects the level sensor in the presence of significant high-frequency components. Instead of this circuit, the dbx Type II system uses a different set of initial pre- and post-emphasis curves. These differences make the two systems incompatible, and a tape encoded with Type II should not be decoded with Type I, and vice versa, since tracking errors will occur in the expander section. Type II phonograph records and FM radio broadcasts have become quite rare, but a version of the

Figure 9.5 dbx noise reduction system specifications.

At +4 dBM signal level, hiss and high-frquency modulation noise are reduced by 10dB on signals whose dominant energy is below 500Hz.
At −16 dBm signal level, there is an additional 10 dB of noise reduction.
At −36 dBm signal level, there is an additional 20 dB of noise reduction.
At −56 dBm signal level, there is an additional 30 dB of noise reduction.
At −56 dBm signal level, there is an additional 40 dB of noise reduction.

dbx Type II system has been adopted by the television industry as the NAB (National Association of Broadcasters) standard for MTS (Multi-channel Television Stereo) broadcasts. Figure 9-7 shows a comparison of dbx Type I and dbx Type II pre- and post-emphasis curves.

It is extremely important that the tape recorder(s) in use be aligned properly for input/output characteristics, as well as for record and playback equalization, to prevent tracking errors as discussed earlier. Since the dbx noise reduction system affects the entire frequency bandwidth and dynamic range simultaneously, gain loss errors can be corrected by adjusting the playback level after decoding the signal. This will raise the noise floor by the same number of decibels. However, if the error was one of equalization rather than overall level, the equalization error would be doubled, while the rest of the bandwidth would be properly reproduced. This type of tracking error would be quite noticeable, and extremely difficult to correct later on.

Figure 9.6 The dbx compressor raises low-level signals. The farther they are below the rotation point, the greater the increase in level.

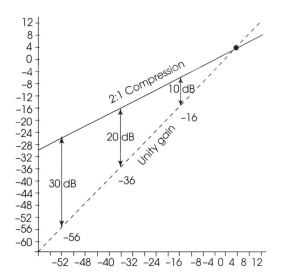

Dolby Type "A" Noise Reduction System

Unlike the dbx noise reduction system, the Dolby Type "A" system does not react to high-level program material. In the former system, high as well as low levels are compressed prior to recording with corresponding expansion on playback. The Dolby "A" system takes into account the fact that compression and expansion actions are most likely to be audible in high-level program material, and that noise reduction is unnecessary once the level is sufficient to mask background noise. Accordingly, the noise reduction system operates on low-level signals only. Figure 9-8a is a simplified description of the basic principle of the Dolby "A" system in the record mode. To understand how the system works, consider a very low-level input signal, with a value of about 2.5 millivolts (about -50dB). This input signal is routed, a) by way of a compressor in the side chain, and b) directly to a combining amplifier. The compressor's gain-before-threshold is 2.16. Therefore, low-level signal voltages through the compressor are multiplied by a factor of 2.16. The signal that travels the direct path is multiplied by 1 (unity gain). At the summing amplifier, the two signals combine to produce an output voltage that is 3.16 (2.16 + 1) times the voltage input. Therefore, the gain of the system is:

$$NdB = 20 \log \frac{Output}{Input} = 20 \log \frac{3.16 \times 2.5mv}{2.15mv} = 20 \log 3.16 =$$

$$20 \times 0.49968 = 10dB$$

In other words, the low-level signal has been amplified by 10dB.

As the input level is increased beyond the compressor's threshold, its gain is reduced, and the compressor's output contributes less and less to the summing amplifier. By the time the input signal has increased to -10dB, the system has become a simple unity-gain amplifier because the compressor's contribution to the summing amplifier is now negligible compared to the direct path.

In playback mode, the compressor is placed in a feedback path as seen in Figure 9-8b. Its output is combined subtractively with the direct signal, which means it is combined with the boosted low-level signal that was recorded on the tape. If the feedback path were not there, the system's output would equal the input, since the combining amplifier itself is simply a unity-gain device. However, the feedback loop reduces the overall system gain by the amount of gain through the compressor in this case. The principles of feedback require a textbook of their own, but we may get a fair idea of the nature of this type of circuit if we can assume, for the moment, that the output of the combing amplifier is 2.5 millivolts. The compressor's gain-before-threshold remains at 2.16; consequently, its output is 2.16 x 2.5 millivolts. This output, combined subtractively with the direct input signal that was boosted earlier to 3.16 x 2.5 millivolts, gives us a system output of (3.16 - 2.16) x 2.5 millivolts = 1 x 2.5 millivolts = 2.5 millivolts. The playback mode system has thus attenuated the recorded signal back to 2.5 millivolts, that is, to its original value.

This brief analysis leaves many questions about feedback unanswered. For example, we have assumed that the playback mode output is what we want it to be (2.5 mv). We then used this value in the feedback path of the compressor, after which it was subtracted from the unity-gain signal. Despite this superficial explanation of the feedback system, it should show that in the playback mode, the Dolby Type "A" noise reduction system restores the low-level signals to their original values by lowering the system gain, and therefore the accompanying noise level, by 10dB. In record mode, higher level signals routed through the side chain are compressed, and as the signal level increases, the compressor's gain decreases, contributing less and less to total output. By the time the program level has risen to -10dB, the compressor's output is negligible with respect to the direct path signal.

Figure 9.7 Pre- and post-emphasis curves for the dbx Type I and Type II noise reduction systems.

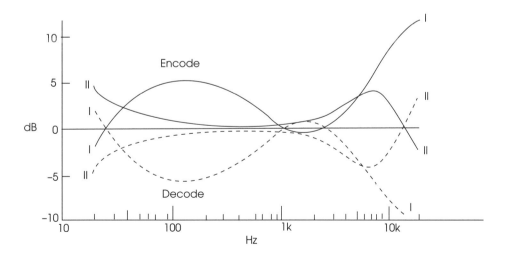

Figure 9-9 is a graph of the Dolby system transfer characteristics in both the record and playback modes. The graph summarizes the noise reduction action. In the record mode, low-level signals are boosted by 10dB, while higher level signals are passed at unity gain. In the playback mode, the boosted low-level signals are attenuated to the original level, while once again, the higher signal levels are passed at unity gain. Note that even though the side chain remains a compressor, its subtractive combination with the direct signal produces an expansion characteristic. Closer examination of the graph shows that the compression/expansion takes place over a relatively small segment of the entire dynamic range. Therefore, much of the signal is unaffected by the companding action.

To further enhance the system and reduce the audible effects of companding, the audio bandwidth is split into four sections. Each

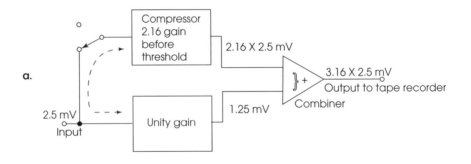

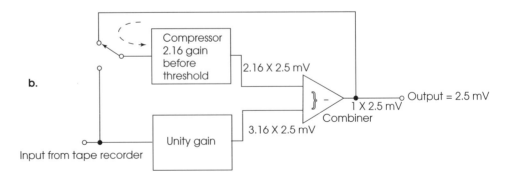

Figure 9.8 Block diagram of the Dolby Type "A" noise reduction system. Low-level signals are raised before recording (a), and then restored to normal before playback (b).

section has its own compression/expansion chain and is thus unaffected by the other. Above 5kHz, the companding action increases gradually until at 15kHz and above, there is 15dB of noise reduction. The four frequency bands are derived from routing the full bandwidth signal through a set of passive filters. These filters are:

1. 80Hz low-pass filter.
2. 80Hz - 3kHz band-pass filter.
3. 3kHz high-pass filter.
4. 9kHz high-pass filter.

The outputs of these filters are then sent to the individual compander inputs. It is the overlap of the top two bands that creates the extra 5dB of noise reduction. The selection of these bands takes into

Figure 9.9 Dolby noise reduction system transfer characteristics.

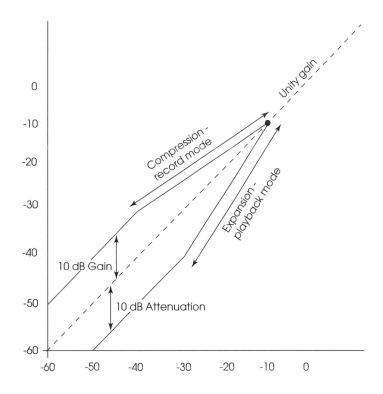

account that fact that at low to moderate program levels, most of the music may lie within the 80-Hz to 3-kHz band, with much less energy found at higher and lower frequencies. Therefore, full noise reduction capabilities may continue to operate in the high frequency (above 3-Hz) ranges, keeping tape hiss at a minimum, while the music is unaffected by compansion.

Two simplified versions of the Dolby noise reduction system have become very popular in many consumer compact cassette tape recorders. These systems, known as Dolby "B" and Dolby "C," are single-band devices. Dolby "B" and "C" only affect the high-frequency portion of the audio spectrum. They use a 5kHz high-pass filter only, in place of the multiple band/filter system incorporated in Dolby "A." Their purpose is to reduce the most audible (high-frequency) portion of tape noise. Commonly found only as built-in circuits, these systems are usually not used professionally. Dolby "B" gives 10dB of noise reduction above 5kHz, and using two "B" circuits serially, Dolby "C" provides 15dB of noise reduction.

Since the Dolby system's companding action affects only a small portion of the program's dynamic range, tracking errors will be confined to levels below -10dB. Since the program material may be above or below that level, these errors cannot be corrected later on. The Dolby system is more tolerant of equalization errors since the tracking error is confined to a small portion of the total dynamic range, but gain errors must be eliminated before playing back the tape through a Dolby expander. To remove these gain-related problems with several magnetic tape operating levels in use, it is important that all Dolby-encoded tapes contain a Dolby level tone at the beginning for Dolby alignment purposes. This tone may be at any convenient magnetic reference level, so long as it matches the level at which the recording was made. When the tape is played back at a later date, the tone is used to verify the input level to the Dolby expander. If the tone does not line up with the Dolby level marker on the Dolby meter panel, the tape recorder output level (or the Dolby input level control) must be adjusted before playing the tape. It is extremely important for this tone to be placed at the head of the tape and that the equalization standard (NAB or IEC) be set properly, because many studios and broadcast facilities use different standards and various elevated operating levels. The Dolby alignment tone is a warble tone that is quite unique and is difficult to mistake for anything else.

Telcom C-4 Noise Reduction System

Another high-quality noise reduction system, although no longer available, is the Telcom C-4 noise reduction system, which combines some of the noise reduction methods found in both the dbx and Dolby systems. The system uses compression and expansion ratios of 1.5:1 and 1:1.5, respectively, as opposed to the 2-to-1 relationship found in the previously discussed systems. The companding action was constant over the entire dynamic range and therefore extended the dynamic range by a factor of 1.5. A tape recorder system with a possible dynamic range of 70dB will have its signal-to-noise ratio improved to 105dB when supplemented by the Telcom system. In addition to this full-range companding, the Telcom system separated the audio bandwidth into four frequency bands. The system used a rapid attack time, coupled with a slow release time for minimum audibility. The attack and release times vary, and are optimized for each band in the system. The filters used to divide the frequency bands are:

1. 215Hz low-pass filter.
2. 215Hz - 1450Hz band-pass filter.
3. 1450Hz - 4800Hz band-pass filter.
4. 4.8kHz high-pass filter.

One of the advantages of this system was its freedom from tracking error problems. Gain errors can be corrected on playback, like in the dbx system, while equalization errors were restricted to their particular audio band, as in the Dolby system. A reference identification tone allowed the engineer to calibrate the system and the tape together for proper playback. Telcom C4 was also used for other transmission lines such as cable, radio links, and satellite systems. A special version of the Telcom C4 was available and recommended in these applications. This modified system used a higher 1:2.5 compression ratio, along with a corresponding 2.5:1 expansion rate.

Dolby Spectral Recording System

The most recent addition to the list of noise reduction systems is the Dolby Spectral Recording noise reduction system, commonly called Dolby "SR." It differs from the traditional Dolby "A" system in many ways, but continues to utilize a main signal pathway in conjunction with a side-chain path. The main signal path is without dynamic signal processing, while the side chain includes a complex series of multiple compressors. At the output of the system, the main signal path and the side chain signal path are summed additively or subtractively, depending on whether the encode or decode function is selected. Prior to compression, the high- and low-frequency components of the side-chain signal are boosted in a manner called "spectral skewing." This effectively raises overall levels further above the noise floor and adds gain prior to compression.

The side-chain signal is divided into three multi-level stages, with the high- and middle-level stages subdivided into high-frequency and low-frequency bands. The low-level stage is not divided, but operates in the high-frequency range only. The crossover frequency between the high and low frequencies of the high- and middle-dynamic stages is 800Hz, with the low-level stage using an 800Hz high pass filter. The filters have gradual slopes and this results in an overlap in the 200Hz to 3kHz region. Therefore, this four-octave intelligibility range is processed by both the high-frequency and the low-frequency compressor in the high-level and mid-level stage, while only the high-frequency component of the low-level stage is affected. This gives a total of five compressors in simultaneous action for each encoding or decoding channel of the system. Each compressor uses a combination of fixed and sliding threshold bands and a 2:1 compression ratio. The thresholds for the multi-level dynamic stages are -30dB, -48dB, and -62dB respectively. This creates dynamic processing for the low frequencies from -48dB to -5dB and from -62dB to -5dB for the high frequencies. No dynamic processing occurs outside these limits; full signal boosting or "spectral skewing" is nevertheless still in effect.

The purpose of the Dolby "SR" system is to reduce only the noise that can be heard by the human ear, leaving subsonic and supersonic sounds unaffected. This results in a decrease in modulation noise and intermodulation distortion. Reductions in harmonic distortion in the recording system are also achieved as well as the elimination of virtually all tape hiss. The overall decode and encode curves bear a resemblance to the Robinson-Dadson Equal Loudness Contours discussed and pictured in Chapter 2. Figure 9-10 shows a simplified block diagram of the Dolby "S" system.

The dual-path multi-level staggered Dolby "SR" System is much less susceptible to tracking errors than the Type "A" system, but Dolby calibration is still required. A pink-noise generator is included in the unit and is used to calibrate not only level, but equalization of the recording system as well. The operating-level full bandwidth calibration eliminates tracking errors of both gain and equalization, and ensures that the Dolby "SR" noise reduction System is inaudible under all conditions.

ANALOG NOISE REDUCTION 265

■ SUMMARY

All the noise reduction systems just discussed are based on the compander principle. Compression on recording is followed by expansion on playback. The gain of low-level signals is raised so that they are recorded as high above the noise floor as possible. On playback, the low-level signals are restored to their normal levels, while gain reduction reduces the noise level by the same amount. Representative production models of the Dolby and dbx noise reduction systems described in this chapter are pictured in Figures 9-11a through d.

Due to the significant differences between the various systems, a tape encoded in one system may not be decoded with another system's playback unit. If an encoded tape is played back with no

Figure 9.10 A simplified block diagram of the Dolby SR noise reduction system (Dolby figure).

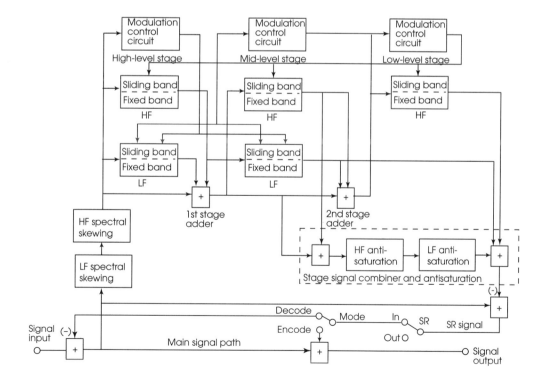

Figure 9.11   a. dbx Type I noise reduction unit designed for rack mounting. (dbx 911; dbx photo); b. Two dbx Type II noise reduction units designed for rack mounting (dbx 941A (encode) and 942A (decode); dbx photo); c. Dolby Spectral Recording and Type A dual channel noise reduction unit (Dolby 363, Dolby photo); d. Multi-channel Dolby SR/A noise reduction rack with SR noise reduction cards mounted. (Dolby SRP series; Dolby photo).

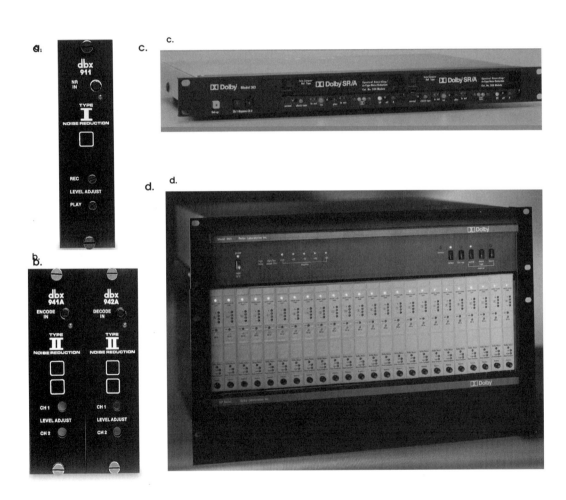

complimentary noise reduction system, or with the wrong system, its dynamic range and frequency response will be distorted. It must also be remembered that compander noise reduction systems are for use with transmission systems only, and signal inputs and outputs must be related. A system such as the Dolby "A" system could not be used to quiet a noisy reverberation chamber. Such a chamber would be creating new sounds that would be inserted between the encoder and the decoder and would cause severe tracking errors.

Noise reduction systems are often designed with an automatic switching function which inserts the system in the input line to the tape recorder during recording, and into the playback or output line at all other times. Thus, the same noise reduction device serves a dual function and can be used for encoding and decoding, though not at the same time. Despite its economic advantage, this arrangement prevents the engineer from properly monitoring the tape while recording.

Several companies now offer what are called "TTM" noise reduction frames. These are rack frames that contain a power supply and switching facilities, along with a number of open slots. The slots are designed to accept noise reduction circuit cards from all major noise reduction system manufacturers. A typical TTM frame will accept dbx Type I and Type II system cards, as well as Dolby "A," Dolby "SR" and Telcom C4 circuits. This allows the engineer to select whichever system he or she prefers, as well as accommodate the various tapes that may be brought into the studio. By simply switching to the appropriate card, the proper noise reduction system is inserted into the designated signal path.

SECTION IV

Analog Audio Recording Systems

■ INTRODUCTION

The tape recorder is rarely thought of as a signal processing device or a transducer, for at all times its output is expected to be a faithful replica of its input. Yet it does indeed process the signal, first from electrical energy to magnetic energy, and later back to electrical energy.

Chapter 10 begins with a brief discussion of the basic principles of magnetic recording, in which tape and tape recorder are viewed as component parts of a total system. The chapter continues with an examination into the makeup and characteristics of modern magnetic tape. Included is a discussion of elementary magnetic principles.

Chapter 11 examines the tape recorder transport, electronic systems, and the entire record/playback process, while Chapter 12 describes the all important alignment procedures that allow the tape and tape recorder to work together as an integrated system.

TEN

Magnetic Tape

Ancient philosophers knew that a certain type of rock, known as lodestone, would mysteriously draw to it particles of iron. This power came to be known as magnetic attraction, and in time was recognized to be a physical, not spiritual, quality. Materials possessing this power are now called magnets.

An artificial magnet can be created by winding a coil of wire around a bar of metal, as shown in Figure 10-1. When a direct current is applied to the coil, a magnetic field is set up and the bar becomes a magnet. Depending on the particular material, the bar may or may not remain magnetized after the current is withdrawn. Hardened steel will retain its magnetic properties long after the applied current is shut off, whereas soft iron is considered a temporary magnet. Although it can be readily magnetized, it does not retain the magnetization in the absence of an applied current.

If a hard ferrous material—that is, an iron-based material with the capacity to acquire permanent magnetic characteristics—is temporarily subjected to a magnetic field and then removed from the field's vicinity, the residual magnetism now stored in the magnet can be measured. This measurement enables us to determine the strength of the force that magnetized the material. The measurement may be made whenever it is convenient to do so, since the material was permanently magnetized at the instant it was withdrawn from the applied magnetic field. It follows then that a series of ferrous particles may be magnetized, each by a different magnetic field strength, and later, measurements of each particle will tell us, as before, the strength of the applied magnetic field at the instant that the particle was magnetized.

These basic principles of magnetism provide the foundation for the magnetic tape recording process. As illustrated in Figure 10-2, the transport system moves the magnetic tape past the recording head where the program to be recorded is magnetically stored on the tape. When the stored program passes the playback head, the playback amplifier produces a signal that corresponds to the input signal, as originally applied to the record amplifier. In practice, the record/playback process is, of course, considerably more involved.

Figure 10.1 An artificial magnet.

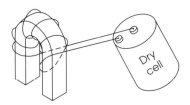

Figure 10.2 The basic tape recorder system.

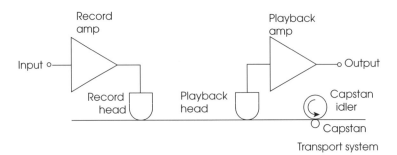

THE MAGNETIC RECORDING SYSTEM

The Tape Transport System

Figure 10-3 shows the essential components of the tape transport system. The tape to be recorded (or played back) is placed on the supply reel side of the transport. The tape is threaded past the tape guides and head assembly to the empty take-up reel. The capstan idler or "pinch roller" presses the tape against the capstan, pulling it past the heads at the required speed, while the take-up motor winds the tape onto the empty reel.

Figure 10.3 The basic tape transport system.

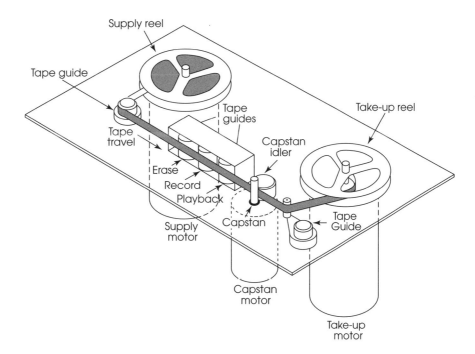

The Record and Playback Heads

The record head itself is, in effect, a temporary magnet. An alternating current, proportional to the audio signal to be recorded, is fed through the coil in the head. The coil is wrapped around a soft iron ring core, in which the applied current creates an intense magnetic field (Figure 10-4). A gap has been cut in the core, and this gap is bridged by magnetic lines of force, known as flux. Flux is the magnetic equivalent of current flow. As the magnetic tape passes the gap, some of the flux passes through the tape, magnetizing it. The flux lines readily penetrate the tape, since this path offers less reluctance to the flux than the air gap itself. Reluctance is the term for opposition to a magnetic force, and is analogous to resistance in an electrical circuit.

The playback head is similar to the record head in construction and operation. A coil of wire is wrapped around a soft iron core. The tape, previously magnetized by the record head, passes by the gap in the playback head. The tape flux penetrates this gap, inducing a current, proportional to the flux, within the coil winding. The playback amplifier converts this induced current into a voltage, which then appears at the tape recorder's outputs.

Figure 10.4 The record head magnetizes the tape as it passes the gap.

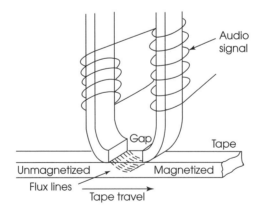

Magnetic Tape

Magnetic tape consists of a polyester-type base material upon which a magnetic coating has been placed. This coating consists of a solution in which gamma ferric oxide (Fe_2O_3) particles are suspended. Depending on the particular application for which the tape is intended, each particle is from 7 to 20 micro-inches in length. The particles are generally one-third to one-sixth as wide as they are long. The particles are mixed in a ball mill with other materials such as a plasticizer, conductants, dispersants, and lubricants. The mixture is composed of 30% solvents, 10% solids (other than the oxide), and 60% oxide particles. When properly mixed, the solution looks like and has the consistency of chocolate syrup and smells like nail polish.

In a continuous process, the oxide formulation is spread on moving base film. Several methods are used to spread the mixture, with reverse-roll coating, gravure coating and knife-edge coating the preferred methods.

Similar to the oxide coating, the back coating of a tape is a thin layer of carbon pigment (rather than oxide particles) suspended in a binder. The resultant surface texture reduces the possibility of slippage as the tape is pulled by the capstan and capstan idler. The back coating also provides more even tape wind, particularly at high speeds. This is critical because uneven wind can cause edge damage with resultant loss of some signal. The coating is generally more abrasion-resistant than uncoated polyester and in addition, its low resistivity minimizes the build-up of static charges. Typical back-coating thicknesses range between 0.05 and 0.15 mils.

Before the formulation has set, the tape is passed through a strong direct-current magnetic field, which physically arranges the particles in a lengthwise orientation with respect to the tape. For video applications, the particles are oriented in a vertical direction. In either case, the particles should be oriented perpendicular to the head gap for the system to work efficiently.

The tape is then dried, which drives off the solvents, and then calendered. The calendering process involves passing the tape between two high-pressure, heated rollers, which smooths the surface of the tape. This is a critical step since the tape's smoothness greatly affects tape-to-head contact. Bumps in the tape cause the tape to lift away from the head for a brief instant; such small surface imperfections in the tape cause dropouts.

After calendering, the tape is stored and "cured" for a brief period of time. This allows the tape to stabilize and the last of the solvents to evaporate. At the end of the storage time, the tape is cut into the proper widths and packed for shipment.

Early tape made use of a paper base, upon which the ferric oxide was distributed. Later developments produced both acetate- and mylar-base films with the thickness of the base ranging from 0.5 mils to 1.55 mils. So-called "extended play" tapes used the thinner base film so that more tape could be wound on a given reel size. However, professional mastering tapes today use a base film thickness of either 1.0 or 1.5 mils, with the latter being preferred for greatest stability. A compound polyester base film with a very high tensile strength that resists breaking, as well as providing low stretch, is currently used for professional audio tape.

A typical studio-grade professional mastering tape that relies on a base film thickness of 1.5 mils may total 2.03 mils in thickness when the oxide formulation and back coating mixture are added. This so-called 1.5 mil tape will have a total length of 2500 feet when wound on a 10-inch NAB (National Association of Broadcasters) reel. Occasionally a reel of 1.0 mil tape is required and usually has a total length of 3600 feet. The record/playback times of these tapes at a speed of 15 inches per second are 33 minutes and 45 minutes respectively.

■ THE RECORDING PROCESS

Important Terms

Coercivity: Measured in oersteds, the coercivity (abbr. H_c) of a tape is an indication of its sensitivity to an applied magnetic field. The published specifications usually list the strength of the magnetic field needed to bring a saturated tape to full erasure. The coercivity of representative studio tapes generally ranges between 280 and 380 oersteds.

Retentivity: Retentivity is a measure of a magnetic tape's flux density after a magnetic field producing full saturation has been withdrawn. Retentivity is expressed in gauss. Gauss is the number of magnetic flux lines per cross-sectional square centimeter of tape.

Remanence: Remanence describes the same condition as retentivity, but is expressed in lines of flux per linear quarter-inch of tape width. This specification is particularly significant, since the playback head's output level is a direct function of the tape's remanence. Figure 10-5 illustrates the measurement of both retentivity and remanence.

Sensitivity: Sensitivity is an indication of a tape's relative output level as compared to some specified reference tape. Thus, given the same input level, a tape with a sensitivity of +2 will produce an output level 2dB higher than the standard reference tape.

Harmonic Distortion: Recorded third harmonic distortion, in addition to being a function of bias level, increases as the recorded level of the tape approaches saturation. Maximum allowable recorded level is usually defined as that level at which the third harmonic distortion reaches 3%.

Headroom: A tape's headroom is the difference between standard operating level (to be explained shortly) and the 3% third harmonic distortion point. Thus, if a tape reaches 3% third harmonic distortion when the applied input level is 9dB above 0vu (when 0vu = standard operating level), it is said to have a 9dB headroom.

Figure 10.5 A comparison of remanence and retentivity.

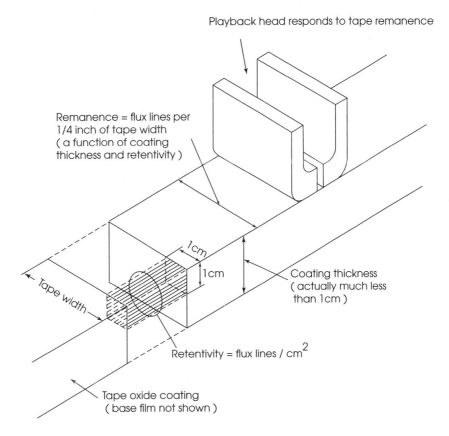

The Transfer Process

When the ferric oxide particles described above pass through the magnetic field that has been created at the record head, they become magnetized. The pattern of magnetism that is spread longitudinally along the length of the magnetic tape is analogous to the signal that created the magnetism. It is the record amplifier that converts the applied signal voltage from the console or other device into the current that flows through the record head. Each magnetic particle consists of one or more domains.

A domain is defined as 10^{18} molecules of gamma ferric oxide, and is the smallest physical unit that can be considered a magnet. Domains are considered to be totally magnetized at all times, but since their magnetic orientation within the oxide particles is random, the net total magnetism of the tape is zero. When an applied magnetic force, as in the record head, produces a weak magnetic field, only those domains whose existing magnetic orientation approximates the direction of the applied field will be forced from their natural "unstressed" state into magnetic alignment with the field. When the field is withdrawn, most—if not all—of the domains will return to their original unstressed, random orientation. In other words, the tape remains non-magnetized.

If a stronger magnetic force field is applied, many more domains will be forced into alignment with it. When the field is withdrawn, some domains will return to their original random or unstressed orientation, while other domains will remain in the orientation forced upon them by the applied magnetic field. Consequently, the tape will remain magnetized. This residual magnetism is not quite as strong as the applied magnetic force (since some of the domains returned to their unstressed orientation when the force was withdrawn), yet the tape definitely remains in a measurable state of magnetization.

If an even stronger magnetic force is applied, all the domains will be forced into alignment with it. When this happens, the tape is said to be saturated, which means that no further magnetic action can take place. In this case, when the magnetic force is withdrawn, very few domains will return to their original unstressed orientation, and the tape is now in its maximum possible magnetized state. Once the applied magnetic field forces the tape into saturation, any further increase in the field strength will have no further effect on the tape. When all domains have been forced into alignment with the applied magnetic field, the tape is incapable of further responding to even greater field strengths.

Transfer Characteristics and Distortion

In order to be of practical use as a program storage medium, a magnetic tape must possess a linear transfer characteristic. Its playback output must be a non-distorted (or linear) replica of the original recorded input signal. A linear transfer curve, that is, one in which input equals output, is shown in Figure 10-6a. Low- and high-level signals must retain their proper relative amplitudes if a usable recording is to be made. When we apply the basic properties of magnetism to recording tape, however, we may find that the weakest audio signals do not get recorded at all. While moderate level signals are recorded satisfactorily, very high-level signals are distorted if the signal strength is sufficient to drive the tape into saturation. In other words, magnetic tape recording seems at the moment to be a non-linear process, as in fact it is if corrective steps are not taken. Before describing these corrective measures, we can illustrate the magnetic properties of recording tape by drawing the so-called hysteresis loop, as shown in Figure 10-6b.

Figure 10.6 a. A linear transfer curve (input = output); b. The hysteresis loop. A graph of the magnetism left on a recording tape by an applied magnetic field. B_r = retentivity; H_c = coercivity.

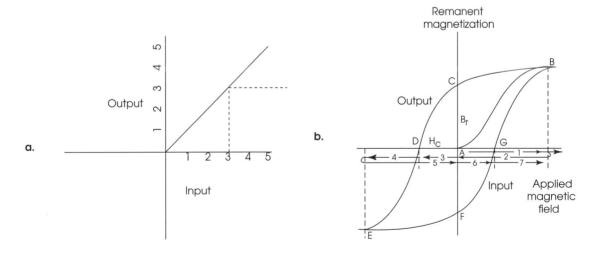

Magnetic force, measured in units called oersteds, is symbolized by the letter H. When a tape passes through a magnetic field, the magnetization that remains on the tape after the magnetic field is withdrawn is called remanent magnetism. This remanent magnetism is measured in gauss, and is symbolized by the letter B. A tape's ability to store magnetic energy is referred to as its retentivity, B_r. To illustrate the magnetic tape's storage capabilities, we shall apply a direct current to the record head. Starting at zero, the amount of current applied is gradually increased in the positive direction toward the right. In Figure 10-6, the applied current creates a gradually increasing positive magnetic force (Arrow 1). The positive magnetization of the tape increases, non-linearly, from zero (A) to positive saturation (B). Once the saturation point has been reached, any additional increase in magnetic force (the dashed line extension of arrow 1) has no further effect on the tape.

When the applied force is gradually decreased to zero by reducing the applied current (Arrow 2), the magnetism left on the tape decreases, not to the zero value at (A), but to (C). The heavy vertical line, B_r, represents the retentivity of the tape, which is the remanant magnetism left on the tape after the applied magnetic force has been reduced to zero. If we cease our experiment at this point, the tape would remain magnetized. However, if a gradually increasing negative force is applied (Arrow 3), the tape's magnetism will be reduced to zero (D) when that force reaches the magnitude represented by the solid horizontal line H_c. H_c is the coercivity of the tape. This is the force required to completely demagnetize the tape, and return all domains to their unstressed state. As the negative force continues to increase (Arrow 4), the tape will become negatively magnetized. Magnetism will increase from zero (D) to negative saturation (E). In either condition of saturation, all domains are magnetically aligned with the applied magnetic field. The terms negative and positive simply refer to the opposite direction taken by the lines of force, and consequently the orientation of the domains. When the negative force is diminished to zero (Arrow 5), the remanant magnetism falls off to (F). Once more, the tape is left in a magnetized state, but the polarity is opposite that shown earlier as B_r. To restore the tape to a non-magnetized state (G), a gradually increasing positive force must be re-applied (Arrow 6). A further increase in the positive direction (Arrow 7) will once again drive the tape to positive saturation (B).

In this rather tedious step-by-step explanation of the properties of the magnetic tape medium, direct current was used so that we could observe the effect of a force of gradually changing magnitude and polarity. In practice, of course, we apply an alternating current, such as a sine wave, to the record head and thence to the tape. In terms of the properties of magnetic tape, however, frequencies within the audio range may be considered as fluctuating direct-current voltages.

To study the effect of recording an alternating current signal on to tape, we may draw a transfer characteristic, such as the one shown in Figure 10-7. This transfer characteristic is simply a graph of retentivity versus applied magnetic force. Although the curve superficially resembles the line between (A) and (B) in Figure 10-6b, it is actually derived mathematically from a series of hysteresis loops. The hysteresis loop shown in Figure 10-6b illustrates the retentivity of a tape that was driven to the saturation point. If the applied force had been somewhat less, however, the hysteresis loop would have been smaller and somewhat differently shaped, thereby giving a different value of retentivity. For any value of applied force, there is a corresponding value of retentivity once that force is withdrawn. For example, if a positive force of H1 is applied, as shown in Figure 10-7, the tape will be left with a retentivity of B_1. If a negative force, H_2, is applied, the retentivity will be B_2, and so on. Therefore, the curve of Figure 10-7 should be understood to be a plot of retentivity for any value of magnetic force applied between negative and positive saturation.

Figure 10.7 A typical transfer characteristic for magnetic tape. An applied magnetic force of H_1 will store a magnetic field of B_1 on the tape.

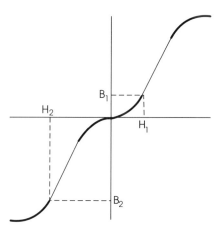

In Figure 10-8, the transfer characteristic is redrawn in an exaggerated form for clarity of explanation. Five waveforms (A-E) are shown, representing various magnitudes of alternating current supplied to the record head. In waveform (A), the progressively longer arrows indicate an increasing magnetic force, first positive, and then negative. The smallest arrows indicate forces that do not permanently affect the tape. The next two sets of arrows indicate forces that do affect the tape slightly, and the largest arrows indicate an applied force operating for the moment within the linear portion of the transfer characteristic. The resulting magnetism left on the tape as a result of waveform (A) may be plotted as shown on the output side of the transfer characteristic. Obviously, the output is quite distorted, both in wave shape and amplitude. The output is redrawn at (A'), and the outputs from the other wave forms (B-E) are shown as (B'-E'). Note that waveform (C) does not produce at all, and that waveform (E') is flat-topped, since the magnitude of the input signal has driven the tape into saturation.

These output waveforms illustrate the non-linear properties of magnetic tape as a storage medium. However, it should be noted that there are two segments of the transfer characteristic that are linear. These segments are identified in Figure 10-8 as well, and any portion of an input waveform that falls within these segments will be linearly reproduced. This may be noted in output waveforms (A') and (D'), where the sections measured by the vertical arrows are seen to be non-distorted reproductions of the equivalent section of the input.

Bias In Figure 10-8, the applied signal is an alternating current varying between zero and some positive, and then negative, maximum amplitude. As noted earlier, only when the amplitude forces the tape's magnetization into the linear portion of the transfer characteristic will the output be a faithful reproduction of the input. If an additional direct current were also applied, the audio signal would now alternate above and below that direct-current value instead of around the zero point as before. In Figure 10-9, a positive direct current, called a bias current, is applied to the record head. The magnitude of this bias current places it, as shown, midway along the positive linear portion of the transfer characteristic. Now, when the audio signal is also applied, the resultant magnetization will vary about this bias level, and the output will fall within this linear segment.

MAGNETIC TAPE 283

Figure 10.8 The effects of a non-linear transfer characteristic. Various levels of input signals (A through E) produce distorted (non-linear) output waveshapes (A' through E').

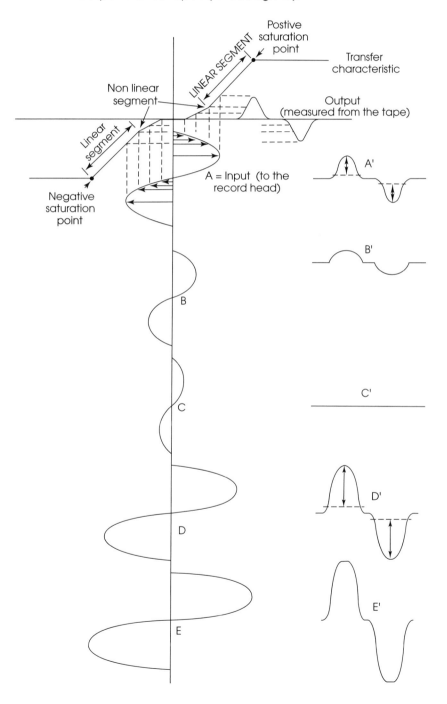

Figure 10.9 Achieving a linear transfer characteristic by using D.C. bias. The D.C. bias shifts the applied signal into one of the linear portions of the transfer characteristic.

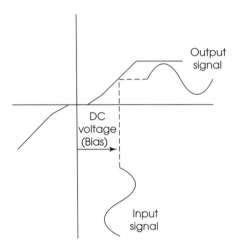

Despite the obvious improvement in linearity, direct-current biasing does not allow the tape to be used to its full potential. Only a small part of the entire transfer characteristic is being used. Consequently, the full magnetic capabilities of the tape are unutilized. Furthermore, input levels must be considerably restricted so that signal values do not fall outside the positive linear segment of the transfer curve. As a further disadvantage, this type of biasing tends to saturate the record head core and magnetize it permanently, introducing noise onto the tape and perhaps partially erasing it. Despite these limitations, direct current biasing is nevertheless a marked improvement over recording without any bias.

Many precursors of the modern magnetic tape recorder used direct-current bias, but it was found that using a very high-frequency alternating current as a source of bias provided a more satisfactory method of overcoming the non-linearity of magnetic tape. Alternating current bias, typically a frequency five to ten times the highest frequency to be recorded, is well beyond our hearing limits. In fact, it may indeed be too high to be reproduced by some playback heads. The bias is, however, actually recorded on the tape, and may be heard as a high-pitched sound if a segment of recorded tape is very slowly rocked back and forth, as in editing, across the playback head. From the shape of the transfer curve, it might be expected that the bias frequency would be recorded in a distorted form, much like any of the other frequencies illustrated in Figure 10-7. However, it has been found that at very high frequencies, the transfer characteristic becomes nearly linear over most of the range between positive and negative saturation. The degree of linearity increases with the amplitude of the bias, as shown by the curves in Figure 10-10. In simple terms, it would seem that the rapid alternations of the bias frequency overcome the magnetic medium's inertia to change in applied force.

Remember that the non-linear transfer characteristic, first seen in Figure 10-7, was mathematically derived from a series of hysteresis loops, and as such, it represents the effects of gradually changing values of applied magnetic force. In magnetic terms, any audio frequency may be considered to be a gradually changing force. By contrast, the bias frequency is a rapidly changing force. The tape is in a constant state of continuous magnetic agitation, and its resistance to change has been overcome.

Of course, the transfer to tape of a bias frequency, while no doubt academically intriguing, is of questionable artistic interest. We are,

Figure 10.10 The apparent linearizing properties of high-frequency A.C. bias. A = typical non-linear transfer characteristic of magnetic tape (no bias signal applied). B = apparent transfer characteristic in the presence of an applied high-frequency bias voltage.

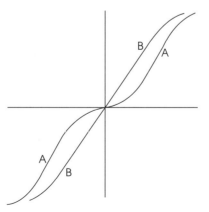

after all, concerned with recording music and speech, and not with the behavior of some frequency many times beyond the range of human hearing. However, A.C. bias allows us to store on tape a faithful reproduction of the applied audio signal at an amplitude that takes full advantage of the full storage capabilities of the tape. Rather than simply applying the audio signal directly to the record head, the signal is first combined with the bias current, which is supplied by a bias oscillator built into the electronics of the tape recorder. The bias frequency may be as low as 150kHz or as high as 450kHz. The mixture of the two will appear as shown in Figure 10-11. The figure shows that the audio signal has become an envelope on the larger amplitude bias signal. This simple addition of bias plus audio is fed to the record head. The bias frequency is linearly recorded on the tape, along with the relatively slight amplitude changes caused by the audio envelope.

When an audio signal is recorded without bias, the output contains a large segment of third harmonic distortion. This means that the output waveform will contain a component that is three times the frequency of the applied audio signal. As bias is added, this distortion is reduced as the bias linearizes the transfer characteristic. This linearizing effect is a function not only of frequency, but of amplitude as well. As the bias amplitude is increased from zero, the transfer curve becomes more and more linear, resulting in less and

Figure 10.11 A typical transfer characteristic for magnetic tape. An applied magnetic force of H_1 will store a magnetic field of B_1 on the tape.

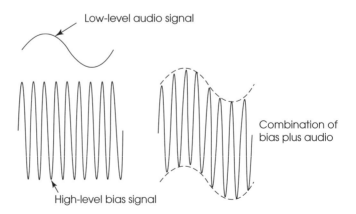

less third harmonic distortion. Conversely, as bias is decreased below optimum, distortion increases.

This may lead us to think that the optimum bias level is the level required to reduce third harmonic distortion to a minimum. Unfortunately, it is not that simple. In a typical situation, it may be found that at some point before minimum distortion is reached, the high-frequency response of the audio signal begins to fall off significantly. Part of this loss is frequency dependent and occurs as the wavelength on tape of the applied audio signal becomes equal to the record head gap space. This loss, called scanning loss, is supplemented by the fact that as the tape leaves the record head, a secondary half-gap is created and causes some self erasure of higher frequencies. Both of these phenomena are affected by tape speed because this varies the physical length of a wave on a given amount of tape (discussed in Chapter 2). The remainder of this loss is a function of the fact that the sensitivity of the tape changes with the amount of applied bias. However, tape sensitivity varies over the audio frequency band so that when the proper amount of bias for maximum low-frequencies output is found, that level does not coincide with the amount of bias required for maximum high-frequency output. For example, if the bias level is gradually increased from zero, while recording a 15kHz tone, that frequency will be recorded on tape at a gradually increasing amplitude until the optimum bias for that 15kHz tone is reached. Further increase in the bias level will result in a decrease in the amplitude of the recorded 15kHz signal. If we now record a low-frequency signal, we will discover that the bias level must be increased several dB beyond what was required for maximum output at 15kHz in order to achieve the maximum possible output for the low frequency.

To summarize, bias is a high-frequency (typically ten times the highest audio frequency to be recorded), high-amplitude signal that is mixed with the applied audio signal and then sent to the record head. Its purpose is to overcome the non-linear transfer characteristics of magnetic recording. Without it, the audio program stored on the tape will be a distorted replica of the original input signal. Bias affects overall output level, high- and low-frequency relative sensitivity, and third harmonic distortion. The amount of bias required is affected by record head gap space, tape coercivity and oxide formulation, as well as tape speed.

■ TAPE DIFFERENCES

The engineer should be aware that the operating parameters of magnetic recording tape will differ appreciably from one manufacturer to another, as well as between different types of tape from the same manufacturer. Different tapes have slightly different oxide thickness and formulations, which affect the overall transfer characteristic of the magnetic domains. These effects are most noticeable in the high-frequency response of the tape, as well as its overall sensitivity.

Although tape recorder alignment will be covered later, (Chapter 12), here it should be understood that if a tape recorder is aligned using a certain kind of tape, it will not necessarily give optimum performance when a different type of tape is substituted; a compromise setting must be sought that balances all of the above factors. There should certainly be consistency between one reel and another of the same type of tape, but in all other cases, the engineer should be prepared to realign his tape recorders when changing types of tape. Even so-called "bias-compatible" tapes will benefit from optimizing the tape recorder/tape interface for each type of tape used. Manufacturers specify the magnetic, physical and electro-acoustic characteristics of their product, each of which will be discussed separately.

■ HIGH OUTPUT TAPES

It would be very difficult and expensive to design a meter that could read magnetic flux level directly from tape. Most professional tape recorders use a level of +4dBm to equal 0vu. When magnetic tape was first put into professional use, a magnetic level that equaled the 0vu on the tape recorder meters was needed as a reference. Many percussive-type sounds may exhibit peak values of 9dB or 10dB above the average level read as 0vu on a loudness-related volume indicator. (This was discussed in Chapter 2.) To compensate for this, the standard operating magnetic level on tape was set so that a reading of zero on the tape recorder VU meters would be 9dB below the 3% third harmonic distortion point. This translated to a magnetic fluxivity level of 185 nanowebers per meter (abbr. nw/m). Today, standard operating level on tape is still defined as 185 nw/m, and for many years this defined the +4dBm = 0vu point.

As a result of advances in oxide formulation techniques, tapes with higher sensitivity and greater headroom are now in use. The sensitivity improvement is a function of the greater retentivity of these high output tapes. For a given oxide coating thickness, this greater flux density results in a higher remanence value and therefore, a higher output level. In addition, along with greater retentivity came higher coercivity, which is expressed as greater available headroom. Because of this, the signal may be applied to the tape at a higher level. This takes advantage of headroom improvement by raising the program material above the residual noise floor by the amount of the increase. The tape may also be recorded at standard operating level with the increased headroom allowing occasional high peaks to be recorded with less distortion. These elevated operating levels have become standardized, and this, as well as the record equalization changes afforded by the increased sensitivity, will be discussed in Chapter 12 on Alignment.

THE PRINT-THROUGH PHENOMENON

Because magnetic tape is stored on reels, each segment is wound between two other segments. The tape's magnetic field may be sufficient to partially magnetize these adjacent segments, resulting in "print-through." Print-through can be defined as the unintentional transfer of a magnetic signal to adjacent layers of tape. It is usually 55dB to 65dB in level below the printing signal, and appears as pre- and post-echoes of the recorded signal. On many recordings, the program itself will mask the print-through, especially the post echoes. However, print-through may be noticeable at the beginning and end of a recording, and during sudden changes in dynamic level, where a quiet passage is not loud enough to mask the echo of a loud passage directly preceding or following it.

Pre-print is much worse than post-print, and it is therefore advisable to store tape "tails out," or without rewinding after playing. This way, the worst case print-through comes as a post-echo and is most likely to be masked by the program itself.

Print-through occurs as soon as the tape is wound on the reel just after passing the record/playback head. Temperature and humidity play a small, but insignificant, role in the amount of print-through found on a master tape. Modern back coatings assist in reducing print-through, but the elevated operating levels used on high-coercivity tapes increase the problem.

Print-through may be reduced by fast forwarding and rewinding the tape prior to playback. This "exercising" of the tape ever so slightly rearranges the alignment of adjacent layers and seems to partially erase the print-through without affecting the original signal. Any of the noise-reduction systems discussed in Chapter 9 will alleviate the problem by reducing print-through, along with tape hiss, to below an audible level. Print-through is not a problem in digital systems, as will be explained in later chapters.

ELEVEN

The Analog Tape Recorder

The analog magnetic tape recorder, as we know it today, has been in use in its present basic form since the early 1950s. The basic transport design shown earlier in Figure 10-3 is as applicable today as it was then.

■ TRACK FORMAT STANDARDS

Early tape recorders were monophonic devices, and the audio signal took up almost the entire tape width. These elementary machines utilized a tape that was ¼ inch in width. The record and playback heads placed a magnetic stripe that was approximately the full width of the tape along the length of the tape. This full-track format was recorded in one direction only. Therefore, if the supply reel and take-up reel were switched at the end of one pass so that the tape could be played in the opposite direction, the program material would be heard backwards.

Soon thereafter, the tape heads were divided into two parts so that the magnetic stripe of recorded energy would take up half as much space as before. Using one set of record electronics, this allowed the engineer to put twice as much information on the tape for a given length. Track 1, the upper track, would be recorded from left to right as before. At the end of the reel, the tape would be placed back on the supply side, effectively inverting the tape and reversing the positions of tracks 1 and 2. The top track (track 2) was then recorded. This became known as recording on "both sides" of the tape, even though both signals are on the same oxide surface, just separate and going in opposite directions. Even today, it is not uncommon to hear someone say, "When you get to the end of the tape, turn it over and play the other side."

The next logical progression was to add a second set of record-and-play electronics so that both tracks could be recorded or played together. Since the two channels of material were recorded on the same tape at the same time, they were always synchronized together on playback. This half-track stereo format remains the professional standard in use for two-channel recording today. Note that it is the record/playback head configuration that determines the track format. A blank tape has no particular format until a recording is made on it.

Consumer high-fidelity manufacturers then divided the tape once again so that a stereo program could be recorded on "each side" of the tape. This is known as a quarter-track format. Quarter-track stereo displaces the tracks to try to keep cross-talk (interference between adjacent channels) to a minimum. Tracks 1 and 3 carry the stereo program in one direction, while tracks 2 and 4 carry it in the other. However, each time the size of the magnetic stripe is reduced, less flux density passes the head for a given signal. This results in a lower remanence value for each track, moving the signal closer to the residual noise level of the tape. This also acoustically lowers the signal-to-noise ratio of the track by 3dB, effectively reducing the possible musical dynamic range that can be put on the tape.

Wider tapes were developed so that more tracks could be put on a single tape without reducing track width. Four tracks on half-inch tape retains the half-track format. In fact, some professional studios have returned to the full-track format by using two channels of information going the same direction on half-inch tape. Figure 11-1 shows most of the available analog magnetic tape formats in use today. The corresponding record/playback head configuration is also shown. Note that there are guard bands between each track to prevent cross-talk, and that even the full-track format on ¼-inch tape is slightly smaller than the full width of the tape. Notice also that professional 4-track, 8-track, and 16-track tapes are all half-track format. Developments in tape sensitivity and lower noise, due to better oxides, led to the 2-inch tape, 24-track format. This format is a hybrid that utilizes quarter-track format track widths with greater spacing between tracks. Attempts were made to establish a 3-inch tape, 32-track format, but problems with tape guidance and the advent of multiple-machine synchronization soon put a halt to such efforts.

Figure 11.1 Common analog tape track formats.

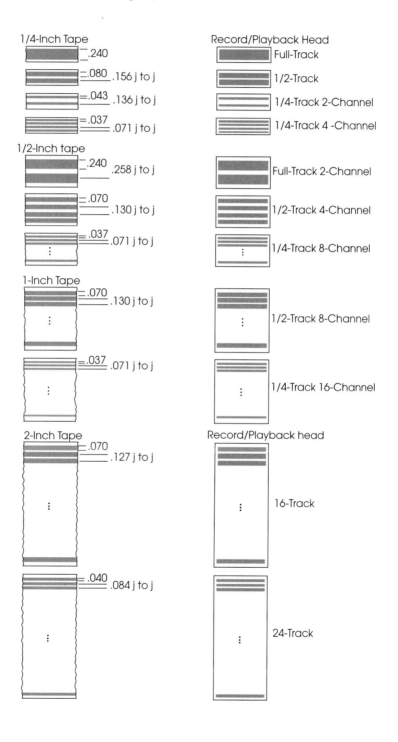

Many so-called semi-professional tape recorders use the smaller quarter-track format for home multi-track recording. These smaller formats are usually coupled with machines that operate at a reduced input/output signal level, such as 0 VU = -10dBm, and unbalanced lines. Many of these machines are quite good when equipped with noise reduction, and the lowered cost has made it possible for many musicians to experiment with multi-tracking at home prior to making a recording in a professional studio.

■ TAPE SPEED

Most reel-to-reel tape recorders found in the professional studio offer several tape speeds. The primary rates of speed are 15 inches per second (15 ips), meaning 15 linear inches of tape pass the head every second, and 30 ips. This is equivalent to 38 cm/sec and 76 cm/sec in countries using the metric standard. Speeds of 7½ ips or 3¾ ips are seldom used as they tend to accentuate dropouts on the tape. The lower speeds also will not allow the engineer to record high-frequency signals at full saturation levels on the tape. This will be discussed later in the chapter.

■ THE RECORDING PROCESS

The record head is the transducer that converts the applied audio signal (the record current) into magnetic lines of force (flux), which in turn penetrate and magnetize the tape. In an effort to achieve a flat frequency response, it might seem that the record current should be kept constant regardless of the frequency of the applied audio signal. As we shall see, however, there are several good reasons why this is not the case.

If we were to analyze any piece of music from an acoustic power point of view, we might find that much of the energy lies within the middle-frequency range. For example, if a series of sound-level meters were tuned to respond to various segments of the audio bandwidth, we might find the situation illustrated in Figure 11-2. The meters assigned to the frequency limits, such as the extreme lows and highs, would show readings considerably less than the meters tuned to the mid-range frequencies. This is because there is often less going on at very high or very low frequencies. Accordingly, if we were to record the music as is, the mid-range frequencies would be the first to approach the saturation level of the tape. Note that the saturation line slopes downward at the high-frequency end of the spectrum. This represents the tape's tendency towards earlier saturation at higher frequencies. From the illustration, we can see that if the level is kept down so that the mid-range is just below saturation, the low and high frequencies are well below saturation. Of course, this places these frequencies that much closer to the residual noise level, or noise floor, of the tape.

Record Equalization If these low and high frequencies were boosted somewhat, just before the record head, they would be recorded at a higher-than-normal amplitude, well above the residual noise level. The resultant recording would now have an unnatural amount of bass and treble, but complimentary circuitry just after the playback head could

Figure 11.2 Energy distribution curve of a musical program. The exact shape of the curve will vary according to the nature of the program.

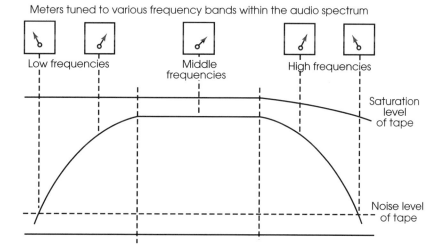

restore the signal to its normal balance, and in so doing, lower the apparent noise level. As the playback circuit restored the frequencies to normal balance by lowering the lows and highs, the recorded noise would likewise be lowered. This is an example of complimentary static noise reduction, as discussed in Chapter 9.

It should be clearly understood that tape recorder equalization has nothing to do with changing the apparent frequency balance of the program as heard by the listener. All equalization described within this chapter is intended solely as compensation for the various limitations of the analog recording medium and to improve the faithfulness of the recording to the original program. When a properly adjusted tape recorder is in use, the listener should be unaware of tape recorder equalization. This is in marked contrast to "program equalization" done at the recording console or on outboard equalizers to audibly change the frequency balance according to the tastes of the engineer or producer. The boosting of low and high frequencies just before the applied audio signal reaches the record head is known as "record pre-emphasis," and is often simply referred to as "record equalization."

Due to the properties of magnetic recording tape, high frequencies will cause saturation sooner, that is, at a lower level, than mid-range or low frequencies. Accordingly, high frequencies may not be boosted quite as much as low frequencies. There is also some tendency toward self-erasure at high frequencies. These facts, along with the differences in available oxide formulations (whose differences affect high frequencies more than lows), are the reason that record equalization adjustments on a tape recorder adjust high frequencies only. Low frequencies are more affected by the bias settings, and because of this, low-frequency boost is not user-adjustable.

■ THE PLAYBACK PROCESS

The playback circuit must compensate for the record pre-emphasis and restore the signal to its normal balance. In addition, equalization must be included to compensate for the output characteristics of the playback head itself. The playback head is the transducer that responds to the magnetic flux stored on the tape. It produces an output voltage that is directly proportional to the rate of change of the magnetic flux. Since this rate of change doubles every time the frequency doubles, we find that as the frequency increases, the output voltage rises according to the following formula:

$$e = N \left(\frac{d\theta}{dT} \right)$$

where:
    e = output voltage
    N = number of coil turns of wire in the playback head

$\frac{d\theta}{dT}$ = rate of change of flux (analogous to frequency)

Since a doubling of voltage equals a 6dB increase, we may say that, for a constant-level recorded signal, the playback output will increase at the rate of 6dB per octave. This rise will continue until the wavelength of the recorded signal is equal to twice the length of the playback head's gap. It is at this point that the rate of change of flux within the gap is greatest, and so the output voltage is at its maximum. It is the gap itself that measures the flux level. As the word implies, the gap is a physical air space across which the tape passes. The distance across the gap has been called "gap width" by some, and "gap length" by others, resulting in some mild confusion. For our purposes, the dimension will be called simply "gap space" and will be labeled G.

To understand the effect of the gap on the output level, wavelengths for several frequencies are shown in Figure 11-3. It should be understood that these wavelengths pass across the gap at a constant velocity, for example, 15 ips. Although it is convenient to think of these wavelengths in terms of some frequencies, it should be kept in mind that each wavelength represents a certain length of tape, and depending on the tape speed, the frequency reproduced by this wavelength will vary. For example, at 15 ips, a 1-mil wavelength will produce a 15-kHz tone. If we play the tape at half speed (7.5 ips), the same 1-mil wavelength now produces a tone of 7.5 kHz. In Figure 11-3, wavelength "A" corresponds to a relatively low frequency. The flux level is changing slowly, which means that the rate of change within the gap is relatively slow, and the output voltage is therefore small. As the frequency rises, so does the rate of change within the gap, until the frequency corresponding to wavelength "B" is reached. At this wavelength ($\lambda = 2G$), the rate of change is at its greatest, because as a positive peak crosses one side of the gap, a negative peak crosses the other, and vice versa.

Beyond this wavelength, the output level falls off sharply, and at half this wavelength ($\lambda = G$), output is zero, because as one positive peak crosses one side of the gap, another crosses the other, for a net change of zero. Across the gap there is no rate of change, and therefore no output voltage. This is illustrated by wavelength "C" in Figure 11-3. At each successive halving of the wavelength, output will again be zero, while in between these points, output will sharply rise and fall, as shown in Figure 11-4. On a professional-quality playback head, gap space is typically around 0.2 mils. Therefore, a frequency whose wavelength is 0.2 mils would give zero output as just described.

Remember that on tape, the wavelength is the physical distance taken up by one cycle of the frequency in question. Every time the tape speed is doubled, for example from 7.5 ips to 15 ips, twice as much tape is used, and the wavelength for any frequency is therefore twice as long. Consequently, the frequency at which maximum output occurs is likewise doubled. This is why high-frequency response is extended with each doubling of tape speed. At 15 ips, 0.2 mils corresponds to a frequency of 75,000Hz, certainly well past the audio range, so zero at this frequency would be of no concern. Maximum output would occur at a frequency with a 0.1-mil wavelength, or 37,500Hz. At 7.5 ips, this same wavelength would yield 18,750Hz, still sufficiently high for most purposes, but now within the audible range. If this gap loss was the only limitation on upper-frequency response, we might safely ignore it when operating at speeds of 7.5 ips or greater. At 7.5 ips, the rapid fall-off in output level, from maximum to zero, occurs within the octave that begins at $\lambda = 2G$ (18,750Hz at 7.5 ips) and poses no real practical problem.

Figure 11.3 Various wavelengths within the playback head gap space.

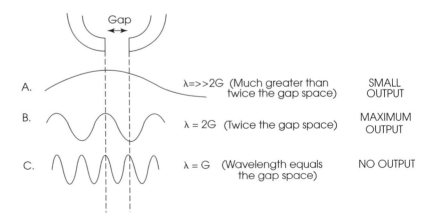

There are, however, other limitations within the playback head and in the tape itself which account for losses that affect the upper audio frequencies even at 15 or 30 ips. The cumulative effect of these losses, added to the gap loss effect, may noticeably restrict the upper audio frequency response. These losses are due to a variety of causes. On the tape, higher frequencies have a tendency towards self-erasure, and bias current may also reduce high-frequency response somewhat. Within the heads, eddy currents further contribute to high-frequency roll-off. The slightest separation between head and tape will also cause noticeable high-frequency dropout. Further losses will occur if the record head and the playback head are not in correct mechanical alignment with respect to each other and to the tape path. Some of these losses will be discussed in detail in the chapter on tape recorder alignment. These losses combine to prevent the tape from recording a full-level, high-frequency signal at speeds below 15 ips.

Playback Equalization

To compensate for the 6-dB/8va rise in the output voltage which occurs over most of the audio range, a playback equalization circuit with a 6-dB/8va cut is employed. At the highest audio frequencies, this 6-dB/8va cut must be replaced with a high-frequency boost to help overcome the losses just described. Of course, once $\lambda = G$, there

Figure 11.4 Output characteristic of a playback head when reproducing a constant flux tape. The output level rises at 6-dB/8va until the wavelength equals twice the gap space. The output falls off rapidly to zero at $\lambda = G$.

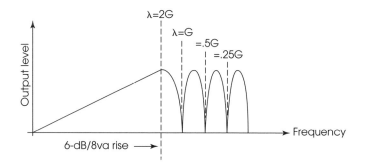

is zero output from the playback head and no amount of equalization will help. However, proper playback equalization should keep the frequency response flat within the entire practical range.

The National Association of Broadcasters (NAB) Standard Reproducing Characteristic is shown in Figure 11-5. It should be understood that this post-emphasis equalization is in addition to the 6-dB/8va cut just mentioned. The combination of the NAB characteristic and the 6-dB/8va cut is shown in Figure 11-6. The low-end roll-off within the NAB standard is to compensate for the low-frequency boost in the record circuit, and it restores these frequencies to their normal balance. Since the high frequencies were also boosted in the record circuit, it might seem that a complementary roll-off would be required in the playback circuit, as indeed would be the case if the record and playback losses described earlier were not a factor. However, the high end pre-emphasis (in the record circuit) only partially overcomes these losses, and so, additional boost, rather than roll-off, is required in the playback circuit.

As a consequence of record/playback losses, there is no published record equalization standard. At lower frequencies, boost is nearly

Figure 11.5 The NAB Standard Reproducing Characteristic (7.5, 15 ips).

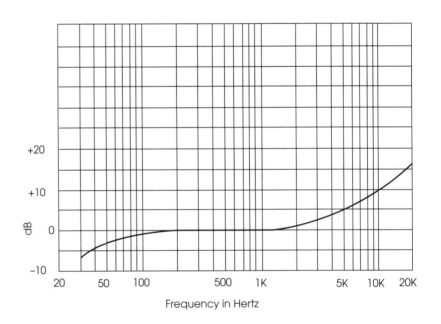

the opposite of the NAB Reproducing Characteristic. However, the cumulative effect of the various high-frequency losses occurring within the record/playback chain cannot be predicted, since they are dependent on bias level setting, recorded level, type of tape, etc. Therefore, proper record equalization is simply that which will yield a flat frequency response when the playback circuit has been aligned according to the NAB standard.

Playback Equalization Standards   As just mentioned, an equalization standard based on the NAB Reproducing Standard is used to compensate for anomalies found in the magnetic recording/playback process. However, tapes and recording equipment used and produced in Europe will often conform to the Consultative Committee for International Radio, or CCIR, Standard. This is also sometimes referred to as the International Electrotechnical Commission (IEC) Standard. In the United States, the NAB reproducing curves are used for tape speeds of 7.5 ips and

Figure 11.6   The NAB Standard Reproducing Characteristic combined with a 6-dB/8va roll-off. Transition frequencies are 50Hz (7.5 ips) and 3180Hz (15 ips).

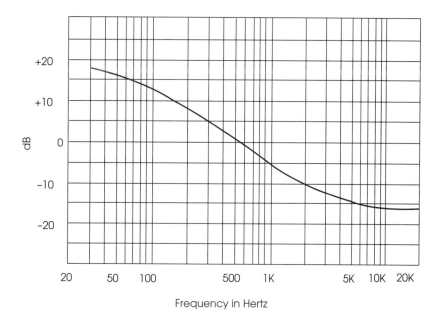

15 ips, while in Europe, the IEC or CCIR reproducing curves are used for these same speeds. The 7.5 ips and 15 ips CCIR curves differ slightly. In both Europe and the United States, another reproducing curve, known as the Audio Engineering Society (AES) reproducing curve, is used for 30 ips. These standards differ in relation to the point at which high-frequency boost and low-frequency roll-off occur. The points at which the reproducing curve flattens out, known as transition points, are defined as time constants. Actual frequencies can then be determined from the time constants based on the following formula.

$$f = \frac{1}{2\pi\tau}$$

where:
f = the transition frequency
τ = the time constant

The IEC curves have a somewhat greater high-frequency post-emphasis than that prescribed by the NAB Characteristic, yet there is no low-frequency roll-off since no low-frequency record pre-emphasis is used. The AES curve uses less high-frequency post-emphasis than either the NAB or CCIR Reproducing Standards at the high end, and has no low-frequency roll-off. Figure 11-7 illustrates the IEC curves, while Figure 11-8 shows the 30-ips AES curve. The following chart gives the time constants and transition frequencies for the various standards.

Speed	Standard	Time Constant	High Transition	Low Transition
7.5 ips	NAB	3180&50 msecs	3180Hz	50Hz
15 ips	NAB	3180&50 msecs	3180Hz	50Hz
7.5 ips	IEC	70 msecs	2275Hz	—
15 ips	IEC	35 msecs	4550Hz	—
30 ips	AES	17.5 msecs	9100Hz	—

Figure 11-9 summarizes the frequency characteristics and corrective equalization found within the modern tape recorder.

Figure 11.7 The IEC Standard Reproducing Characteristic combined with a 6-dB/8va roll-off. Transition frequencies are 2275Hz (7.5 ips) and 4450Hz (15 ips).

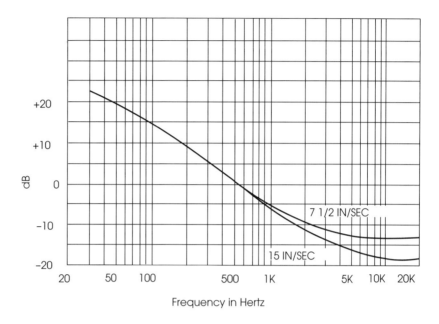

Figure 11.8 The AES Standard Reproducing Characteristic combined with a 6-dB/8va roll-off and a transition frequency of 9100Hz.

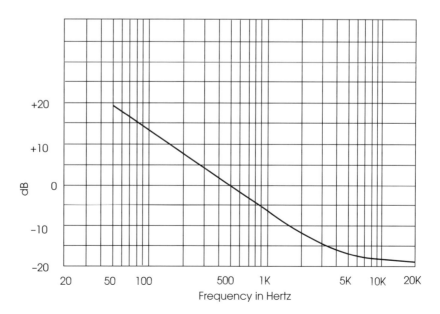

Figure 11.9 Summary of the frequency characteristics and corrective equalization within a tape recorder: a. A flat frequency response is applied to the input; b. Record equalization boosts low and high frequencies; c. Tape is recorded with boosted low and high frequencies; d. Playback head exhibits 6-dB/8va rise in output level, with rapid fall-off beyond $\lambda = 2G$; e. Recorded response (C) as reproduced by playback head with 6-dB/8va rise. High-end boost of response C may partially counteract rapid fall-off beyond $\lambda = 2G$; f. 6-dB/8va roll-off to compensate for playback head characteristic; g. Resultant response after 6-dB/8va roll-off; h. NAB equalization to restore flat response; i. Resultant flat frequency response corresponding to input response a.

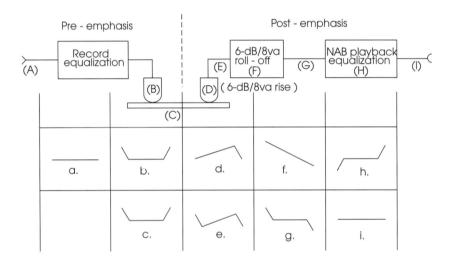

■ OTHER FEATURES AND CHARACTERISTICS

The Erase Head The function of the erase head is to remove any remanent magnetization from previous use of the tape. The need for an efficient erase mechanism is not unique to magnetic recording. Notebook pages, blackboards and magnetic tape are alike in that they have to be erased after each use if they are to be re-used. To simply record over an already recorded tape would be just as confusing to the ears as writing over a previously used page of notes would be to the eyes.

In the tape path, the erase head immediately precedes the record head. Generally, the tape recorder's high-frequency bias oscillator also supplies a very strong alternating current to the erase head. This "erase current" completely saturates the tape, and of course, there is no audio signal mixed with it, as in the record head circuit. This heavy erase current forces the tape's magnetic particles out of their previous magnetic orientation and into saturation. The high frequency of the erase current alternately drives the tape from positive to negative saturation; as each segment of the tape moves out of the magnetic field around the erase head, the effect of the field on the segment becomes less and less. The hysteresis loop for the segment grows smaller and smaller, the remanent flux, consequently diminishes eventually to zero, and the tape segment leaves the vicinity of the erase head completely demagnetized.

Of course, all of this happens very quickly as the tape travels past the erase head on its way to the record head. By the time the tape arrives at the record head, all traces of previously recorded information have been removed and the tape is ready to accept the new signal.

The Tape Transport System

For optimum performance of the recorder's electronics system, it is essential that the magnetic tape maintain constant contact with the head over which it passes. The slightest up-and-down movement or variation in tape-to-head pressure will seriously degrade the tape recorder's performance. Even a slight speed error will become a serious problem if the tape thus recorded is played back on a different machine.

In addition to the supply, take-up and capstan motors briefly mentioned earlier, the complete tape transport system includes a series of tape path guides, located on either side of, as well as within, the head assembly. What follows is a description of the various components that comprise the complete transport system. Although construction details may vary from one tape recorder model to another, most of the components mentioned may be found on every studio grade reel-to-reel tape recorder, whether analog or digital.

Figure 11-10 is a close-up photo of the capstan/pressure roller assembly. As shown, the machine is stopped and the pressure roller has swung away from the spinning capstan. When the play button is depressed, the roller swings down against the capstan, which pinches the tape between them and pulls it past the heads at a constant speed.

Figure 11.10 The capstan/capstan idler system. The capstan idler is often called a pinch roller (Sony APR-5003).

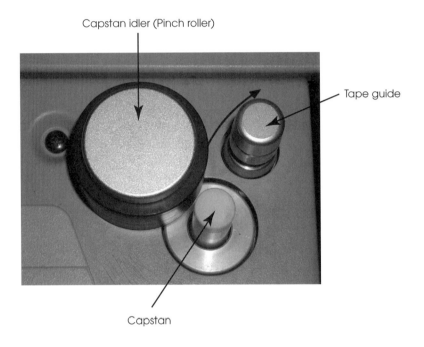

In Figure 11-11, the top of a typical ¼-inch tape recorder is seen. When the machine is in the play or record mode, both motors attempt to wind the tape onto their respective reels. Because the capstan motor is pulling the tape in the direction of the take-up motor, the tape instead winds off the supply reel and onto the take-up reel as intended. The reverse torque applied to the supply reel prevents the tape from exiting the supply reel too quickly, and the constant back tension promotes good tape-to-head contact. If this torque were excessive, the tape might be stretched as the capstan and supply motors attempt to pull the tape in opposite directions.

The reels are kept under tension by a current applied to both the take-up and supply reel motors. The greater that current, the greater the tape tension. With a constant applied current, the actual mechanical tension on the tape will vary with the amount of tape left on the reel. As less tape remains on the supply reel, the hold-back tension on the tape increases. Similarly, the take-up tension on the take-up reel decreases as it fills up with tape. Overall, the tension must be sufficient to maintain good tape-to-head contact, yet not so great as to cause tape stretching or to affect the speed of the tape past the

Figure 11.11 Top view of a two-channel quarter-inch analog tape recorder (Sony APR-5002, Peabody Recording photo).

heads. Although a full roll of tape on either reel requires more tension than an almost empty reel, tension is usually adjusted for some mid-point value between the optimum tension for full and empty reels. Many current microprocessor-based machines actually sample the tension at determined intervals and, by varying the current applied to the respective reels, can optimize the tension continuously as the tape pack moves from one reel to another.

Most studio recording is done with reels of tape wound on NAB hubs. The inner diameter of these hubs is 4½ inches. Nevertheless, it is often necessary to use smaller plastic reels. These may have an inner diameter of only 2¼ inches, and normal tension adjustment made with the NAB hub would place an undue amount of tension on tape being wound on such a reel. Many older tape recorders have a reel size switch (or switches) which reduce the tension for smaller hub diameter reels. Multi-track tape recorders, all designed for NAB hubs, are notorious for being extremely sensitive to reel tension. The inertia of a large reel of 2-inch tape can put great demands on the supply and take-up reel motors. Older multi-track machines often need to be re-tensioned (a time-consuming process) when changing from 10½-inch NAB reels to 14-inch NAB reels. These larger reels, which hold twice as much tape as the 10½-inch reel, are sometimes used when more recording time per reel is necessary. However, the modern microprocessor-controlled transport successfully copes with these changes in tension requirements with a minimum of operator adjustment.

Supply Side Idler(s) These idlers provide guidance for tape height, and also help to reduce wow and flutter (discussed later). The idlers turn with tape motion and help keep the tape moving smoothly. These rollers may also contain the mechanically-driven tape counter mechanism. This mechanism may provide a readout in an hour/minute/second format. Even though the counter may be an electronic device, the physical tape motion is what is measured. Therefore, absolute tape positions are not possible since any tape slippage on the idler can create minor timing errors. Absolute tape addressing can only be carried out when a timing code is placed on the tape itself. This type of system will be discussed in greater detail in Chapter 15 on SMPTE (often pronounced "simp-tee") time code. A tape motion sensor may be placed between these supply side idlers as well.

Take-up Side Idler(s) Located just beyond the capstan/pinch roller assembly, the take-up idler serves as a guide for proper tape height, and in some machines contains a micro switch that senses tape presence and functions as a transport on/off switch. When no tape is threaded, or the tension is incorrect, the machine will not enter any mode that would cause the tape to move. This prevents tape spillage or breakage and, in the event of tape breakage while in any moving mode, will cause the transport to stop. Similarly, when either reel runs out of tape, the switch stops both reels from spinning.

Braking Systems Some tape recorders employ a mechanical braking system on both reel motors. In the stop mode, a clutch system may allow the reels to spin freely in one direction only, allowing tape to be wound up on either reel, yet preventing it from accidentally spilling off. In the play, fast forward, and rewind modes, the brakes may be applied only to the reel motor from which the tape is being unwound as a precaution against tape spillage. Once the machine stops, the brakes are applied to both reel motors.

In some recorders, mechanical brakes have been completely eliminated, with all braking being done electronically. In the stop mode, reels are kept under tension by a current applied to both motors. Since both reels are attempting to take up tape, the tape may slowly creep across the heads while the machine is stopped if the braking currents are not properly balanced.

However, most tape recorders today apply a combination of mechanical brakes along with electronic reel tensioning. This allows for electronic braking during wind and rewind (which is more gentle to the tape) and total stoppage with no creeping when in the stop mode.

Edit Switch In some machines, when the edit button is pressed, the capstan/pinch roller pulls tape across the heads in the usual manner, but the take-up reel motor is disabled so that the tape spills off the machine rather than winding onto the take-up reel. On some machines, this is called the "dump mode," which allows the unwanted part of the tape (when editing) to spill to the floor. The edit switch may merely release the mechanical brakes while keeping a minimal amount of tension on the reel motors. This allows the engineer to easily move the tape across the heads by hand. When combined with the play switch, this may cause the transport to enter the dump mode.

Fast Forward and Rewind Switches When either the fast forward or the rewind switch is pressed, the pressure roller releases the tape, and the appropriate reel motor begins spinning very quickly, winding or rewinding the tape at high speed. In either mode, tape lifters hold the tape away from the heads. This protects against excessive head wear, and prevents the sudden increase in flux change across the playback head gap from reaching the playback electronics. Many tape recorders have a "library wind" or "shuttle speed" that restricts the maximum wind or rewind speed to a fixed inches-per-second rate. This produces better tape pack for storage and prevents edge damage. Typical rates are between 150 ips and 180 ips. The edit button may defeat the tape lifters in the shuttle mode so that pauses between recorded material can be found by ear. However, care should be taken to make sure that the playback monitor volume is lowered first to avoid loudspeaker and ear damage.

Wow and Flutter

Proper uniform tape speed is a function of the capstan motor, pressure roller (capstan idler) tension, and the torque applied to the reel motors. Two common irregularities in tape speed are termed "wow" and "flutter." The terms closely describe the type of sound that is heard if the irregularity becomes excessive.

"Wow" is a slowly varying pitch change, which may be caused by a variety of factors. Chief among them may be: transport misalignment, particularly tension discrepancies between the take-up and supply reels; an out-of-round capstan or idler; or a worn bearing that may cause a low-frequency fluctuation in the tape speed that results in an audible wow, noticeable especially on sustained tones.

"Flutter," as the word implies, is a very rapid speed fluctuation, producing a fluttering kind of sound. It is usually traced to some sort of tape vibration, not unlike that of a string on a violin which has been set into motion by pulling the bow across it. For example, as the tape moves across the transport system, it passes over, and is supported by, tape guides at various locations. Between these guides, the unsupported tape may be forced into mechanical vibration, and if a record or playback head detects this vibration, or flutter, the listener hears a high-frequency modulation of the music program.

Scrape Flutter Idlers

To keep the flutter frequency as high as possible—well beyond the audio range—so-called "scrape flutter idlers" may be installed between the heads. The scrape flutter idler reduces the length of unsupported tape, thus raising the resultant vibration, or flutter, frequency. The supply side idler also acts as a flutter idler, isolating the supply reel from the head assembly, and reducing the length of unsupported tape on that side of the transport. Figure 11-12 shows the head block of a ¼inch tape recorder. Note the scrape flutter idler between the record and play heads. On some machines, an idler of this type may also be found between the erase and record heads.

Variable-Speed Operation

The capstan on most tape recorders is driven by a direct-current servo motor. Here, a tachometer registers the motor speed (and hence the tape speed) and compares it to a reference signal supplied from either a built-in oscillator or a crystal frequency. An error signal equivalent to any difference between the two signals continuously corrects the capstan motor, keeping it precisely on speed.

For variable-speed operation, the built-in reference frequency is adjusted by the operator with a variable potentiometer. As the refer-

Figure 11.12 Detail view of the head assembly showing heads and scrape flutter idler (Sony APR-5002 head block).

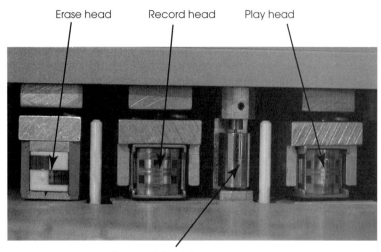

ence frequency is changed, tape speed varies continuously over a wide range as the error signal brings the capstan motor into synchronization with the applied reference signal. Many tape recorders today have this as a standard feature, and either the positive or negative percentage of change, or the change in plus or minus semitones, is rotated electronically on the front panel.

Tape Transport Remote Control

Most multi-track tape recorders are equipped for remote-control operation, which allows the engineer to control tape transport from his position at the console. This means that the multi-track tape recorder, with its attendant cooling fans and transport mechanics, can be located away from critical listening areas. A typical multi-track tape recorder remote control is shown in figure 11-13. Most remote controllers will provide access to input/output switching, safe/ready status (to be discussed in Section VI), as well as the play, rewind, and other transport functions. Many also include an autolocator function, which can store a number of tape locations. A readout will often give the engineer the tape's current location, while a

Figure 11.13 A typical remote controller for a multi-track tape recorder (Sony PCM-3324, Peabody Recording photo).

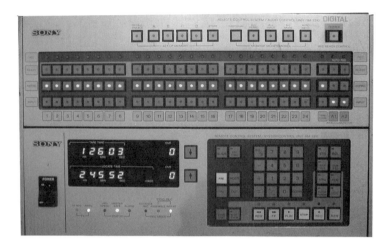

keyboard will let him call a new location in an hour, minute and second format. These data are then stored in a memory and assigned a number. To call up that location, the engineer simply enters that number followed by a search command. When the auto-locator is engaged, the transport will enter the appropriate wind or rewind mode and move to the new location. Once there, the transport will either stop or begin playing the tape, depending on the status required and pre-programmed by the engineer.

TWELVE
Analog Tape Recorder Maintenance and Alignment

The vitality of an inspired musical performance can be all but extinguished if the equipment used to record it is not properly maintained. The performance of the analog tape recorder, like that of any sophisticated machine, depends on regular maintenance. In this chapter, we look at the various procedures engineers employ to optimize the performance of the analog tape recorder, from simple cleaning to precise electrical and mechanical adjustments.

■ CLEANING AND DEMAGNETIZING

Before beginning any alignment procedure, all tape guides and heads should be thoroughly cleaned. Head cleaning fluid is readily available, and should be applied with a cotton swab to all surfaces with which the tape comes in contact. Since some head cleaners may damage the rubber capstan idler, it should be cleaned with denatured alcohol instead.

Heads and tape guides should be routinely demagnetized at regular intervals, since after a few hours of use, these surfaces can become slightly magnetized. Even a small amount of magnetization may cause erasure of some higher frequencies. In severe cases, a tape passing over a magnetized head or tape guide will be partially erased throughout the frequency band. With the tape recorder turned off, a head demagnetizer should be brought slowly toward the heads and slowly moved away again. The demagnetizer must be well clear of the heads when it is turned on and off, or the sudden change in magnetic flux will remagnetize the heads.

■ TEST TAPES

To completely evaluate a tape recorder's level and frequency-response performance, a test recording must be made and played back. Since the recording is evaluated by observing the machine's own playback meter or by externally measuring the machine's playback output, the playback circuit must first be properly aligned. Adjustments to the record circuit should be made only after proper playback calibration.

The skilled engineer uses precision test tapes to properly align the tape recorder's playback electronics. These tapes, made under controlled laboratory conditions, contain a series of test tones at levels corresponding to the selected reproduction curve (NAB, IEC, etc.; see Chapter 11) at a standard reference fluxivity measured in nanowebers per meter. A nanoweber (abbreviated nw) is 1×10^{-12} weber. The specification refers to the number of flux lines per unit of tape width. These flux lines will be detected by the playback head and converted into an output signal voltage. Fluxivity is an indication of the magnetic field strength of the test signal recorded on the tape. This should not be confused with remanence, which was described earlier as the magnetism left on a tape after a saturation-producing force has been withdrawn. Consequently, for a given playback head, a greater fluxivity on the tape produces a higher playback output level.

■ OPERATING LEVELS

There are three basic operating levels in use today. To understand how they were derived, we begin with a look at the advances in the dynamic range of magnetic tape.

The tape that was developed shortly after World War II had an approximate dynamic range of 68dB. This was the level in decibels between the noise floor of the tape (the point at which the signal would be masked by the tape's residual noise), and the point of maximum permissible amplitude (the point near saturation where the tape reached 3% third harmonic distortion). A point several dB below saturation was arbitrarily chosen to equal 0 VU (= +4dBm). This point below saturation was chosen so that there would be a safety margin above 0 VU before distortion made the tape unusable.

This point corresponds to a level of 185 nw/m at a frequency of 700Hz and has become known as "standard operating level." Later developments in the magnetic oxide itself, along with the manufacturer's ability to put thicker oxide coatings on the tape, both lowered the noise floor and raised the saturation point. When the headroom above 0 VU reached 12dB due to oxide improvements, a choice could be made with regard to the trade-off between headroom, print-through and low noise.

By keeping the 0 VU point at 185 nw/m, 12dB of headroom could be maintained while keeping the same noise level. By raising the zero point 3dB, noise floor could effectively be lowered by 3dB while maintaining 9dB of headroom. This elevated operating level has a fluxivity of 260 nw/m and has become known as a +3 EOL (Elevated Operating Level). Further developments in magnetic tape created a +6 EOL at 370 nw/m. This lowered the noise floor an additional 3dB while retaining 9dB of headroom for peaks and sudden transients. Note that magnetic fluxivity corresponds to the 20log formula used for voltage doubling (2 X 185 nw/m = 370 nw/m = +6 EOL). Print-through, however, increases in proportion to the rise in operating level. The greater the level on tape, the more audible print-through becomes. 260 nw/m or +3 EOL has become the most used standard at this point in time. However, many multi-track studios prefer the lower noise of +6 EOL (370 nw/m), and in many archival situations all standard operating level of 185 nw/m is still used. Some confusion may arise when the engineer finds himself with an alignment tape labeled "Standard Operating Level of 200 nw/m." This will usually refer to a test tone of 1,000Hz rather than 700Hz. In actuality, since fluxivity changes with frequency, these levels are the same and only the frequency of measurement is different.

■ ELECTRONIC ADJUSTMENTS

For purposes of clarity, adjustments to the electronic and mechanical systems are discussed separately. There is, however, a certain amount of interaction between these systems. As each is adjusted, the performance of the other should be verified.

Playback Electronics Alignment

On most tape recorders, the electronics system parameters can be adjusted with a screwdriver. On many of the newer machines, however, electronic alignment takes place by changing values that are stored in a microprocessor memory. These values, when selected, are often changed with a thumbwheel or other rotary control. In other systems, values may also be changed by holding down a push button until the correct reading is attained on the VU meter. A calibration panel of this type is shown in Figure 12-1.

There is also, usually on the front of the machine near the VU meter, a switch or series of switches that lets the engineer select the origin of the signal delivered to the output jacks of the tape recorder, as well as what is read on the associated channel's meter. These switches select "input" "reproduce," or "sync." When "input" is selected, the signal being sent to the recorder is sent through the input level controls, metered on the VU meters and delivered to the output jacks at the rear of the machine. When "reproduce" is selected, the output of the playback head is sent through the reproduce level controls, metered, and sent to the output jacks. "Sync," which stands for synchronization, will be discussed in detail in Section VI, "The Synchronization Systems." These input/repro/sync switches are independent of the record function of the tape recorder, and allow the engineer to monitor either the input of the machine or the playback signal from the tape.

Figure 12.1 An alignment panel for a software-driven alignment system for a two-channel, three-speed tape recorder (Sony APR-5003).

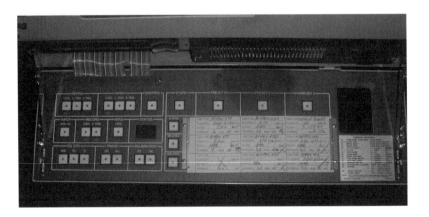

Another set of switches located in the same area is used to select the record or the safe mode. When "safe" is selected, no recording is possible. "Safe" locks out the record and erase functions of the machine, so that pushing the play and record button will only cause the machine to play. This lock-out prevents any accidental erasure of a master tape. When "ready" or "record" is selected, the tape recorder is ready to go into the record mode; pushing the record and play buttons simultaneously will energize the bias and erase circuits and send the applied signal to the record head. When adjusting the tape recorder's playback electronics, the machine should be placed in "reproduce" and "safe." Reproduce is selected so that the applied test tape can be monitored both visually and aurally, while the safe mode prevents accidental erasure of a valuable and expensive test and alignment tape. It is assumed that the following procedures will be performed for each channel of playback electronics.

Generally, the first tone on the test tape is 700Hz or 1,000Hz, and is used for setting the proper playback output level. If using a standard operating level alignment tape to set a machine for standard level, the tone should read 0 VU. However, if the machine is to be aligned to an elevated level such as +3 EOL or +6 EOL using the standard tape, the VU meter should read -3 VU or -6 VU respectively. (If 185 nw/m = -6 VU, 0 VU = 370 nw/m.) 0 VU will usually produce a level of +4dBm at the balanced (three-conductor) output of the tape recorder, but could, in the case of some small-format machines, produce a level of -10dBm with an unbalanced (two-conductor) output. If the engineer elects to use an elevated-level test tape, the metered reading should be adjusted accordingly.

Immediately following the level set tone, the test tape will usually provide a series of high-frequency tones. The first one is usually specified for azimuth adjustment. Azimuth is the angular relationship between the head gap and the tape path (discussed in detail later in this chapter). It is at this point that the playback head azimuth should be adjusted. If the azimuth is not correct at this point, all further adjustments will be incorrect by the amount of the azimuth error. The remaining high-frequency tones are monitored while adjusting the machine's high-frequency playback equalization control. The tones have been recorded in accordance with the NAB Standard Reproduce Characteristic (see Figures 11-5 and 11-6), so that when the machine is properly aligned, the tones will read the same as the first reference tone (0 VU, -3 VU, or -6 VU depending on the operating level) on the playback meters.

Most tape recorders contain an additional equalization control to adjust the low-frequency response of the playback circuit. However, the so-called "fringing phenomenon"—an apparent low-frequency boost—must be taken into account when evaluating the machine's low-frequency performance. Many test tapes are made in a full-track format, which, as you may recall from Chapter 10, means that the reference tones are recorded across the entire width of the tape. In this way, the tape can be used with any head format available for that tape width. At long wavelengths (low frequencies), however, a playback head tends to respond to flux recorded over an area greater than its normal track width. Accordingly, the machine's low-frequency response appears to be higher than normal. If the test tape had been recorded over just the width of the track being measured, this fringing effect would not occur. The amount of fringing varies from one playback head to another, and so no attempt was made to compensate for it when preparing the test tape. Most manufacturers indicate in the machine's documentation how much fringing is to be expected with that particular machine's head configuration and at the various tape speeds available on that transport. When the actual amount of fringing is unknown, however, low-frequency playback equalization adjustments may be deferred until after the record circuit adjustments are complete. Low-frequency tones at a standard reference level are then recorded and the low-frequency equalization is then adjusted for flat response. Since the tones are being recorded at the proper width for the playback head in use, there will be no fringing, and the low-frequency response may be adjusted to read the same as the beginning reference tone. (This procedure assumes that the low-frequency record equalization is correct.)

After the playback response has been set, no further adjustment should be made to the playback circuit or level controls. This is, however, only an adjustment for the first or primary tape speed to be used. For machines with multiple speeds, the manufacturer usually includes a set of adjustments for each speed. For instance, if the primary speed of 15 ips had been adjusted to the NAB reproducing characteristic, the engineer would now proceed to equalize the playback of the other available speeds, but without adjusting the overall playback output levels. Typically, an AES Standard Reproducing Characteristic test tape would now be used to adjust the 30-ips high- and low-frequency responses. On some machines, both speeds would have to be set up for the same operating level, while on others, independent playback level controls would allow the engineer to select different operating levels for different speeds. For instance, a tape recorder could be set to play at an elevated operating level of +3 EOL at 15 ips using some form of noise reduction, and could also be adjusted for an operating level of 370 nw/m (+6 EOL) without noise reduction at 30 ips for lowest noise since, as discussed in Chapter 9, all noise reduction systems can cause problems on some program material.

Since the NAB Playback Characteristic is standard in the United States for speeds of 7.5 and 15 ips, few studios are likely to have an IEC (or CCIR) test tape available. Nevertheless, many U.S. facilities regularly receive tapes from abroad that may have been recorded using the IEC Characteristic. Assuming a tape recorder's playback response has been properly aligned for IEC playback, an NAB test tape played on such a machine should display the response shown in Figure 12-2. Therefore, when it is necessary to adjust a machine for IEC playback, an NAB test tape may be used, and the playback adjusted to match the response shown in the figure. Many modern machines now offer both NAB and IEC equalization circuits, and the engineer can select between them at the touch of a switch. This means, however, that at some point the machine's circuits will have to be adjusted for all these standards. A 3-speed dual-equalization tape recorder will require twelve different equalization adjustments, and the studio or technician must own the corresponding test tapes (three speeds times two equalization standards times high- and low-frequency response adjustments)!

Record Electronics Alignment

Now that the playback electronics of the tape recorder have been adjusted to the prevailing standard, the record electronics need to be adjusted to match. We will assume for the purpose of this discussion that the machine has been calibrated to NAB Standard Reproduction Characteristics. By calibrating the record electronics to

Figure 12.2 Tape recorder frequency response when an NAB test tape is played on a machine properly aligned to the IEC (CCIR) reproducing characteristic.

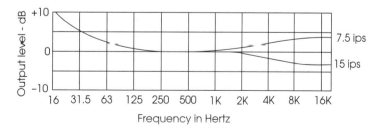

match the playback characteristics just set, we are assured that all recordings made on this tape recorder correspond to the NAB standard. Once again, it is assumed that the alignment procedures will be performed for each channel of the tape recorder.

To begin aligning the record function of the machine, the tape recorder is threaded with a fresh reel of tape of the type that will be used for later recording, and the machine is switched to the ready mode. An externally supplied oscillator is used to first calibrate the input/output characteristics of the tape recorder to the selected operating level. While monitoring input, a 1,000Hz sine wave with a level of +4dBm (-10dBm for some machines) is fed to the channel to be aligned and the input level control is adjusted so that a reading of 0 VU is obtained on the meter. The machine is then set in motion in record mode and the channel is switched to reproduce mode. The record gain control (not the input level control) is adjusted so that the meter reads 0 VU. This raises or lowers the gain of the record circuit to match the operating level selected during playback alignment. If the machine's playback level has been raised to a +6 EOL, then the record circuit gain must be raised to match. Some older tape recorders will not have enough gain in the record circuit to be used at elevated operating levels. If this is the case, lower operating levels will have to be used with the corresponding rise in noise level. This would necessitate realignment of the playback side before further record adjustments could be made. The record head azimuth would now be adjusted; as mentioned earlier, this will be discussed fully under mechanical adjustments.

Once the initial record level and azimuth adjustments have been made, the bias is adjusted. Bias is set using a short wavelength signal to find the bias peak. This peak is the point at which an additional amount of bias would cause a fall in the high-frequency response of the tape. Recommended alignment frequencies are 20kHz at 30 ips, 10kHz at 15 ips, and 5kHz at 7.5 ips. As mentioned in Chapter 10, proper bias is a trade-off between maximum output level and lowest distortion. The two points do not usually coincide. Figure 12-3 shows the relationship between output and distortion. Note that minimum distortion is obtained when additional bias current is applied beyond the bias peak. Bias is set by applying the proper frequency signal to the recorder's inputs, and while monitoring reproduce, adding bias until the bias peak is found. As more bias current is added, the recorded output level drops by an amount based on the size of the record head gap space. At this point, the tape's sensitivity is fairly uniform across the audio bandwidth, and third harmonic distortion is quite low. The exact amount of overbias

Figure 12.3 Tape output versus distortion. Note that the maximum output level (MOL) does not line up with the minimum distortion point.

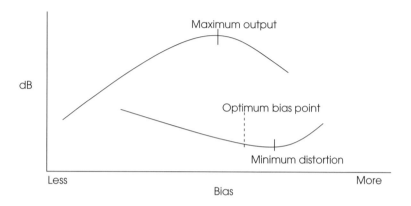

is usually specified in the tape recorder's operation manual. The following chart summarizes these figures for a tape speed of 15 ips.

Record Head Gap Space	Amount of Overbias
1 mil	1.5dB
.5 mil	2.5dB
.25 mil	3.0dB

These figures change with tape speed, head configuration and tape type. Therefore, the prudent engineer will not fail to consult the respective manufacturer's specifications. Figure 12-4 shows the electrical performance characteristics of two types of magnetic recording tape. In Figure 12-4b, it will be noted that a further increase in bias level will lower the distortion even more; however, the tape's sensitivity also falls off and modulation noise begins to increase again.

After the bias has been properly set, the record level should be adjusted again using a frequency of 1,000Hz. A change in bias level will cause a change in the input/output calibration of the tape recorder, and this must be corrected before proceeding with the record equalization adjustments.

The record equalization control is used to adjust the record circuit response at high frequencies. The setting will vary depending

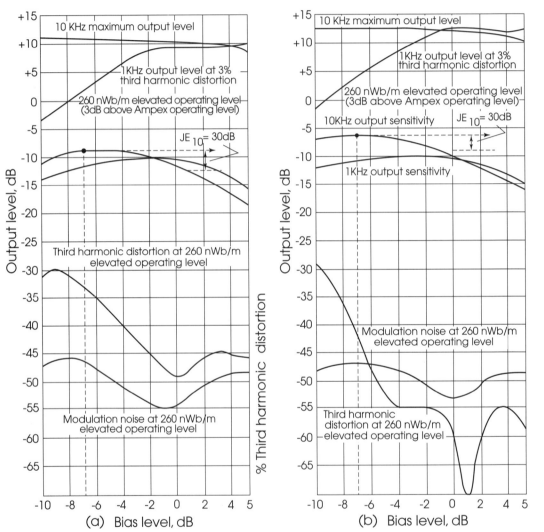

Figure 12.4 Electrical performance characteristics of two types of magnetic recording tape.

on the characteristics of the tape being used, the bias level and the design of the record head itself. A high-frequency tone (typically 15kHz to 20kHz) is recorded, and the equalization adjusted as before, until the machine's output level matches the input level at this frequency. Once the high-frequency response has been adjusted, the signal generator or oscillator output frequency should be varied over the entire audio bandwidth to verify that the system response is within the required specifications. It is important to have available a signal oscillator with a continuously variable frequency adjustment, rather than one with just a few fixed frequencies. Although the latter is adequate for routine signal tracing and spot checks of performance, it may not reveal response errors lying between the switch-selected frequencies. It is also possible to verify the machine's frequency alignment by applying pink noise to the input and metering the output with a real-time frequency analyzer.

■ MECHANICAL ADJUSTMENTS

For the tape recorder system to function properly, the tape must remain in good physical contact with all three heads. The slightest misalignment may cause errors in level and response that may be incorrectly attributed to the electronics systems. Figure 12-5 illustrates the following five mechanical adjustments that may be required in the alignment or realignment of a head:

Tape-to-Head Contact (Rack): The tape must remain in firm contact with the head at all times. Since the head is convex, a slight wrap around this surface will help maintain good contact. Excessive rack on one head, however, may cause the tape to be lifted from a neighboring head.

Tilt or Zenith: The pressure of the head against the tape must be evenly distributed across the entire tape width. If the head (or the tape) is not truly vertical, contact pressure may be greater at one edge than at the other, resulting in a skewing of the tape away from the centerline of the head.

Height: Head height must be properly adjusted so that each track is properly aligned with its corresponding head gap. A slight discrepancy between head heights in the same head block may result in incomplete erasure prior to recording, inter-channel cross-talk and increased noise level. Height inaccuracies between tape recorders create non-standard playback characteristics that can cause a tape that sounded fine in one studio to seriously mistrack in another.

Wrap or Tangency: As a part of the tape-to-head criteria, the tape must remain tangental to the head gap for optimum performance. An error in wrap angle adjustment may cause the tape to enter or leave the gap area prematurely. It is also important that the tape trail off the head as smoothly as possible to prevent the formation of a secondary head gap that can cause self-erasure at high frequencies.

Azimuth: Azimuth adjustments are particularly critical, especially in the case of short wavelength signals. If the head gap is not exactly perpendicular to the direction of tape travel, segments of the same recorded wavelength will enter and leave the gap at different increments of time, as shown in Figure 12-6. In the drawing, lines A and B represent an identical signal recorded on two tracks (or, for that matter, A and B may be the edges of any one track). The sine waves represent the instantaneous amplitude of the signal(s). As shown, the head azimuth is misaligned; as a positive maxima passes over a line in the gap, a negative maxima passes over the same line.

Whether the just mentioned illustration represents a single-track width, or two tracks that may be eventually combined, total output will be zero. In practice, a single track will rarely, if ever, cancel out completely, although two adjacent tracks that are out of azimuth will be severely attenuated when they are combined. In the case of some multi-track tape recorders, all head alignment adjustments have been made by the manufacturer and are permanently fixed. However, the heads of most machines with four or fewer tracks can be adjusted by the user. In such cases, the head is held in place and adjusted with a series of screws. A properly installed head should not require repeated mechanical adjustments, with the exception of azimuth, which should be checked periodically as a matter of routine.

Use of the cathode ray oscilloscope, described in Chapter 2, is the preferred method of verifying azimuth alignment. Various oscilloscope displays were shown in Figure 2-4 to illustrate the phase and coherency relationships between two signals. Here, Figure 12-7 shows the display when a test tape is monitored on the oscilloscope. If the head azimuth is correctly aligned, an in-phase display will be seen, since there is zero phase shift between the two inputs. If the head is slightly off-azimuth, however, there will be a phase shift displayed on the test instrument. Figure 12-7a indicates a properly aligned head, while the other oscilloscope displays indicate various amounts of phase shift, and therefore azimuth error. The error in degrees refers to the phase shift, not the angle of the head!

Figure 12.5 Exaggerated misalignment of the tape head: a. Contact: tape does not make good contact with the head; b. Tilt: tape makes uneven contact with head; c. Height: head height is misadjusted; d. Tangency: tape does not make good contact with head gap; e. Azimuth: tape head is not perpendicular to tape path.

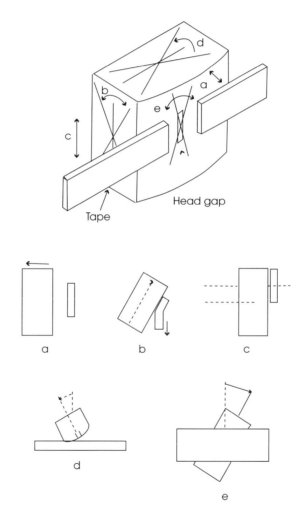

Figure 12.6 An out-of-azimuth head gap. Different segments of the recorded wavelength cross the gap at the same time. A and B represent an identical signal recorded on two tracks.

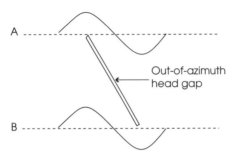

Figure 12.7 Cathode ray oscilloscope displays for various amounts of azimuth error: a. Properly aligned head. No azimuth error; b. Azimuth error of about 45°; c. Azimuth error of about 90°; d. Azimuth error of about 135°; e. Azimuth error of about 180°.

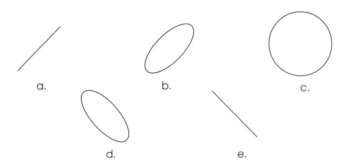

If an oscilloscope is not available, it is possible to set the azimuth relatively accurately by combining the outputs of the two outermost tracks out of phase. If one of the tracks is put electrically out of phase (reversing the positive and negative in a balanced line), and the combined outputs are monitored at a high level, the azimuth adjustment screw can be adjusted until monitor output is at a minimum.

When checking the alignment of an adjustable multi-track head, the azimuth of two adjacent tracks may be checked first, and then successively more distant tracks, until the two outside tracks are compared. Ideally, a single setting will bring all tracks into perfect alignment, but in practice, this is often not the case. Within the head itself, the gaps may be slightly misaligned, making complete azimuth alignment of all tracks impossible.

As already mentioned, azimuth alignment becomes more critical as frequency increases. It is often a good idea, therefore, to make coarse azimuth adjustments at 1,000Hz to 5,000Hz, and to then play successively higher tones on the test tape while making fine adjustments. When the azimuth is properly set, it should remain aligned over the entire audio bandwidth. If the scope display goes in and out of phase as the frequency is varied, the azimuth is improperly set.

Once the playback head is set, the record head azimuth may be adjusted by observing the playback head output on the oscilloscope while recording a tone and adjusting the azimuth screw on the record head. On machines with sync (record head) output capabilities, the test tape may be played while direct readings are made from the record head output.

SECTION V

Digital Audio Systems

■ INTRODUCTION

Despite many impressive advances in recording technology over the past century, at least one basic principle has remained unchanged. Ever since the first recording was etched onto a wax cylinder, the recorded format, if not the quality, has closely resembled the waveform of the original sound source.

In air, the sound source creates a series of pressure variations from which a recognizable waveform may be drawn. In the record groove, the stylus traces the same waveform, now permanently etched into the vinyl surface. Even on tape, the alternating magnetic field is a faithful reproduction of the acoustical sound wave. In each case, we may say that the stored information is directly analogous to the original source of the sound. It is an analog recording.

This has all changed, and now it is difficult to discuss audio without the word "digital" appearing in the conversation. The Compact Disc is inexorably replacing the vinyl phonograph record, and the DAT (rotary head Digital Audio Tape recorder) will undoubtedly have the same effect on the Compact Cassette.

This quantum leap forward in sound storage technology has placed an extra burden of care on the recording engineer. Small flaws in the recording are no longer covered by the surface noise of the vinyl or the hiss of the tape, but stand out for all to hear on the finished product.

How does this system work? What are its advantages? Chapter 13 takes a comprehensive look at the process we call digital recording and examines these questions. Then, Chapter 14 discusses the available hardware for storing, editing and playback of the digital information.

THIRTEEN

Digital Audio

No matter how simple or complex a waveform, any instant at which we examine its amplitude, we will find that the amplitude is different from any other adjacent instant in time. As the intervals between examinations become smaller and smaller, the change in amplitude becomes less and less. Nevertheless, the change is always continuous, and an infinite number of examinations will simply reveal an infinite number of amplitude variations. Ideally, the recorded waveform will be a faithful replica of the original, and for our purposes here, we may consider this to be so, until the signal is sent to magnetic tape. At this point, a variety of distortion factors may be added, as shown in Figure 13-1. In effect, the distortion is an analog noise combined with the analog music as the tape is recorded. Later on, the playback head faithfully reproduces the signal, which is now a somewhat distorted version of the original source. As discussed in Chapter 10, the distortion may come from a variety of sources, such as improper bias, tape saturation and so forth.

The problem, though certainly not the solution, is obvious. The playback head is incapable of distinguishing between the wanted and the unwanted components of the analog signal. For us to get around this difficulty, we must devise a method of recording in which the playback system will simply not recognize (and therefore, will not produce) the unwanted noise components of the recorded signal.

A useful analogy may be made by considering an early broadcast medium and a spoken message which must be sent over a noisy transmission system. The analog receiver picks up a transmitted message along with enough noise to make the message difficult to understand. To get around the problem, a sender may encode the analog message into a series of dots and dashes (A = .-, B = -..., etc.), and transmit these. At the receiving end, the listener (who has previously studied up on Morse code) mentally decodes the data stream of dots and dashes penetrating the transmission noise, then types out the message. Although our human decoder may get a serious headache, anyone else who reads the message will be unaware of the noise problem, which was simply not reproduced. Of course, the utility of the Morse code goes no further than the transmission of the printed word. It is of no use in the recording or reproducing of an actual sound source, since the small handful of dot/dash combinations falls far short of meeting the requirements for reproducing even the simplest sound wave.

Instead, we need an encoding system that is not only more versatile than Morse code, but also one that will not require operator assistance during the encode/decode process. Such a system should faithfully decode and reproduce the recorded musical signal, while remaining insensitive to any (analog) noise components that are a function of the storage medium. Obviously, analog noise components that are a function of the recording session itself will also be faithfully reproduced. In fact, if our new system is really more effective in eliminating analog tape noise, we may find that some studio noises are much easier to hear since they are no longer masked by tape hiss.

Figure 13.1 The analog recording process combines signal (a) and noise (b) to create a distorted version (c) of the original sound source.

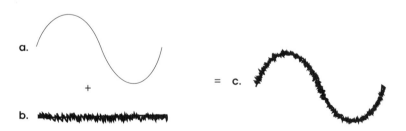

■ DIGITAL DESIGN BASICS

We can begin our design of such a system by examining a familiar analog waveform, such as the one seen in Figure 13-2a. When viewed continuously, even this very simple waveform may be seen to possess an infinite number of discrete amplitudes, as discussed earlier in this chapter. Now, let us view the waveform through a series of equally spaced "windows," and assume that we have no idea of what is happening between the windows. This is illustrated in Figure 13-2b and c, where we note that the more windows there are within a given distance, the more often we are able to see—that is, to sample—the waveform. Since we are really dealing in terms of time, we may say that the closer the windows, the higher the sampling rate, and thus the more accurate our view of the waveform.

Assuming that the waveform in Figure 13-2c represents the highest audio frequency of interest, we may be surprised to discover that a sampling rate that is merely twice that frequency will be sufficient to give us all the information we need in order to recover (decode) the original waveform. As long as we know the rate at which we are sampling, we can always determine the frequency of the sampled waveform.

Sampling Frequency

To illustrate this important point, we freeze, or "sample and hold," the value seen through the first window until we reach the second window. Then, sample and hold that value until the third window is reached, and so on. The sequential combination of these windows forms a square wave of the same frequency as the original sine wave. Since a square wave may be defined as a sine wave plus an infinite number of odd harmonics, we need only pass the square wave through a suitably designed low-pass filter to remove the harmonics to recover the original sine wave.

This may be quickly demonstrated with a square-wave generator, a low-pass filter and an oscilloscope. Select any convenient generator frequency, f, and set the filter's cut-off frequency to 2f. With the filter in the circuit, the square wave output of the generator will be displayed as a sine wave of the same frequency on the oscilloscope.

The highest audio frequency that can be accurately sampled and the requisite sampling rate are often referred to as the "Nyquist frequency" and the "Nyquist rate," after Henry Nyquist of Bell Laboratories, who first noted this 2:1 relationship. Nyquist's theorem states that in order to accurately represent any waveform, we must have at least two complete samples of that waveform. In other words, when sampling a broadband signal, the sampling rate must be twice the highest frequency we let pass through the system. Thus, a Nyquist rate of 50kHz, for example, will allow us to accurately sample any audio signal below a Nyquist frequency of 25kHz.

Aliasing If we attempt to sample a frequency that is higher than the Nyquist frequency, we generate an enharmonic pitch called an "alias" frequency. The alias frequency is equal to the sampling frequency minus the input signal that is above the Nyquist frequency. Therefore, if we attempt to sample a 32kHz signal with a 50kHz sampling rate, an 18kHz alias frequency will be produced. This 18kHz signal

Figure 13.2 A simple waveform (a) may be thought of as an infinite series of discrete amplitudes. We may sample the waveform periodically through a series of equally-spaced windows (b, c). In the time intervals between samples (shaded areas), we have no information about the waveform.

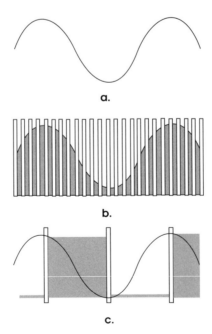

bears no musical relationship whatsoever to the 32kHz signal. In fact, the 32kHz signal is probably an overtone of a 16kHz signal. It is unlikely that we could perceive this overtone in any event. However, after we send the squared-off waveform through the low-pass filter to remove the odd harmonics, the 32kHz signal will be removed, but the aliased 18kHz signal will still be there. Another way of visualizing this phenomenon is to picture the signal above the Nyquist frequency folding down into the audible bandwidth by the same amount that it is above. To prevent this unrelated frequency (which is quite audible) from occurring, the signal to be sampled is first fed through a low-pass filter to remove any high-frequency components that might produce these aliasing effects. This input filter is often called the anti-aliasing filter.

Because of the practical design limitations of filters and the possibility that the sampling point may occur at the same time as the zero crossing for high frequencies, a sampling rate is chosen that is somewhat higher than twice the highest frequency that needs to be represented. This ensures that no information is distorted, and that a flat frequency response is maintained across the entire audio bandwidth.

Quantization

In Figure 13-2, we saw that the sample-and-hold process replaced the continuously variable sine wave with a square wave. In other words, a relatively simple waveform (Figure 13-2c) with only two levels has been substituted for the infinite number of levels that were previously seen (13-2a). This process of representing an analog waveform with a series of discrete levels (two, in this case) is known as "quantization." The square wave is said to be a quantized waveform with two quantization levels. It should be noted here that no analog information gets lost due to the sample-and-hold/quantization processes. Because of the anti-aliasing (low-pass) filtering previously described, there is no information present above the Nyquist frequency, and so quantization cannot lose what was not there in the first place.

Our next step is to devise a method of encoding the quantized amplitude information found at each sampling window. Although it is easy enough to define the maximum amplitude peaks as, for example, +100 and -100, with intermediate levels falling between these two extremes, recording these levels on tape is merely a needlessly complex variation of the traditional analog recording process. Instead, we must seek out a method of recording this information in a way that makes it immune to tape-induced analog noise. To accomplish this, we must first convert our quantized analog amplitudes into a binary-coded digital system.

■ DIGITAL ENCODING: THE BINARY SYSTEM

Even without further elaboration, we may intuitively realize that a coded signal of any form is no longer an analog of the original audio signal. Therefore, if we can successfully devise a system that will encode, record, playback, and decode our signal, we can free ourselves from the storage medium's noise, which is neither encoded nor decoded. What could be more elegant than using a series of on and off pulses to represent our audio signal waveform?

A binary numbering system is simply one with a base of 2, rather than our familiar base-10 decimal system. By the time most of us have passed through high school, our familiar decimal system of counting has become an almost intuitive process, and it takes some concentration to "un-learn" the decimal system, where ten symbols (0-9) are used to express any quantity. One might suspect that if our ancestors had evolved with eight fingers instead of ten, we would be using an octal system (0-7) instead. Or, if those eight fingers were on each hand, we may have been saddled with a hexa-decimal (6 + 10) system, which would require six more symbols beyond the familiar 0 - 9. The table shown in Figure 13-3 tabulates several of these alternative numbering systems, including the apparently quite primitive binary system, in which only two symbols (0 and 1) are used. In each case, we note that once we run out of symbols, we simply begin a new column and start all over.

At first glance, the binary system may appear the least promising of the lot. For example, an easily-read decimal number such as 217 becomes a rather awkward 110111001. We may well ask ourselves, just what is the point of requiring eight bits (Blinary digiTS) to express a quantity that is certainly a lot easier to read and understand in the decimal system? While the binary system is indeed something of a chore for most decimal-trained readers, it is quite another matter for an electronic system. Here, the disadvantage of a long succession of bits is far outweighed by the fact that the status of each bit may be unambiguously conveyed by a simple two-position switch. When the switch is closed, the bit is 1, and when it is open the bit is 0. Alternatively, the presence of a voltage signifies 1, while the absence indicates 0.

Figure 13.3 A comparison of various numbering systems, using binary, octal, decimal, and hexadecimal bases. The binary system is the foundation of digital recording technology, while the octal and hexadecimal systems are regularly found in contemporary electronic systems design.

Quantity	Binary (base 2)	Octal (8)	Decimal (10)	Hexadecimal (16)
	0	0	0	0
I	1	1	1	1
II	10	2	2	2
III	11	3	3	3
IIII	100	4	4	4
					101	5	5	5
I ||||	110	6	6	6
II ||||	111	7	7	7
III ||||	1000	10	8	8
IIII ||||	1001	11	9	9
|||| ||||	1010	12	10	A
I |||| ||||	1011	13	11	B
II |||| ||||	1100	14	12	C
III |||| ||||	1101	15	13	D
IIII |||| ||||	1110	16	14	E
|||| |||| ||||	1111	17	15	F
I |||| |||| ||||	10000	20	16	10
II |||| |||| ||||	10001	21	17	11

example: 16+0+0+0+1 = 2X8+1= 10+7= 16+1=17(decimal)

Coding Systems

Taking this a step further, the simple pulse code shown in Figure 13-4a may be used to express any binary number. Even if the waveform is severely distorted in transmission, the information is still retrievable, provided the waveform manages to cross a defined threshold, as shown in Figure 13-4b. A binary 1 is indicated whenever the signal is above the threshold, and a binary 0 is signified when it is not. All other characteristics (actual recorded level, distortion, etc.) are irrelevant and have no effect on the transmitted or recorded information.

The waveform may be recorded as a series of positive and negative polarity voltages that alternatively saturate the tape in opposite magnetic directions. This ability of magnetic tape was discussed in

Chapter 10. However, in such a brute-force recording system, "waveform fidelity" does not have the significance it had in the analog system, and consequently there is no need for a bias signal. This form of encoding is known as Pulse Code Modulation, or PCM, encoding. Other methods such as Pulse Width Modulation (PWM), Pulse Number Modulation (PNM) and Pulse Position Modulation (PPM) have been tried, but PCM is more easily coded to reduce the number of individual pulses necessary to represent a series of binary numbers. These coding or modulation schemes allow much greater amounts of data to be stored over a given tape area. Because of this, PCM has become the method of choice for representing the binary ones and zeroes. Although there are now other systems such as DSD (Direct Stream Digital), a detailed discussion of a PCM encoding/decoding system will illustrate all the basic principles of any digital coding system.

As a practical matter, the waveform just described needs some modification to make it a useful recording system. For example, a

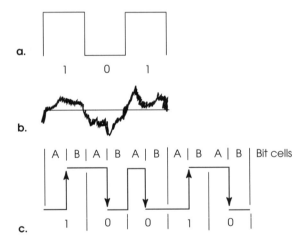

Figure 13.4 A simple square-wave-like series of pulses (a) may be used to transmit binary-coded data. Even when the waveform is severely distorted (b), the binary-coded data are preserved. The data code may be modified (c), so that 1s and 0s are defined by positive- and negative-going transitions instead.

long series of ones or zeroes will simply create a DC (direct current) component which will give no indication of when one bit stops and the next begins. Therefore, as a refinement in the recording of binary information, the waveform shown in Figure 13-4c may be used. In this particular modulation scheme, a 1 is indicated by a positive-going transition in the waveform, while a 0 is defined by a negative-going transition. When two or more identical bits appear in succession, there is a level change at the beginning of each successive identical bit. When a 1 is followed by a 0, or vice versa, there is no level change. This coding scheme allows the appropriate level transitions to take place for all possible combinations of ones and zeroes. For study purposes, in Figure 13-4c, each bit may be divided into two bit cells, A and B, with the transition that defines the bit as a 1 or 0 occurring between the cells.

In the practical digital tape recorder, an even more sophisticated coding scheme may be employed. Without some form of modulation scheme, the time information that is derived from the spacings of the individual bit cells would be lost, unless some external source of timing was employed. A modulation scheme called the "Non-return to Zero," or NRZ, code is used by the video-based digital recording systems. This code is very simple to use, but requires some external source of synchronization to determine the timing between successive ones or zeroes. This is provided by the control track pulses on the video recorder itself. Other codes such as the HDM-1 (High-Density Modulation) code, the EFM (Eight-to-Fourteen Modulation) code, and the Miller code are used with various different storage media. These will be discussed in greater detail in Chapter 14, "Digital Data Storage Systems."

The Digital Word

Returning now to our analog source, at each sampling window, we must convert the signal's amplitude into a binary number. In Figure 13-5, we may see that the more bits we have at our disposal, the better we may approximate (or quantize) the actual analog level.

A quantity expressed as a binary number is referred to as a "digital word." Thus, 101 is a three-bit word, and 11010 is a five-bit word. A one-bit word permits us to use only two discrete levels (0,1), whereas a two-bit word yields four (00, 01, 10, 11) levels, and so on. In other words, an N-bit binary word will give us 2^N discrete levels.

Barring an infinite number of bits, the quantized waveform that we will encode is only an approximation of the analog input waveform, as was just shown in Figure 13-5. In the more detailed drawing

Figure 13.5 The more quantization levels that are available, the more accurate the quantized waveform will be. In each example given here, the sine wave represents the analog waveform, and the stepped waveform, its quantized equivalent.

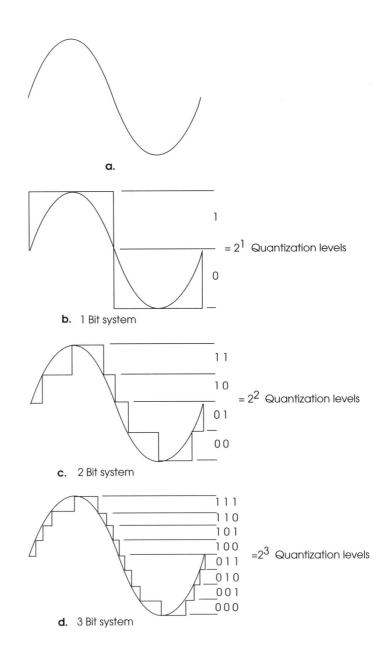

shown in Figure 13-6, note that all analog levels within any sampling interval (that is, between any A-B pair) are represented by the quantization level Q that appears within this interval. Thus, a single quantization level now represents all the analog levels that occurred within the sampling interval. Obviously, the more bits that are available, the higher will be the resolution of the system. There will, however, always be an analog amplitude that will fall between the available levels. This error is called quantization error and is limited to one-half the distance between two available levels. Most systems use 16-bit quantization that generates 2^{16} or 65,535 discrete levels. This number of bits generally lowers the noise caused by quantization error to an almost indistinguishable level.

Figure 13.6 Detailed drawing of the quantization process. The analog waveform (a) becomes the quantized waveform (b). The resultant waveform error may produce quantization noise.

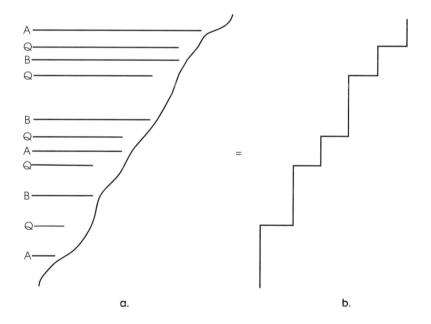

The maximum signal level that can be encoded in the digital word is determined by the number of bits used and the maximum quantization error. This signal-to-error ratio is roughly equivalent to the more familiar signal-to-noise ratio of an analog system. Generally, the dynamic range of the system in dB is given by the formula:

$$6.02(N) + 1.76 = \text{System Dynamic Range in dB}$$

where N = the number of bits in the system.

Therefore, the signal-to-noise ratio or dynamic range of a 16-bit system is:

$$6.02 \times 16 + 1.76 \text{ or } 97.98 \text{dB}.$$

This approximately 98dB range is considerably more than the maximum possible 76dB signal-to-noise ratio of most analog tape systems, as discussed in Chapter 10. Even a 14-bit digital system with 86dB of dynamic range exceeds most analog systems by 10dB.

■ THE RECORD PROCESS

Now that we have a basic understanding of how an analog waveform can be represented with a series of digital bits, we can look at the actual process of converting and storing a continuous analog audio signal. Figure 13-7 shows a block diagram of a typical PCM digital encoding system. The following discussions will describe each of these sections.

The first item that the signal passes through is usually an amplifier of some sort. This amplifier (a) raises (or lowers) the input signal to a level that is usable by the digital encoder. Digital processors are available with a wide range of input capabilities, ranging from unbalanced high impedance -10dBm inputs to balanced low impedance +4dBm inputs.

Dither Following this amplifier in many systems is a dither generator (b). As mentioned earlier, there will always be an analog amplitude that will be exactly halfway between two possible levels. This last or least significant bit cannot decide whether to be a 1 or a 0, and, as a

Figure 13.7 A simplified diagram of a two-channel analog-to-digital encoder: a. Input amplifier; b. Low-pass filter; c. Dither generator; d. Sample-and-hold circuit; e. Analog-to-digital converter; f. Multiplexer for parallel-to-serial data conversion; g. Error processing, and h. Record modulation.

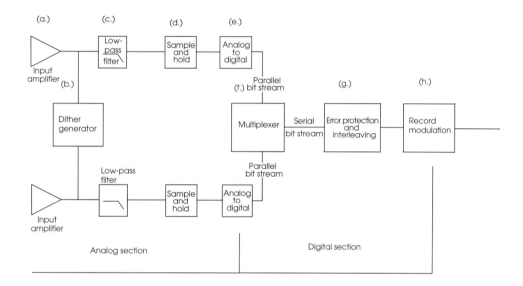

result, continually toggles between the two. This is the quantization noise mentioned earlier, and even though it is 98dB below the maximum signal, it can be audible since it is not being masked by tape noise. It can be especially evident during the fade-out at the end of a song or selection, and can be heard as a graininess in the signal at the very softest levels. This is often called granulation noise. To remove this noise, an analog noise signal called dither is added to the signal prior to quantization. The dither signal is usually a type of band-limited white noise; the amount of dither that is added is usually between one-quarter to one-third the value of the least significant bit. This amounts to about one to two dB of actual level. Dither was first developed by Bell Labs during the 1960s when they were experimenting with digital video. Dither actually lowers the noise of the digital system by pushing the toggling bit that is causing quantizing error to the next available level.

The Anti-Aliasing Filter

The analog input signal with dither is then sent through a low-pass filter (c) to remove any frequency components that are above the Nyquist frequency. If these audio frequencies are not removed at this point, they will cause the aliasing effects that were described earlier in this chapter. The design of this filter is critical, and various types of filters have been designed to give as flat a frequency response as possible with minimum phase distortion. Ideally, an infinitely steep cut-off would be preferred, but this brick-wall filter usually creates more problems than it solves. The typical anti-aliasing filter found in most professional digital recorders uses a steep 24-dB/8va slope that begins just below the Nyquist frequency. With a sampling rate of 48kHz, the filter slope gives a flat response with little phase distortion up to 20kHz. This allows a safety area of 4kHz prior to the Nyquist frequency.

The Sample-and-Hold Circuit

The filtered signal next passes to a sample-and-hold circuit (d). At regular intervals, as defined by the sampling rate, the analog signal is sampled, and this value is held until the next sample is taken. The sampling rates currently in use are 48kHz and 44.1kHz. The former is used by reel-to-reel digital recorders, while the latter is used by the rotary-head video-based systems. The compact disc plays back at a sampling rate of 44.1kHz, and the DAT will play back at 44.1kHz and 48kHz, but only record at the 48kHz rate when operating in the normal mode. A few video-based systems sample at a 48.056kHz rate, and this difference will be discussed in the next chapter. It is important in PCM systems that pulse spacing remain constant, and that the sampled value be held fully until the next sample occurs. If the sample were not held, the analog voltages would change between samples, and the following A-to-D converter would create false digital words. Figure 13-6 shows an analog waveform being sampled at a fixed rate and the resultant staircase waveform after the sample-and-hold circuit.

The Analog-to-Digital Converter

Each sampled analog signal is now converted into a 16-bit digital word based on its voltage level, and these words appear at the sixteen outputs of the A/D converter. This analog voltage to digital word circuit (e) must decide which of the available quantization levels most closely approximates the analog input voltage for each sample, and derive a binary number from that chosen level. If the sampling rate is 48kHz, this occurs 48,000 times per second, or once every 20.416 microseconds. The circuit must be extremely fast and accurate since any under- or over-level assigning will create errors in the digital word.

The Multiplexer   Tape recorders (analog or digital) may be thought of as serial devices. This means each recorded channel is only capable of processing a single data sequence. Depending on the type of recorder, this may be a continuously varying analog signal, or a serial bit stream of ones and zeroes. However, the A/D converter has a parallel output. All sixteen bits appear simultaneously at the sixteen output pins of the converter. In order to record the 16-bit word on tape, the bits must be sent to the recorder one at a time (serially). The multiplexer (f) accomplishes this by converting the sixteen parallel bits into a serial output, usually utilizing a shift register. This bit stream is then sent to the next stage of the digital encoder.

At the multiplexer stage, any necessary data coding is also done. Many systems require the use of an address bit to identify the locations of the data on the storage medium. A synchronization bit is also needed in some systems so that the beginning of each digital word can be identified. Other identification bits or "flags" can signify the presence of pre-emphasis, represent index numbers, or even be used for absolute timing such as time code numbers or reel numbers. These bits are often called "preambles" and are similar in nature to the "P" and "Q" codes added to the compact disc's data stream to identify track and index numbers.

Error Protection and Interleaving   Digital tape is subject to the same coating imperfections found in analog tape formulations and, of course, is just as susceptible to airborne dust particles and dirt build-ups along the tape path. Consequently, there are the inevitable dropouts in the serial data stream. Although a reasonable number of dropouts can be tolerated and may even go unnoticed on an analog tape, a digital tape dropout may be impossible to ignore. Furthermore, there may even be drop-ins. This occurs when tiny magnetic particles become dislodged from the oxide surface at one location and become re-located elsewhere on the tape.

In the digital domain with the digital signal recorded as a series of ones and zeroes, a dropout or drop-in creates an error in the code, whereby the decoded signal is likely to contain an annoying noise or even an interval of silence. Fortunately, the digital format permits various error detection and correction codes to be recorded (g) along with the audio signal. Typically, the recorded data stream for each audio channel consists of four components: a preamble (sync bit and identification code), the audio data itself, a cyclic redundancy check (CRC) code, and a parity check code. In some digital recording formats, the parity check code may be recorded at a separate location or on a different area of the tape. In fact, the entire data stream may be recorded several times and at different locations on the tape. Some systems may have as many as six data streams for each channel of audio information.

The CRC Code

The CRC (Cyclic Redundancy Check) code is used as an error detection system. To understand the general theory of operation, consider an encoding system in which the audio data stream is examined at regular intervals, called "data blocks." Each data block is mathematically divided by a fixed number. Although the division is, of course, a binary operation, we will use the more comfortable decimal system equivalent for ease of explanation.

Each binary number is to be divided by a fixed divisor. The quotient is ignored, and the remainder is noted and recorded (in binary form) at the end of the data block. This is the cyclic redundancy check code for this segment of the data stream. Later on, when the audio data stream is reproduced, it is again divided by that same number, and the remainder is compared to the reproduced CRC code. If the two numbers are not the same, it is an indication of an error. Either the data block, or the CRC itself, contains the error. Of course, the CRC code is not capable of correcting the error. It is merely an error detection system. To make the necessary correction, an additional code is required.

The Parity Block

Returning to the actual binary format to be recorded, let us examine a series of the three simplified 5-bit data blocks shown here:

```
01100   Data Block 1
10111   Data Block 2
10101   Data Block 3
-----
01110   Parity Block
```

The parity block code is found by adding the data blocks and ignoring all arithmetic carries. Thus, although binary $1 + 1 = 10$, we discard the carried 1, and so we have $1 + 1 = 0$. In decimal notation, we might similarly discover that now $9 + 8 = 7$, rather than 17. Note that in our parity system, an odd number of ones in any column yields a parity bit of one, and an even number of ones makes the parity bit zero. As mentioned earlier, the parity check code is added to the serial bit stream and becomes part of the digital word for that particular sample. We will see how this parity block is used to correct errors detected by the CRC in the discussion about playback.

Interleaving

As a further safeguard against data errors, most systems use an interleaving format, in which successive digital words are scattered so that no two successive words follow each other. Often, the data stream is first divided into odd and even word sequences. Thus, a 12-bit word sequence is generated as 1, 3, 5, 7, 9, 11, 2, 4, 6, 8, 10, 12. This sequence may be further cross-interleaved to produce 1, 5, 9, 3, 7, 11, 2, 6, 10, 4, 8, 12. One of the more popular coding schemes is the Reed-Solomon code. The Cross-Interleave Reed-Solomon Code (CIRC) is the standard for data recorded for the compact disc.

This interleaving means that a drop-out or other physical defect that could cause a data error in more than one word is dispersed across a larger area. When the data stream is restored to its proper sequence, it will be easier to restore the missing erroneous data since the error detection/correction parity scheme makes the assumption that only one block of data is wrong. CIRC prevents burst errors (large blocks of lost data) from creating non-correctable areas of digital data in the serial bit stream.

The Record Circuit

The digital bit stream is now almost ready to be recorded on magnetic tape or other storage media. It is at this point that the record modulation mentioned earlier (Nonreturn-to-Zero, Eight-to-Fourteen, HDM-1) takes place (h). The derived modulating voltages are sent to the record circuit where they are converted to magnetic reversals and recorded on the magnetic tape as positive and negative saturations. The reversals on tape represent the binary ones and zeroes of the digital signal.

THE PLAYBACK PROCESS

Now that we have successfully digitized and stored the analog signal on some form of storage medium, be it magnetic or optical, we must now retrieve the digital information and reconstruct our original continuous analog waveform. The conversion, in most ways, is simply a reversal of the record process. A block diagram of the D-to-A process found in most PCM systems is shown in Figure 13-8.

The Playback Circuit (Demodulation) The first step in the playback process (a) is to convert the magnetic saturation reversals back to recognizable binary ones and zeroes. The output of the playback head is usually very low in level and some processing is needed to retrieve all the data. In actuality, the waveform has become rounded and the sharp transitions that were

Figure 13.8 A simplified diagram of a two-channel digital-to-analog decoder: a. Playback demodulation; b. Error processing (detection and correction); c. De-multiplexer for serial-to-parallel data conversion; d. Digital-to-analog converter; e. Sample-and-hold circuit for wave reconstruction; f. Low-pass or "smoothing" filter; g. Output amplifier.

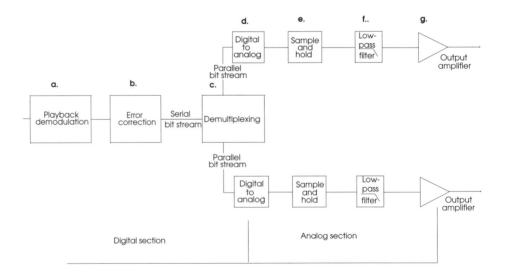

produced at the output of the encoder have been lost. A wave-shaping circuit restores the signal to more angular wave-like transitions, thereby allowing the ones and zeroes to be recreated accurately. Because we have completely recovered all the binary digits in the same order in which they were recorded, no data have been lost or distorted. The digital data have been retrieved without tape hiss, print-through, tape distortion, wow and flutter, or any of the other ills that plague analog magnetic tape.

Error Correction The serial bit stream we have just produced is now de-interleaved using a reverse of the original interleaving scheme. Dropouts that occurred on the tape itself may have affected a number of words, but after de-interleaving, these errors are most likely not successive in nature. This fact makes it easier for the digital decoder to detect and correct any data errors (b).

Earlier we created a series of three simplified 5-bit data blocks, and from them we generated a parity block which was recorded along with the digital data word. Now assume that during playback, the CRC code determines that a data error is present in Block 2. The three data blocks are now combined with the recorded parity block to produce a new parity block. The new parity block is combined with the incorrect data block, and this operation corrects the information that was in error.

```
         01100   Data Block 1
         01001   Data Block 2 (Incorrect)
         10101   Data Block 3
         01110   Recorded Parity Block
         -----
         11110   New Parity Block
         01001   Incorrect Data Block 2
         -----
         10111   Corrected Data Block 2
```

Note that if an error occurs within the CRC code, indicating an error where none actually exists, the new parity block will not change the value of the data block. For example:

```
         01100   Data Block 1
         10111   Data Block 2
         10101   Data Block 3 (Falsely assumed to be incorrect)
```

```
01110   Recorded Parity Block
─────
00000   New Parity Block
10101   "Incorrect" Data Block 3
─────
10101   "Corrected" Data Block 3 (No change)
```

As can be seen from the above example, block errors can be easily corrected when detected by the CRC code. Block code error correction systems are very effective. A further refinement of this technique uses a "hamming" code to locate the error within the block. In effect, a parity bit is generated for each section of the 16-bit word, and the above process occurs in miniature.

Other types of error-correcting schemes are used by various manufacturers. Some use the mathematical approach as above, while others use convolutional or predictive methods. Many of these methods can apply to the individual bit or to the whole word, or both. As digital technology continues to develop, we will see more and more sophisticated error detection and correction schemes in use.

Error Replacement and Concealment

What if, for some reason or other, an error is detected but cannot be corrected by the available parity blocks, hamming codes, etc.? The decoder's first response will be to look elsewhere for redundant data. In the preamble of the digital word, there is an address code that will tell the D/A converter where to look for the redundant data. If this is not possible, other methods must be employed.

One of these methods is "interpolation." Since audio is a continuous and related signal, it is relatively easy in many cases to logically determine the missing data. If, for instance, I were counting from 1 to 10 and there were a silence where the word "five" would normally have been, it would not take you long to figure out which word was missing. Similarly, if a string of voltages were being measured as 1.0v, 1.1v, 1.2v, 1.3v, 0v, 1.5v, it would seem logical that instead of the fifth reading being zero volts, it should be 1.4 volts. This is called linear interpolation, and it takes place when the digital converter makes up data that would logically fit in a sequence between two known values.

If two or more successive values are missing even after de-interleaving, the decoder will usually hold the previous value to conceal the missing or erroneous data. A combination of linear and lateral interpolation may be used, depending on the number of successive errors that are found to be non-correctable.

In the event that the above methods of correcting, replacing or concealing have been unable to produce a suitable data word to correct an error, the digital decoder will usually then mute. At this point, the data word in error is changed to all zeroes, and the output of the decoder is muted. Even in this most severe case, our ears may not perceive the data as missing. An absence of sound is much less noticeable than a click or sputter from invalid data. It would take one thousand successive muted digital words to create a mute of about 20 milliseconds in the analog output. The period of twenty milliseconds is recognized as the shortest noticeable integration time for the human ear. We may hear some distortion or roughness in the sound before that, but it is doubtful that a space in the music will be heard.

De-multiplexing

After the serial data stream has been restored to its original form, it is "de-multiplexed" (c). That is, it is restored to its parallel state before it is sent to the actual digital-to-analog converter. This restores the original sample timing by clocking out a parallel digital word of the same sampling rate as the original waveform.

The Digital-to-Analog Converter

The digital-to-analog converter (d) takes the parallel bit stream from the de-multiplexer and converts the 16-bit binary word back to an analog voltage value. These voltage values are equivalent to the levels of the originally sampled pulses derived from the input analog waveform.

The Sample-and-Hold Circuit

This series of voltage pulses is then put into another sample-and-hold circuit (e) to reproduce the staircase waveform. The correct voltage value is held until the next sample is decoded. This effectively filters out any false readings between samples because the initial value is held until the D/A converter outputs another.

One effect of the sample-and-hold function is a high-frequency roll-off phenomenon called "aperture loss." A perfect D/A converter would output a series of pure pulses with nothing in between and timed exactly to the original samples. In this imperfect world, however, switching pulses, miscellaneous transients, etc., can occur between the samples. While the sample-and-hold function filters these, it also creates a high frequency roll-off that is infinite at the sampling frequency, but that has only fallen by about 4dB at the Nyquist frequency. (A Fourier transform of a pulse generates a sin x/x function with its first null point at the sampling frequency.)

Two methods are commonly applied to return the frequency response to a flat state. The first is to add some high-frequency pre-emphasis to the original input signal, similar to that used in some analog noise reduction units. This is sometimes switchable, and an identification flag is usually added to the preamble to indicate this to the decoder. The second method is to tailor the output filter to compensate for the fall-off of the highs. In either case, the response of the overall system remains flat.

The Output Filter The reconstructed, quantized signal is now passed through a low-pass filter (f) to remove all unwanted high-frequency components which are a function of the square wave-like quantization waveform. This filter is sometimes called the "reconstruction filter" or "smoothing filter," and usually has a very sharp roll-off (24-dB/8va), similar to the anti-aliasing filter found at the beginning of the system. This filter actually performs a superposition of the sin x/x function at each sample point, and as a result, no data are lost. The resultant output signal is a faithful reproduction of the original input source. The signal then often goes through an output amplifier (g) to raise the signal to system level.

■ THE DIGITAL ADVANTAGE

In the following section, various specifications of the digital tape recording process are reviewed, and comparisons are made to analog recording systems.

Tape Noise An immediate advantage of the digital recording process is its immunity to tape hiss. Although analog noise may indeed distort the waveform of the digitally-encoded signal, the decoding process is insensitive to this form of distortion. At any instant, the D/A output is either a "1" or a "0." There is no such thing as a "1.003" or a "0.0025." In fact, the ones and zeroes may be thought of simply as HI and LOW logic states whose actual voltage levels are irrelevant. Therefore, even a severe deviation from the defined level is of no practical consequence. The decoder looks for positive- and negative-going transitions, and is not concerned with the levels at either end of the transition. The analog noise component introduced by the tape medium is thus incapable of affecting the decoded signal output.

Dynamic Range

Dynamic range is no longer a function of the tape medium. Since the digital tape is always fully magnetized in one direction or another, saturation distortion is meaningless. The dynamic range is only limited by the number of bits that are available for each digital word. As already noted, the de facto standard is 16 bits, which gives a dynamic range of almost 98dB.

Cross-Talk

In analog tape recorders, the cross-talk specification is usually a function of head design limitations, and is typically within the -50 to -55 dB range. Cross-talk within a record or playback head is an analog phenomenon, and so has no significance to digital recording. It suffers the same fate as analog tape hiss, and simply does not survive the decoding process.

However, in the absence of tape-head cross-talk, the electronic cross-talk within the analog sections of the tape recorder is now measurable, with typical specifications within the -80 to -95 dB range.

Print-Through

Print-through from adjacent layers of tape manifests itself as analog distortion of the signal waveform. The printed signal is below the switching threshold of the HI/LOW logic states and as such, is not recognized by the decoder. As before, waveform distortion has no effect on the decoded output, and therefore the print-through signal is not present in the analog audio output.

Harmonic Distortion

The distortion characteristics of the digital format are significantly different from those associated with an analog system. In analog recording, distortion rises gradually as the tape's saturation level is approached. Generally, an output level is specified, and at some point the third-harmonic distortion reaches a three-percent level. Beyond this point, distortion rapidly rises, effectively defining the upper limit of the analog system's dynamic range.

As described earlier, the dynamic range of the digital recording system is not a function of the tape medium, and distortion does not increase with recorded level. Instead, distortion (or lack of it) remains constant as the maximum output level is approached. Maximum output level depends on the number of digital bits available. Once this ceiling is reached (all bits are ones), further level increases simply cause errors in the bit stream, resulting in instant and severe distortion. Error correction systems may cause mutes to occur to mask the distortion, but this solution is not much better than the distortion itself. Therefore, the transition between minimum and maximum distortion is almost instantaneous, and the engineer must take care not to let levels stray into the red. Unlike analog tape, which may forgive an occasional peak or two, digital distortion is immediately and painfully obvious.

Since the signal-to-quantizing-error ratio in a digital system remains constant until all the bits are used, it is common practice to set a "0" or system calibration point 10 to 15 dB below the maximum ceiling. Most systems use peak reading or "PPM" indicators to meter incoming analog signals. The system level of +4dBm is usually set at -15dB. Various digital metering systems will be discussed in Chapter 14.

Wow and Flutter   In the digital playback system, the serial bit stream is converted into a parallel output by the de-multiplexer, as mentioned earlier in this chapter. Since each parallel output is held until the next 16-bit serial word has been received, there is—in effect—a built-in "wow-and-flutter" filter in the decoding process. The start-and-stop nature of the data flow is such that any wow and flutter within the tape transport system has no effect on the decoded output.

Tape Duplication   Because the digital process is immune to the various distortion-producing phenomena described above, multiple-generation copies of a tape will be indistinguishable from the original master. There is very little that can go amiss, since only a simple series of Hi/Low logic states is being copied. Due to the nature of some storage media, the signal may lose the transition sharpness of the leading edge of the waveform, so some reshaping of the signal prior to re-recording is often done so that the transitions between the logic states remain as instantaneous as possible.

FOURTEEN

Digital Audio Recording Systems

In Chapter 13 we examined the process for converting a continuous analog waveform into a binary bit stream that accurately represents the original signal. Once this bit stream is generated, it needs to be stored so that it may later be retrieved and then converted back into an analog signal. This signal can be sent to amplifiers and loudspeakers, mixing consoles or other analog equipment.

Today there is much equipment that is capable of processing the digital signal entirely in the digital domain, and there are more devices of this type on the horizon. These digital signal processors may accept the digital signal from a variety of sources. However, any digital signal must be converted back to its equivalent analog waveform, either as an electrical signal or as an acoustic wave, before our ears can perceive the nature of the sound.

The accuracy of the digital storage medium is an important link in the digital recording/playback chain, and the engineer who wishes to accurately make use of this technology should understand the various storage media available for his use.

■ TRACKS VERSUS CHANNELS

In analog recording, the terms "tracks" and "channels" are used almost synonymously. However, digital specifications sometimes list two or more digital tracks per audio channel. To understand this distinction, we must first examine some of the specifications for a hypothetical digital tape recorder. These specifications are tabulated in Figure 14-1.

Reviewing the data tabulated at 76 cm/sec (30 ips), we find that with a 16-bit word transmitted during each sampling interval, each bit has a word length of 9.5×10^{-5} cm, and the bit stream is recorded on each tape track at a rate of 800 kbit/sec. If this bit stream represents an alternating 10101010.... series, the recorded frequency will be 400kHz. At 76 cm/sec, this represents a wavelength of 1.9μm. At 38 cm/sec the wavelength is 0.95μm. Given existing longitudinal tape recorder technology, recorded wavelengths of less than 1.5μm are considered unreliable, and therefore the wavelength at 38 cm/sec is unacceptable. However, if the bit stream for each audio channel is distributed across two digital tracks on the tape recorder, the bit rate per track becomes 400 kbits/sec. Therefore, f_{max} = 200kHz, and the minimum wavelength is once again within the acceptable limits. This will be explored further as the various types of digital recorders and track formats are discussed.

■ SAMPLING-RATE CONVERTERS

For any digital signal to be decoded to its analog equivalent, the specifications of the digital bit stream must be matched to the specifications of the digital playback machine. Different sampling rates are used by several of the various systems, and in order to transfer a signal from one machine to another, the sampling rates must match. The predominant sampling rates are tabulated in Figure 14-2.

Leaving aside all considerations of track layout and tape speed, the output bit stream from one recorder/player must match the sampling rate of the device to which it is being sent. If the rates are different, a sample-rate converter is needed to accomplish this task. The converter takes the incoming signal and, by re-clocking the signal, modifies the sampling rate to match the required configuration.

Many sampling-rate converters also allow signal interchange between various digital input/output signal formats. The most commonly found formats are the AES/EBU (Audio Engineering Society/European Broadcasting Union) balanced interface format (requiring XLR connectors), the S/PDIF (Sony-Philips digital interface) unbalanced format (requiring unbalanced RCA connectors and also known as the IES-958 standard), and the SDIF-2 (Sony digital interface 2) format requiring BNC connectors.

Figure 14.1 Data specifications for a tape recorder with a sampling frequency of 48kHz.

Specification	High	Low	Units
Sampling rate f_s	48	48	kHz
Tape speed V	76	38	cm/sec
Wavelength λ_{fs}	1.52×10^{-3}	$.076 \times 10^{-3}$	cm
Bit length BL = 1/16	9.5×10^{-5}	4.75×10^{-5}	cm/bit
Bit rate BR = V/BL	800	800	kbits/sec/track
f_{max} BR/2	400	400	kHz
λ_{min} V/f_{max}	1.9	0.95	μm

Figure 14.2 Predominant sampling rates and uses.

Sampling rate	Principle use
32kHz	Digital audio broadcasts (EBU), long-play DAT, and some reduced-bandwidth multi-channel audio recordings on digital video systems such as DV-Cam.
44.056kHz	Used with NTSC color video systems running at a SMPTE frame rate of 29.97 fps.
44.1kHz	Compact disc, DAT, DTRS and ADAT, and U-Matic-based monochrome storage systems (PCM-1630).
48kHz	Open-reel digital audio tape recorders, DAT, an option for DVD-V, and audio for digital video systems.
96kHz	Standard for audio on the DVD-V format.
192kHz	Standard for 2-channel audio on the DVD-A format.

FIXED-HEAD DIGITAL RECORDERS

Advances in thin-film-head technology have permitted the design of recording heads for longitudinal recording that meet the frequency requirements of digital recording. This has resulted in the development of two systems of multi-channel digital tape recording that are capable of all the synchronous recording techniques required for recording and mixdown of popular music. The Digital Audio Stationary Head System (DASH) and the Professional Digital System (PD) have replaced earlier types of digital multi-track tape recorders developed by other manufacturers. The uses and functions of these two systems are similar, and an engineer familiar with the operation of a conventional analog multi-track tape recorder should have no trouble adapting to these machines. It should be noted, however, that the actual methods of data storage, error detection and correction, channel codings, and track layouts are different; each format will be discussed separately. Both systems formerly offered two-track longitudinal digital recorders as well as multi-track machines, but the popularity and ease of DAT use has caused them to disappear.

In our discussions of analog recorders, reference was made to the erase, record, and reproduce heads. In the digital domain, these heads are often referred to as the erase, write, and read heads, respectively. Because it is very difficult for a head to simultaneously perform both a read and a write function when handling high-density digital data streams, there is a second read and/or write head found in the headblock of a digital multi-track recorder. This extra head is necessary for the recorder to be able to perform synchronization functions.

Electronic cross-fading is used to smooth the transitions between old and new data during punch-ins and -outs. As in analog multi-track recording, old data are erased and replaced with new data in synchronization with the material on other tracks. To keep these transitions as smooth as possible and to reduce data errors, data buffers hold the old data at the punch-in point and cross-fade that with the new data before sending the signal to the write head. At the punch-out point, the reverse occurs and the new data are buffered and cross-faded with the old. This electronic cross-fading and data buffering is also used to prevent data errors caused by overlapping areas of data that occur when the tape is spliced with a razor blade. Electronic cross-fading is illustrated in Figure 14-3.

Figure 14.3 (a) shows the physical head positioning required for replacing old data; (b) illustrates the cross-fading of old and new data.

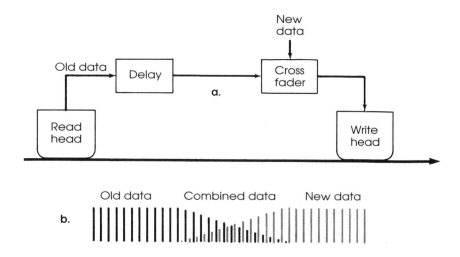

The DASH System

The Digital Audio Stationary Head, or DASH, system is one of the two professionally accepted longitudinal digital multi-track recording systems. Most DASH multi-track tape recorders have twenty-four channels of digital audio, two analog auxiliary cue tracks, a SMPTE time-code trac,k and a control track or CTL. This total of 28 tracks is arrayed vertically from 1 to 28 on ½-inch tape. Each digital track is 0.24 mm wide; the gap spacing within the head for each track is 0.20 mm. The analog tracks are slightly separated from the digital tracks to prevent the required analog bias from affecting the digital tracks. The transport operates at 30 ips and provides 60 minutes of recording using 9000 feet of tape, which is packed on 14-inch precision metal reels. The DASH system uses 16-bit linear pulse code modulation (PCM) and has switchable sampling frequencies of 48kHz or 44.1kHz. At the lower sampling rate, the tape speed is lowered to 27.56 ips. This keeps the recorded wavelengths the same regardless of sampling frequency.

The analog auxiliary tracks are used primarily for cueing purposes. The digital audio channels can only be listened to while the tape recorder is at the proper play speed. There are times, though, when it is advantageous for the engineer to be able to hear the signal while the tape is being slowly moved by hand. The associated analog cue inputs receive audio signals during the recording of the basic tracks. When using a typical in-line recording console, these cue inputs can be fed from the stereo monitor bus outs or from a spare auxiliary bus.

The control track (CTL) is used to control the speed of the transport during reproduce operations. It is recorded during the basic tracking sessions and should extend continuously all the way through the song. Because of this, the prudent engineer may pre-stripe the tape to be used prior to the beginning of the session. Most DASH multi-track recorders have an "Advance Record" function to handle this task. The control track consists of a synchronizing pattern that contains a sector-identifying mark, a control word that defines the sampling rate in use, the address for data location, and a Cyclic Redundancy Check Code (CRCC) word for error protection. It is possible to synchronize two multi-track DASH tape recorders together using the control tracks.

The SMPTE time-code track stores non-drop-frame longitudinal time-code that is used for synchronization with other machines, as well as transport control of auto-locate and cue functions.

A 48-track version of this system is also available, in which tracks 25 through 48 are interleaved with tracks 1 through 24. Thus, tracks 1 through 24, recorded on a 24-track DASH machine, can be played on the 48-track version if desired. Likewise, the first 24 tracks recorded on the 48-track machine will be available for playback on the 24-track digital tape recorder. The larger channel version DASH 48-track digital tape recorder retains the four analog tracks (2 auxiliary, 1 time-code, 1 CTL) of the original format. This interleaved channel configuration was selected to maintain machine-to-machine compatibility within the DASH operating community.

The analog audio is converted to digital audio through the A-to-D converters, and the data are used to form data blocks. Each data block contains a sync word which is used to identify the block, 12 data words which contain the audio information, four parity words for error checking, and a Cyclic Redundancy Check Code (CRCC) word. The data blocks correspond to the sector-identifying mark of the control track. As mentioned in Chapter 13, various modulation codes must be used with binary data for more compact high-density data storage. The DASH system makes use of a system called High

Density Modulation-1, or HDM-1, for maximum packing density and to reduce the bandwidth requirements for the digital data.

As previously mentioned, it is very difficult to construct a head that will read and write data on different tracks simultaneously. However, when overdubbing new material in-sync on analog tape recorders, this is exactly what is done. To overcome this limitation inherent in digital multi-track heads, the DASH system utilizes an additional write head. The first write head is used for normal recording; it is followed by the read head. This allows off-tape monitoring during the recording of the basic tracks. In "sync" mode, the signal from the read head is delayed; the new data are written by the second write head and applied to the tape synchronous to the data on the previously recorded tracks.

The Professional Digital (PD) System

The other reel-to-reel stationary-head digital recording system is the PRO-DIGI (Professional Digital) system, commonly referred to as the PD format. Although the PD system is no longer manufactured, there are still some PD systems in operation. But as parts for repairs become harder to acquire, the format is disappearing, leaving the DASH system as the last reel-to-reel digital system. The PD system differs from the DASH system in several ways, and it is useful to examine these differences.

At first glance, the most notable difference between the DASH format multi-track tape recorders and the PD format machines is the number of available recording channels. PD tape recorders allow thirty-two channels of information to be recorded and replayed. Along with the thirty-two channels of digital audio, there are two auxiliary cue tracks, two digital auxiliary tracks, and one SMPTE time-code track. The SMPTE time-code track, used for auto-locating functions and synchronization, is also an integral part of the block-locating system within the encoding/decoding process. In addition to these tracks, there are eight additional digital tracks for a total of forty-five tracks on 1-inch-wide tape. The eight additional digital tracks are divided among the primary digital tracks so that for every eight channels of information, ten tracks are used. These extra tracks store the check words of the associated digital-audio tracks for the error-protection systems. Each digital track is 0.29 mm wide and the heads have a gap spacing of 0.27 mm. The tracks are not, as in the DASH system, arranged numerically from top to bottom, but are shuffled so that no two consecutively-numbered tracks are adjacent to one another. Similar to the DASH format, however, the auxiliary tracks are placed on the extreme edges of the tape where they act as guard bands, protecting the digital tracks from edge damage. Figure 14-4 shows the track layout for both the PD and DASH (24- and 48-track) formats.

Figure 14.4 Track formats for digital reel-to-reel multi-track tape recorders: a. the 32-channel ProDigi format, b. the 24-channel DASH format; and c. the 48-channel double density DASH format.

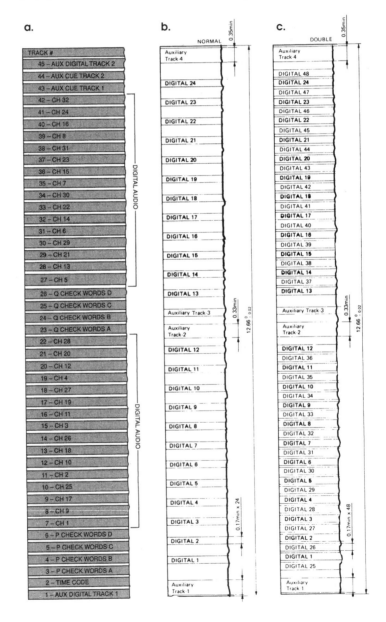

The transport operates at 30 ips. Sixty minutes of recording time are available when using 14-inch reels. The recording system uses 16-bit Pulse Code Modulation (PCM) encoding for the digital audio tracks and Pulse Width Modulation (PWM) for the auxiliary tracks. Sampling frequencies of 48kHz and 44.1kHz are supported, but in the PD format, tape speed is not affected by a change in sampling frequencies. This means that at the lower sampling frequency, less data are stored on a given length of tape, which results in a change of wavelength on tape for any given input frequency. Speed control is achieved by locking the capstan to a reference frequency. A control track is therefore not required.

As in the DASH format, data from the A-to-D converters are used to assemble data blocks. Within the PD format, the data block contains a 16-bit sync word followed by twelve data words, in turn followed by a 16-bit CRCC word, for a total of 224 bits within each data block. This is slightly less than the 288-bit data block used by the DASH system which, as mentioned earlier, contains four 16-bit check words prior to the CRCC word in each block. The check words used by the PD format are stored on the additional dedicated digital tracks. It is also important to note that the check word tracks are displaced vertically from their associated audio tracks, so that the check words are not lost with the audio data in the event of a large tape dropout. The PD system uses a Four-to-Six Modulation (4/6M) scheme to achieve the high packing density necessary for digital multi-track tape recorders.

To allow overdubbing, the PD system uses two read heads and one write head. The basic tracks, which have already been recorded, are played by the secondary read head (which precedes the write head), and delayed by the appropriate interval to ensure that the players will be "in sync" as the new information is written by the single write head. Figure 14-5 shows the head configurations for both the PD and DASH systems.

ROTARY-HEAD RECORDING SYSTEMS

In our previous example of a hypothetical digital recorder, bit rates of 800 kbits/sec were being recorded and, as mentioned earlier, this represents a frequency of 400kHz for each channel of recorded material. A stereo signal that has been converted to a digital bit stream, with additional bits added to each word for preambles and parity, can easily generate frequencies in the 1MHz to 2MHz range. Since most video recorders have a frequency bandwidth of around 4MHz or better, they can be considered an ideal recorder for storing digital data streams.

A digital audio processor is used to convert the stereo signal into a digital data stream that can be recorded as video data on various types of video recorders. The two channels of the stereo signal are combined or "multiplexed" into a continuous stream that contains both the left- and right-channel audio information. Vertical and horizontal synchronization pulses are also generated and added to the

Figure 14.5 Head position for the (a) DASH multi-track system and (b) the ProDigi multi-track system.

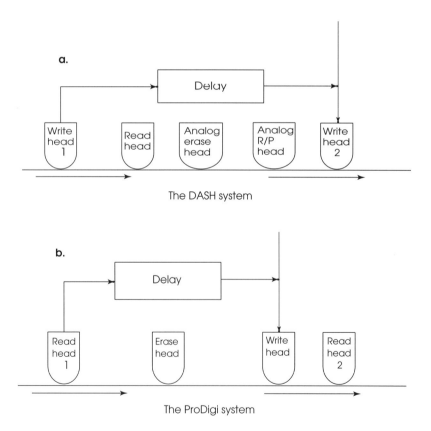

signal so that it will appear to be a video signal and thus be recorded correctly on the video recorder. Video recording, however, is a non-continuous process, in contrast to the more familiar analog recording. This is because after each video field and associated pass of the electron gun on the television screen, there is an interval of time while the beam flies back to the upper left-hand corner of the screen. This is called the "vertical blanking interval." The digital audio processor contains buffer memories, which hold the signal during this interval so that the continuous serial data stream from the A-to-D converter can be stored in the non-continuous video fields and frames of the video system. Professional digital recorders that use a video-based storage medium usually have SMPTE timecode recorded on one of the audio tracks of the video recorder (discussed in Chapter 15). This allows for synchronization and editing and will be discussed later in this chapter.

U-Format Recorders

Originally, most of these professional video-based systems stored their data on what is called U-Format video tape. U-Format tape, often called U-Matic, is a video cassette containing video tape that is ¾-inch in width. Tape speed is 3.75 ips and recording duration can vary from 10 minutes to as long as 75 minutes. The recording system uses a technique called helical scanning, in which the tape is wrapped around a rotating-head cylinder. The cylinder contains two video heads, each capable of recording one field of each video frame. Each head produces a track width of 3.4 mils; the tape is wrapped around the head at an angle which causes the tracks to be recorded diagonally across the tape. This yields a track length per field of 6.7 inches. Separate fixed heads record analog audio signals longitudinally along the bottom of the video tape. A CTL or control track, which controls the spot at which the signal switches from head to head, is recorded at the top of the tape. For a further discussion on video recording and frame rates, refer to Chapter 15.

The high-frequency response necessary for video and digital audio is achieved by spinning the video head cylinder at a rate of 1800 rpm while moving the tape past the spinning heads at a relatively slow speed of 3.75 inches per second (ips). The resulting tape speed (often called "writing speed") is roughly 400 ips, allowing data to be recorded in the megahertz range. U-Format recorders also place a guard band between each field of the video or pseudo-video signal. This prevents the signal in one frame or field from bleeding into the next, thus minimizing tracking error.

The predominant frame rate for these professional digital rotary-head systems is 30 frames per second, which is the frame rate for NTSC monochrome video. The video scan rate, which determines the number of frames per second, also controls how the digital audio data are stored in the video frame. The video scan rate of a professional U-Format video tape recorder allows a sampling rate of 44.1kHz with 16-bit quantization when used in the NTSC monochrome mode of 30 frames per second.

Professional digital-audio processors and their associated U-Format recorders are still used in the mastering and manufacture of the compact disc, although other formats, such as Exabyte and the pre-mastered CD (PMCD) are making major inroads.

Beta- and VHS-Format Recorders

Home digital-audio processors designed for consumer video cassette decks have been used successfully to record some memorable performances. They are not often seen nowadays and there are no compatible processors currently in production. These processors differ from the professional versions in several very important ways, including tape speed and width. Helical-scan recording is again employed, but at a reduced tape speed of .787 ips for Beta II and .657 ips for VHS LP. Even though the head drum rotates at 1800 rpm, as in the U-Format, the reduced speed results in equivalent tape speeds of 270 ips and 225 ips, respectively. Track widths are also reduced to less than half that of the larger format and guard bands have been eliminated to allow more frames of video information in a given tape length.

While U-Format recorders are capable of recording at a frame rate of 30 frames per second (fps) in the NTSC monochrome video mode, and 29.97 fps in the NTSC color mode, Beta and VHS recorders can only record at the color frame rate. As mentioned earlier, a frame rate of 30 fps allows a sampling rate of 44.1kHz. Sampling at the slightly slower color rate translates to a sampling rate of 44.056kHz. This difference in sampling rate requires that it be sample-converted before a consumer-format PCM digital audio tape can be mastered for compact-disc or DAT release.

DAT Recorders

Probably the most popular recordable digital storage medium is the Digital Audio Tape, or DAT, recorder. The DAT was developed primarily for the consumer marketplace, but because of its stability and accuracy, it found its way into professional recording studios. The desire to develop small, convenient digital recorders suitable for home use led to the creation of two small-format digital recording systems. One used a stationary-head transport system, while the other used a rotating-head transport similar to current video technology. They were originally called S-DAT (for stationary head) and R-DAT (for rotating head), respectively. The superiority of the R-DAT method established it as the sole consumer digital-audio recorder, the R- prefix has been dropped and the unit is now simply called a DAT recorder.

Tape is contained in a plastic cassette, called a "DAT cassette," which is similar to a video cassette and is a high-coercivity metal-particle tape. The cassette dimensions are 73 mm (2⅞") × 54 mm (2⅛") × 10.5 mm (⅜") (W × D × H). Tape width is 3.81 mm. The completely sealed structure prevents dirt and debris (as well as fingerprints) from causing dropouts and errors in the digital signal.

The DAT recorder uses a helical-scanning method of recording similar to that found in U-Format video tape recorders. However, the amount of tape wrapped around the drum during record and playback is considerably less than in the video formats. The tape is in contact with the head cylinder over an angle of 90 degrees, which is one-fourth of the circumference of the drum. This allows the tape to be fast-forwarded or rewound while still wrapped around the head drum with minimal or no tape wear or damage. Conventional video recorders remove the tape from the head cylinder prior to fast-speed shuttling. The rotating cylinder has two heads on a drum with a 30-mm diameter. The narrow drum-to-tape contact angle means that the record or playback signal is applied to the tape only 50 percent of the time. The remaining percent of the time, the signal is interrupted and buffer memories are once again used to convert the continuous data stream to a non-continuous recording format.

The tape moves at a speed of 8.15 mm per second, which is about ¼ ips. This slow longitudinal tape speed, coupled with long tape lengths in the DAT cassette, results in possible recording times of over two hours. The head drum rotates at a speed of 2000 rpm, resulting in an equivalent writing speed of approximately 123 inches per second. The tracks are recorded on tape without guard bands. As mentioned earlier in the discussion of consumer video recorders, the lack of a guard band between tracks can create tracking problems that can produce errors in the digital bit stream. This is avoided in the DAT format by using a method called "azimuth recording." The tracks are tilted away from the relative vertical of the tape by an angle of plus or minus 20 degrees. The first track has positive azimuth, the following one has negative azimuth; this situation continues for the length of the tape. The length of each diagonal track is 23.501 mm, yet only a portion of that length contains audio data. By constantly comparing the azimuth angle of the pilot and sync signals between heads, automatic track finding (ATF) is possible. This ATF-azimuth-type recording allows precise tracking without guard bands or the CTL (tracking control head) signals that are necessary in video recorders. The track layout of the DAT recording system is shown in Figure 14-6.

The DAT recording system uses a double Reed-Solomon code for error detection and correction. The signal is also interleaved, and data are recorded across two tracks. This makes data retrieval possible even if only one head is functioning. The DAT system was designed to be able to record and playback at sampling rates of 48kHz, 44.1kHz and 32kHz. However, not all DAT machines will

Figure 14.6 Track format for the DAT (rotary) showing positions for audio data, subcode information, and ATF (Automatic Track Finding) encoding.

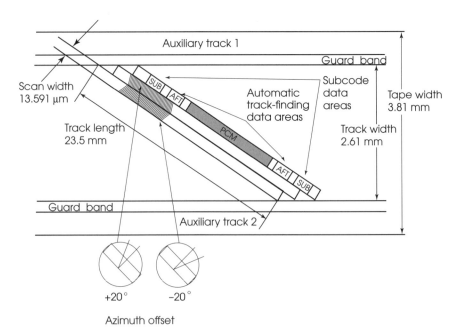

DIGITAL AUDIO RECORDING SYSTEMS

record at all sampling rates. The quantization used is normally 16-bit. There have been several experimental 88.2kHz and 96kHz DAT recorders produced, as well as a model that will store 24-bit word length data. It remains to be seen whether these machines become more prevalent. The DAT will automatically select a sampling rate of 44.1kHz if a 44.056kHz digital signal is applied. This will cause a pitch discrepancy and raise the pitch by 0.1 percent. Figure 14-7 shows a typical DAT recording system.

Multi-Track Rotary-Head Systems

Advances in digital audio coding systems, in conjunction with the advent of ATF (Automatic Track Finding), have made it possible to develop multi-track rotary-head recording systems. As of publishing date, two systems are available. One system uses the VHS-format video system, the other relies on the 8mm video format. Both rotary-head multi-track systems use a high-band version of their corresponding video systems, meaning that they operate at higher video carrier frequencies. Although the VHS system is now the standard consumer video system in use, its data recording capabilities are far below those of professional video recorders. On the other hand, its high-band cousin, S-VHS, offers improved response and lower noise by operating at a higher carrier frequency. The 8mm system was developed principally for hand-held camcorders; the high-band version of that system, known as Hi8, has been used successfully in professional video situations. Information capacity is the key element for high-bandwidth recording systems. Both S-VHS

Figure 14.7 A professional DAT recorder (Tascam DA-40, Tascam photo).

and Hi8 video systems significantly shift the luminance signal up in frequency. The standard 8mm system uses a range from 4.2MHz to 5.4MHZ, while the Hi8 system uses a 5.7MHz to 7.7MHz range. The improvement in video resolution (to more than 400 lines of vertical) translates directly to higher density digital recording capabilities. VHS uses a carrier range of 3.4MHz to 4.4MHz, while S-VHS uses 5.4MHz to 7.0MHz. One of the major differences between Hi8 and S-VHS is that, like DAT, Hi8 uses ATF or azimuth recording and requires no control track, whereas the VHS system needs the timing pulses which the CTL (control) track provides.

■ The Digital Tape Recording System (DTRS)

The Digital Tape Recording System (DTRS), developed by Tascam, is a rotary-head digital multi-track system based on the Hi8 video system. It has eight channels of digital audio plus a subcode channel used for synchronization and time-code data. The helical scanning transport uses two read and two write heads. Data are written to tape as radio frequency (RF) signals much like the U-Matic system. The helical tracks are divided into 8-channel blocks with edit gaps between them. Sampling rates of 44.1kHz and 48kHz are supported and the system uses 16-bit linear PCM quantization. Since the system writes data in a format that is compatible with neither Hi8 nor 8mm video, it is wise to "pre-format" the tape prior to recording. It is also important to note that a tape used (and thus formatted) for video cannot be used for digital audio recording.

In the DTRS, data blocks or packets are read from the tape and assembled into tracks in a buffer memory prior to output. Tascam uses two heads to read and write data into long data packets that are spread out over a large area. The system uses transparent track sharing, which means that some of the data for Channel 1 may end up stored on Channel 8. Data will nevertheless always play from the correct output. The first four packets are encoded and held; the data from the next four are combined with these and then written to tape. Punch-ins and -outs are accomplished using digital cross-fading and memory buffers. The Hi8 system uses a 40mm head drum and has a tape speed of 14.3 mm per second. Tascam, however, has increased the speed to 15.9 mm per second to raise tape-to-head velocity, thereby increasing frequency capabilities at the expense of recording time. Typical Hi8 machines are belt-driven, but for digital audio, higher stability is required and a three-motor direct-drive transport is the norm. The higher bandwidth and greater speed stability allow a smaller minimum recordable wavelength, thereby increasing data density. This yields a recording time of 108 minutes of eight-channel recording on a P6-120 tape. The tape is a high-coercivity (1450 Oersteds) metal-particle tape that gives excellent retentivity and a high output. A DTRS eight-channel recorder is shown in Figure 14-8.

Earlier models allowed 20-bit recording on eight channels, while the current versions allow 24-bit recording on these machines as well. Three manufacturers make add-on processors that permit high-bit recordings to be made and stored on the DTRS system. One system by Rane, called PAQRAT, takes the 20- or 24-bit stereo input signal from the AES/EBU digital input and divides it into four parts. These signals are then recorded on either the first four or last four tracks of the DTRS recorder. The PAQRAT also lets the user send the signal to all eight tracks for redundancy. With two PAQRAT devices, four-channel 20- or 24-bit recording is possible. The PAQRAT is also available for use with the ADAT system, which will be discussed shortly. Another high-bit rate coder called the MR2024T, and made by PrismSound, uses a similar approach but with different coding and channel distribution. Therefore, the two systems are not compatible. The MR2024T allows either six channels of 20-bit audio, or four channels of 24-bit recording, or two channels of 24-bit audio sampled at 96kHz, with the data spread over the DTRS-available tracks. A third system comes included in one of the Apogee high-bit high-sample-rate A-to-D and D-to-A converters.

One of the advantages of this eight-channel modular approach is that several units can be chained together to increase the number of

Figure 14.8 An eight-channel DTRS digital recorder (Tascam DA-98, Tascam photo).

simultaneous recording channels. Three or more units can be daisy-chained together to provide 24, 32 or even 48 channels. Each Hi8 unit in the chain receives a sequential identification number. This allows the system controller to know, for instance, that track 23 is on channel 7 of machine ID 3. Each machine has a built-in chase synchronizer. Slave machines follow the master recorder, which is usually machine number 1 and contains the first eight channels. With the introduction of machines lacking all synchronization control functions, an economical system can be put together using one full-featured master and a number of less-expensive slaves. Up to 16 units can be controlled from one remote for a total of 128 channels! Subframe accurate synchronization with other digital and video media is possible through a synchronization board that also generates and reads SMPTE time-code.

■ Alesis Digital Audio Tape Recording System (ADAT)

The other popular rotary-head digital multi-track system is based on the S-VHS transport system. S-VHS is the high-band version of the VHS video system commonly found today in the consumer marketplace. Its advantage over the standard VHS system is similar to the quality gain of Hi8 over regular 8mm video. The Alesis Digital Audio Tape Recording System (ADAT) records eight channels of digital audio onto S-VHS tape. It uses 64-times oversampling and Delta-Sigma converters. Oversampling is a method of outputting a digital bit stream using a multiple of the original sample rate. This process raises the Nyquist frequency, thereby reducing phase anomalies and quantization noise (see Chapt. 15).

The system uses a standard sampling rate of 48kHz with 16-bit linear quantization. Other sampling rates can be achieved by varying the record/play tape speed. The system requires a control track on the tape and writes eight separate data blocks with each helical scan. Recording time with an ST-120 tape is 40 minutes (three times standard VHS speed). This speed is required to achieve optimum packing density for the eight channels of digital audio. A RAM (Random Access Memory) buffer for each track holds data until clocked out at the designated sample rate with all channels in sync. Prior to recording, the tape must be pre-formatted with control track and sync information. Addition of a proprietary sync block allows a combination of up to 16 units, for a total of 128 channels.

A more recent version of the ADAT allows 20-bit recordings to be made without an external processor. This new version, called ADAT Type II, is backwards-compatible with existing ADAT recordings. The new machines detect whether a tape has been formatted in the 20-

bit Type II format or in the original 16-bit Type I. The 20-bit machine records in either the 16-bit or the 20-bit mode, depending on how the tape is pre-striped by the user prior to recording. It should be noted that a Type I machine cannot play back a Type II tape. At a sampling rate of 44.1kHz, the Type II ADAT can store up to 67 minutes of 8-channel sound on an ST-180 tape.

The ADAT system uses a proprietary multi-channel optical digital interface to connect multiple ADATs in the system with fiber-optic cable. This interconnection permits copying between machines in the digital domain. Copy-and-paste editing can also be accomplished between tracks using RAM memory included with the larger system controller. This RAM can also be used to delay tracks up to 170 milliseconds for synchronization purposes. Track delays and other setups are stored in a data block area at the head of each tape in the system. When a tape is loaded for playback or record, the previously entered control data are recovered and implemented. The large system controller option, referred to as a "BRC," also allows SMPTE synchronization, MIDI time-code and MIDI clock information. MIDI data such as song pointers, tempo maps, and system exclusive data can be generated and stored in the recordable data block at the beginning of each tape as well. An external sync clock-in port allows the system to be locked to SMPTE, an external word clock (at 48kHz), or composite video sync. An ADAT multi-channel tape recorder is shown in Figure 14-9.

Figure 14.9 An eight-channel ADAT Digital Recorder (ADAT M20, ADAT photo).

■ FIXED-HEAD VERSUS ROTARY-HEAD DIGITAL RECORDERS

Open-reel digital tape recorders are now only available in the 24- and 48-track versions of the DASH format. The major advantage of open-reel formats over the helical-scan recorders is the ability to edit with a razor blade. However, open-reel digital tape recorders are very susceptible to errors created by tape-edge damage, dirt and debris, fingerprints, and tape dropouts. Tape-to-head contact is extremely critical and tape head alignment is not possible without sophisticated test and alignment equipment. Moreover, no satisfactory electronic editing systems have yet been devised for open-reel digital recorders.

Rotary-head digital recorders have many advantages over fixed-head versions. Chief among them is that the tape is contained in an enclosed environment, and therefore protected from contamination that can cause dropouts and high error rates. Helical-scan digital recorders tend to be less expensive than their reel-to-reel counterparts since their data are interleaved and recorded as a single bit stream. This type of recording uses simpler record and playback electronics and, with the exception of the DAT, makes use of existing video recorders. Another financial advantage of these systems is the low tape consumption. Tape used for digital recording must be manufactured to very high standards and tolerances in terms of dropouts and uniformity, and can be expensive. In contrast, the slow longitudinal tape speed of helical-scan systems allows lengthy recordings on very little tape.

Editing options for helical-scan systems far exceed what can be achieved with mechanical editing systems. Note, however, that any required editing must be done electronically with the associated extra cost of editing hardware. Moreover, since helical-scan editing systems are only available in the two-channel format, multi-channel recording requires the use of one of the existing open-reel format systems, unless the data are copied from a multi-channel DTRS, ADAT or DASH system into a DAW. A discussion of electronic editing methods and systems is included later in this chapter.

HARD-DISK RECORDERS

Alternatives to open-reel and helical-scan digital recorders, i.e., recorders that do not use traditional magnetic tape, are becoming increasingly more available and affordable. Although in some ways they have some of the functions of a digital audio workstation (DAW), they do not have quite the same extensive capabilities. They do, however, maintain some of the characteristics of the analog and digital reel-to-reel tape recorders. One of the current popular systems uses 9.1-GB drives in lieu of tape, equivalent to nearly three reels of 2-inch, 24-track analog tape on each drive. The remote control for a hard-disk recorder is shown in Figure 14-10. Since the drives will most likely be remotely located, only the remote controller needs to be located in the control room. A further detailed picture of a remote control is shown in Figure 14-11.

Figure 14.10 Remote control for a multi-channel hard-disk recorder (Euphonix R-1; Euphonix photo).

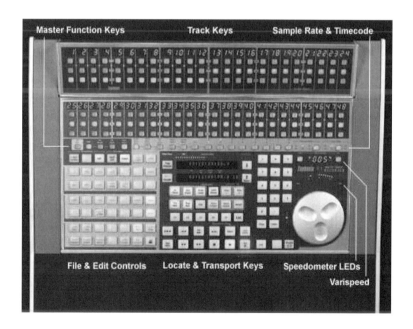

Moreover, hard-disk recorders offer highly detailed directories for stored tracks, along with the ability to do some rudimentary cut, copy, and paste editing. A major advantage of these recorders is that when doing overdubs, if the take requires correction, the hard-disk recorder simply creates another file with the same pointer coordinates as the first overdub, instead of the engineer having to either replace that overdub at the punch-in point through the punch-out point, or using another track. Later, the engineer can "comp" or compile the best section of each overdub take into a final version for playback and mixing.

Personal Computer-Based Recording Systems

With the proliferation of personal computers and the attendant fast-paced developments in software, it did not take very long for the audio industry to make use of this technology. Random-access memory storage systems and other disk-based storage media are

Figure 14.11 Detail of hard-disk recorder remote showing track metering and editing screen.

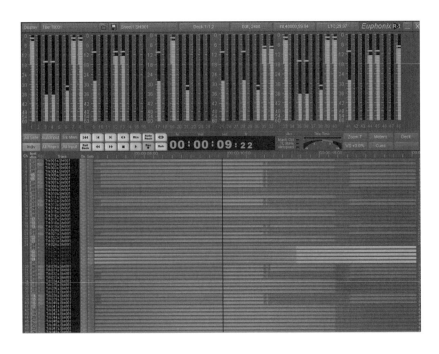

appearing in the control rooms of major recording studios at a very fast rate.

The principles for recording digital audio onto disk-based systems are not so different from those associated with recording digital audio on either helical-scan recorders or longitudinal digital systems. With these systems, however, the digital bit stream is written to either an optical or magnetic storage disk. The limiting factor is simply the storage capabilities of the storage disk itself. Maximum recording time is limited by the storage capacity of the disk(s) divided by the sampling rate and quantization, and again divided by the number of tracks required. In many systems, the engineer can decide how many tracks are necessary for the work to be done and apportion the disk storage system accordingly. A system of this type is often purchased initially with a minimum amount of storage time, and as the fortunes of the studio increase, additional blocks of storage time may be added by supplementing the number of disks in the system. Typically 10MB of storage are required for every minute of stereo digital audio when recorded at a sampling rate of 44.1kHz with 16-bit quantization. If we needed to record sixteen tracks at a sampling rate of 96kHz with 24-bit quantization, approximately 240MB per minute of storage would be required!

Another limiting factor of these very flexible systems is the amount of time necessary for downloading and uploading data. If, for instance, a series of tracks is recorded, overdubbed and mixed down within the system, the whole system will be tied up until the material is downloaded (often in real time) to some permanent storage medium. Conversely, if existing material needs to be loaded into the system for processing, editing or even further overdubs, the time required can be extensive. This can create problems in facilities where one client follows another with a minimum amount of turnaround time. Some dedicated DSP systems allow background recording and playing of files from projects other than the one currently in use. This lets the engineer work on the project at hand while preparing or finishing others.

Most of these systems, as just mentioned, write the digital-audio data to magnetic disks. These disks can be easily erased and prepared for new data after the desired material is downloaded, using a simple format-type command. Other systems use Write-Once-Read-Many (WORM) optical disks. With Direct-Read-After-Write (DRAW) systems, once information is written, it is permanently encoded into the disk. The disk is then encased in a protective envelope that is inserted into the front of the recorder, not unlike a video cassette; the disk therefore is not open to dust and debris. These systems are not to be confused with the CD-R and CD-RW systems, which are primarily storage media. Members of the CD and DVD family do not have the access speed or seek time required for real-time digital editing. A benefit of the permanent systems is that the recorded material cannot accidentally be erased by an errant format command. On the other hand, a large project can use quite a few disks, and once the session is over, you may not wish to ever use the material again. An optical disk can contain a substantially greater amount of data, often as much as a thousand times more, than a magnetic disk. This can give a recording time of over 90 minutes in stereo per disk.

■ DIGITAL EDITING PROCEDURES

Digital Tape Many ways of editing digital tapes have been devised since the advent of digital recordings. The initial problem has been that editing a digital tape requires cutting the data stream, while editing an analog tape means cutting a magnetic pattern that is analogous to the audio waveform. Data stream interruptions can wreak havoc with the error-protection and -correction schemes. One of the great benefits of digital editing is the precision that is possible with electronic editing systems. These systems can be tape-based or personal-computer-based with the new types of digital workstations now available.

■ Punching In/Out One of the easiest ways to append or delete information from a digital tape is to either erase or replace existing data. This is done more often on multi-track recorders than on two-track recorders, but the technique is applicable to both. A discussion of the procedures for punching-in/out or insert editing for multi-track recording is included in Chapter 18.

When punching in on an analog tape recorder, the headphone cue to the musician is played from the record head so that the new material is recorded "in sync" with the old. It is not possible to do this with a digital multi-track recorder because the write head is not capable of reading data. Digital multi-tracks utilize a second read or write head that follows the read head, as previously mentioned. The data from the read head are delayed and then sent to the second write head in synchronization with the new material, or the data are taken from the extra read head and recorded "in-sync" at the write head. Digital cross-fades are used to blend the two data streams together so that the transition from the old material to the new material is not noticeable. Figure 14-3 shows a typical signal flow for recording and punching in on a digital multi-track tape recorder.

Another advantage realized when using digital multi-track recorders is that the punch-in and punch-out points can be memorized by the tape recorder. This lets the engineer rehearse and trim the entry and exit points precisely, so that no old data are lost. Once these points are decided and programmed into the digital recording system, the punch-in and punch-out will be performed automatically, at precisely the same point, and as many times as necessary to correct the musical slips or errors on the selected track.

■ **Razor-Blade Editing**   Razor-blade editing of digital tapes presents some practical difficulties that are not encountered in the traditional analog domain.

The first problem is that the familiar technique of rocking the tape back and forth across the head to find the splice point is now impossible. At the very slow speed involved, the decoding system simply does not function, and the operator cannot listen for the splice point. Many open-reel digital recorders get around this problem by providing a channel or two onto which analog audio information is recorded in synchronization with the digital audio tracks. The analog track(s) may then be monitored in the traditional way during editing.

The next problem is that the razor-blade edit is quite likely to interrupt the normal sequence of sync word (or preamble), audio data, CRCC, and parity information. There is really no guarantee that the razor will sever the tape at precisely the spot necessary to preserve the proper sequence of data flow. Thus, there will surely be a momentary data discontinuity from which it may take the system a relatively long time to recover. As a result, the splice will create a noisy transition as it momentarily disrupts the orderly flow of data. This may be a disruption that can be corrected by some of the error correction schemes mentioned in Chapter 13, but if severe enough, it could create a momentary audio mute.

Various systems have been designed to make razor-blade editing of digital data possible. Generally, discontinuity in the data is sensed and an electronic cross-fade makes a data interpolation to smooth out the transition across the splice. It is important that the ends of the tape be joined as smoothly and as evenly as possible to prevent audio mutes from occurring. One major difference in the splicing block is readily apparent when editing digital tapes with a razor blade: the cut that is made is perfectly vertical, compared to the angled cut associated with analog tape. Since the flux change across the magnetic head gap is cross-faded electronically, a butt splice (90°) is perfectly acceptable, and is actually preferable.

The fact that digital tape is considerably more delicate than analog tape is a further limiting factor. For comparison, Figure 14-12 lists some of the physical characteristics of several well-known analog and digital reel-to-reel tapes. Note that the oxide and the total thickness of the digital tape is considerably less than any of the analog formulations. This thickness reduction is necessary to ensure that the tape conforms to the head contours as completely as possible, which produces the tape-to-head contact necessary for high packing densities and low error rates.

In other words, these digital tapes must be handled with extreme care, and razor-blade editing must be approached with great caution. Most manufacturers recommend wearing white gloves to protect the tape from fingerprints, body oils and other contaminants.

Electronic Editing

One of the more preferred formats for digital editing is based on the electronic editing techniques used in video tape production. These techniques are principally used for two-channel audio recorded on professional U-format video cassettes. It is possible, but not necessarily financially feasible, however, to accomplish this type of editing with two (or more!) multi-track digital recorders as well.

Figure 14.12 Physical dimensions for several representative analog and digital tapes.

	Analog 1.5 mil tape	Analog 1 mil tape	Digital tape
Base film thickness	1.92	0.88	0.79
Oxide coating thickness	0.55	0.50	0.16
Backcoat thickness	0.05	0.05	0.04
Total thickness	2.02	1.43	0.99

Unlike traditional razor-blade editing, in which a collection of tapes may be played back and spliced together on a single tape recorder, the electronic editing process requires two machines: a player and a recorder.

The takes to be assembled into the master tape must be reproduced on the player and dubbed onto the recorder one at a time. For example, consider a final master that will be comprised of sections of three separate takes, as illustrated in Figure 14-13. In the illustration, note that the master will begin with Take 3, followed by a portion of Take 1, and then a smaller segment from Take 4. Finally, the last portion of Take 1 is used to complete the master.

Briefly, there are two types of electronic editing: assembly editing and insert editing. Assembly editing is used when there is no existing control track or SMPTE time code and the material is being assembled onto a fresh video cassette. Insert editing is required when there is already continuous control track and/or SMPTE time code recorded on the tape and the new material must be placed between two segments of existing material. Assembly editing edits the video (digital material) and audio (analog material) signals simultaneously. The control track on the video cassette is edited at the same time so that the head switching lines up properly with the

Figure 14.13 Excerpts from various takes are to be assembled into a final master tape. The procedure begins by recording Take 3 onto the master tape recorder, and then cross-fading and dubbing Take 1. The procedure continues until the entire master tape is assembled.

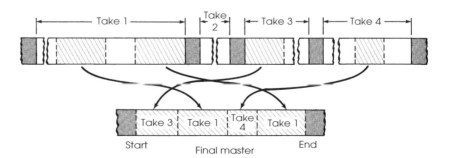

helical-recorded video tracks. Insert editing allows the video signal (digital) and audio track 1 and audio track 2 (SMPTE time code) to be edited independently or simultaneously, and all signals are referenced to the existing control track. Figure 14-14 shows examples of both assembly editing and insert editing on a U-format video cassette.

In digital editing, the assemble mode is generally used only when a song or master tape needs to be extended, because the possibility of a break in the control track signal exists. Normally, the tape that is to be used for the assembled master is "pre-striped" with SMPTE time code (audio track 2) and control track signals prior to the edit session, and before the insert edit mode is used. This ensures continuous time-code and control tracking, which are essential for proper mastering onto CD or DAT products.

We begin by dubbing the beginning of Take 3 (from our previous example) onto the recorder. The transfer should continue for several seconds beyond the actual edit-out point required. Then, Take 1 is cued up on the player and is stopped a few seconds before the edit-in point at which it is to begin replacing Take 3.

■ **The Edit Rehearse Mode**

Obviously, simply transferring Take 1 onto the end of Take 3 will not work. The odds of making an accurate and undetectable transition from one take to the other at the proper spot are too remote to be worth considering. However, all electronic editing systems have some form of a rehearse mode, in which the edit may be simulated rather than actually executed. Once the simulation has been perfected, the actual edit takes place electronically.

The actual editing procedure begins after recording the first sequence by listening to the end of Take 3 from the recorder (not the player) and executing an "edit-out" command at the appropriate moment. This command marks the selected edit point where Take 3 ends, and transfers several seconds of audio from both sides of the edit point into a buffer memory. The contents of this memory may now be reviewed at regular or slow speeds so that the edit point can be modified or confirmed.

Next, the "edit-in" point of Take 1 (the next segment) is selected from the player. Once again, several seconds of program on either side of the proposed edit are loaded into the buffer memories in the editor system and the engineer may adjust or confirm the actual point.

Then the actual edit is played in its entirety. Some systems do this from the buffer memories, while others actually play the previously

Figure 14.14 a. Assembly editing. Note that the control track does not line up with the video or data track. This may cause unstable data transfer; b. Insert editing. Data can be edited (transferred) without interrupting the control track or time code.

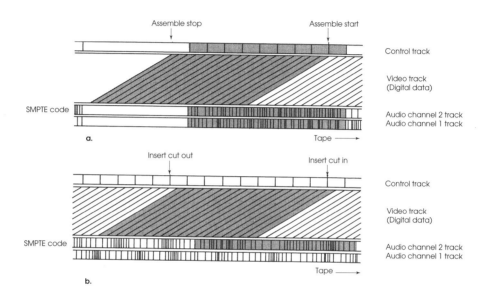

recorded material and switch to the new material at the edit point. As before, the edit point(s) may be adjusted again and again until the transition is perfected. Various cross-fade intervals may be tried to achieve the optimum inaudible edit, similar to varying the angle of cut on a standard splicing block. If the proposed edit simply won't work, the engineer is able to discover this without wasting time making the actual assembled transfer.

If it is necessary to use the traditional tape-rocking-style search for an edit point, most digital editors are able to simulate this effect. The contents of the buffer memory may be "rocked" back and forth at any convenient speed. This is usually done by manipulating a rotary knob, designed to create the illusion that a piece of analog tape is actually being moved back and forth across a playback head. Figure 14-15 shows a typical example of such a system.

Figure 14.15 A typical digital editing controller and processor (Sony DAE-3000; Sony photo).

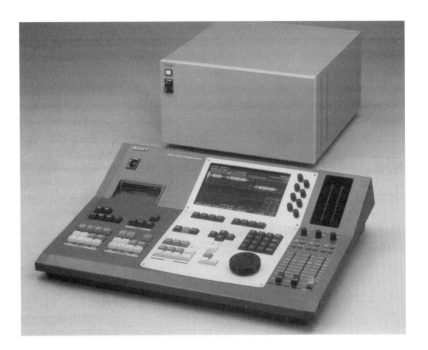

Once the edit points have been located and confirmed, the electronic editing process begins. The master recorder rewinds to a convenient "pre-roll" location well ahead of the edit, and begins playing. As the edit point approaches, the player transport (containing Take 1) starts in synchronization with Take 3 on the recorder. SMPTE time code is used to lock the two machines together. At the appropriate moment, the recorder switches from play mode to record mode, erasing the final moments of Take 3 (thus the importance of rehearsal!) and simultaneously replacing it with the new material from the player (Take 1). Take 1 continues to be dubbed onto the recorder through the next edit-out point, after which the recorder is stopped and the process begins again with the next take. This process continues with succeeding edits and takes until the end of the project is reached.

The electronic editing procedure just described may require some re-thinking of traditional editing techniques, especially for engineers who are long-accustomed to working with a razor blade and some splicing tape. For example, once the electronic edit has taken place, the new material must continue to be dubbed onto the master recorder until the next edit point is reached. There can be no skipping ahead to the next edit, and the editing process takes considerably longer than the razor-blade method.

Some systems require that the engineer review the previous material up to the next edit point before proceeding. Systems with "read-after-write," or so-called "confidence," heads may not require this, but it is still best to verify that the material was dubbed properly before proceeding. An advantage of this assembly edit system is that even if the edit is not correct and some data have been copied incorrectly, the edit can be recreated because the original material on the playback machine is still intact.

■ **Editing With a Hard-Disk System**

It is in the area of electronic editing that the advantage of hard-disk systems becomes readily apparent. The editors that have recently appeared have put electronic editing within the realm of possibility for the home studio. Using already existing personal computers, coupled with high-capability PC-type hard disk drives, digital-audio data can be edited using specially written software, similar to the way we now edit a text file. The audio data are treated as a file and can be moved, joined with new material, replaced, or deleted almost as easily as editing a letter on a word processor.

In most systems, the file can be called up and monitored as the punch-in/out points are determined. The edit point can then be brought up on a screen and visually edited with a mouse or other pointing device. Cross-fade time can be varied, and the edit can easily be previewed and changed prior to actually performing the edit. When the edit command is given, the computer assembles a new file by copying the original material up to the punch-in point. Then the new material is added in with the appropriate cross-fade between the new and the old up to the marked punch-out point. It is possible to completely pre-assemble an edit list for an entire selection before telling the editor to proceed. Even then, the old material is retained in the event that a change needs to be made. The completely edited selection can then be downloaded to a permanent storage medium such as DAT or other digital storage media.

Almost all hard-disk recorders and editors have their own analog-to-digital-to-analog converters, as well as a digital AES/EBU port for transferring data in and out of the system. If necessary, a DAT recorder can be used as a converter, by simply placing the unit in record or standby, and then using the digital outs of the recorder to feed the computer system.

Further developments along these lines have created so-called "Digital Audio Workstations," or DAWs, which combine recording and editing functions with digital mixing and signal processing. There are two basic kinds of systems: ones that use the computers processors to do audio data manipulation, and so-called dedicated DSP systems that have proprietary cards designed strictly as audio processors. Figure 14-16 shows the component parts of a typical DAW. Note that there are outboard SCSI (Small Computer Serial Interface or "SCUZZY) hard drives used for storing audio files, while the computer's internal hard disk is used for program information and Edit Decision Lists (EDLs).

Figure 14.16 The component parts of a PC-based digital audio workstation (DAW).

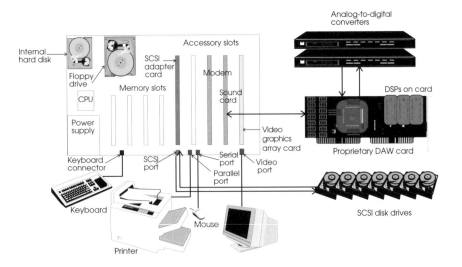

DIGITAL AUDIO RECORDING SYSTEMS

The latter systems are generally able to do more processes in real time. While a system that uses the computer's CPU to process audio, refresh the screen, store the data on the disk, etc., may take a few seconds to calculate and implement a cross-fade, systems with dedicated DSP cards and chips can perform this function in real time as the music is played. The quality of the edit may be the same, but the time factor is not. Figure 14-17 shows a premier DAW that utilizes custom-designed DSP chips for audio processing. In addition to the computer and its peripherals, this DAW has a specially-designed control surface that allows the system to mimic either a console when using the mixer or an editor when using the track ball and mouse. Figure 14-18 shows the external processing rack that contains the A/D and D/A converters, as well as the analog and digital inputs/outputs.

■ **DAW Functions**   Most DAWs let the user do more than just edit. Fig 14-19 shows a typical edit screen from a DAW system that was designed to operate on an Intel-based computer system. In addition to the waveform representation, there are also edit-fade controls, transport controls, monitor mode, and track controls, among others. There is also a P and Q Subcode editor that lets the user add track and index points

Figure 14.17   A modern digital audio workstation with hardware controllers (SADiE DAW; SADiE photo).

Figure 14.18 Some digital audio workstation components. On top: physical analog and digital inputs and outputs; middle: an eight-channel analog-to-digital encoder (converter) and on the bottom, SCSI hard-disk storage drives (SADiE DAW, SADiE photo).

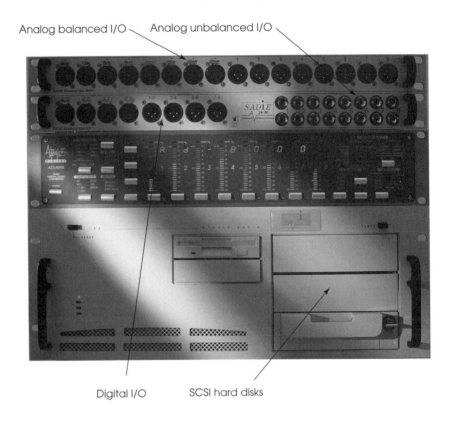

for preparation of Red Book-compliant master compact discs, also referred to as PMCDs (Pre-mastered CDs).

Figure 14-20 shows another view of the same system, which includes a mixer with equalization, auxiliary sends and returns, and a unique surround-panning system with separate subwoofer channel controls. Figure 14-21 is a screen shot of a similar system showing twelve pairs of stereo tracks, for a total of 24 tracks. Many systems of this type, as well as the DAWs with proprietary DSP, are virtually boundless in the number of tracks that they can have. The limiting factor, however, is the amount of computing power available to play out all the tracks at once. Even though a system may allow up to 99

Figure 14.19 A digital audio workstation editing screen (Sek'd Samplitude 2496, APK photo).

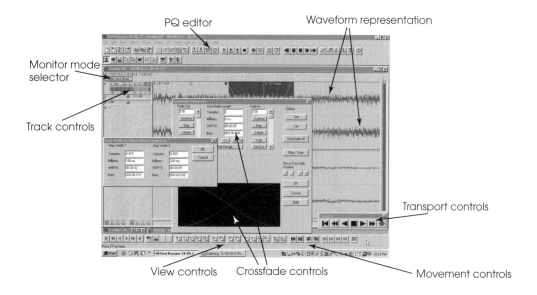

virtual tracks, it may only be capable of outputting eight discrete data streams at any one time. This limitation can often be worked around by grouping some tracks within the system, for instance, a large number of layered vocals, premixing them down to a stereo mix and then "bouncing" the stereo composite to another virtual track. This then only requires two data stream outputs instead of the multiple streams that were used before. Although this saves DSP power on playback, there is no further balance adjustment of the layered vocals other than a simple left/right pan.

Further, there are often dynamics modules within the system that allow compression, gating and other such effects. Such so-called "plug-ins" may be designed specifically for a certain system, while other broad-based plug-ins work with virtually any system, including a simple computer sound card. DSP will often also allow pitch and tempo variations and can be tied together with a MIDI sequencer for maximum output.

These systems are certainly the wave of the future. When a powerful DAW is combined with the state-of-the-art digital console, as

Figure 14.20 DAW editing screen (Sek'd Samplitude 2496, APK photo).

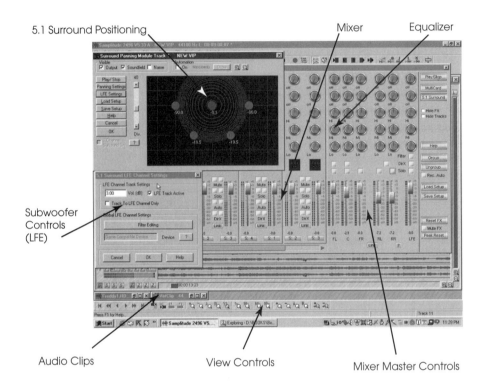

long as you have first-class musicians and engineers, the production possibilities are boundless.

■ OPTICAL DISK STORAGE SYSTEMS

The compact disc (CD) has revolutionized the way we listen to music at home. A 30-centimeter disc was introduced in 1977 by Sony, Mitsubishi and Hitachi. Simultaneously, Philips of the Netherlands was developing a 11.5-centimeter disc. Sony and Philips then

Figure 14.21 DAW 24-track editing screen.

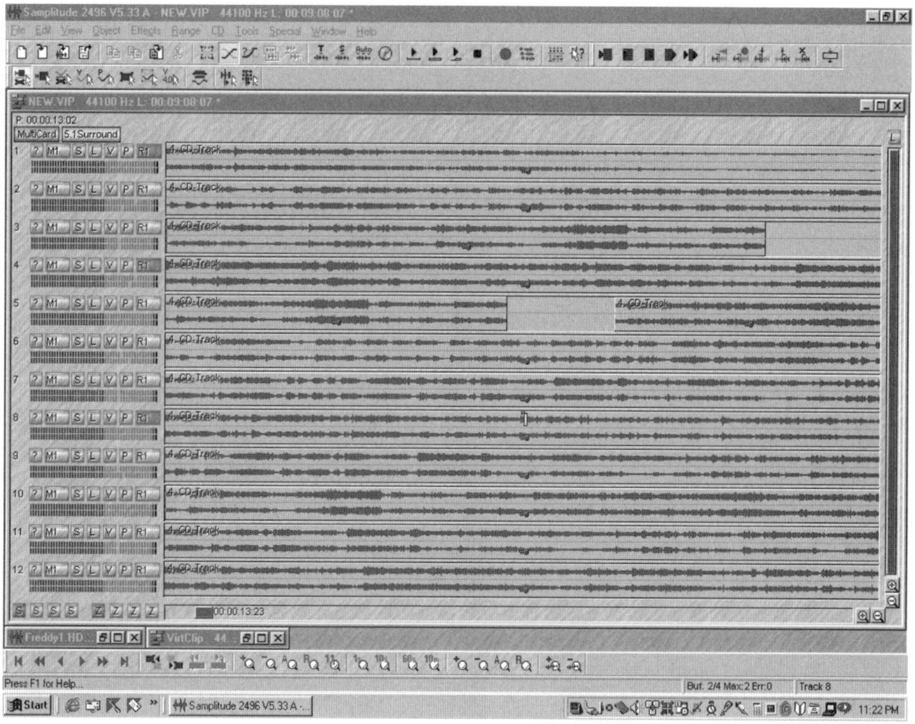

combined their ideas and developed a 12-centimeter disc with a capacity of 74 minutes. The compact disc as we know it was introduced to the record-buying public in October of 1982.

The Compact Disc

The compact disc in its standard form is a read-only two-channel digital audio optical storage medium. Physically, the disc is 120 millimeters in diameter, with a center-hole diameter of 15 millimeters. Thickness is a nominal 1.2 millimeters. Data are stored on an area that is 35.5 millimeters wide, with the remaining space used for the disc clamping area and lead-in area on the inside, and a lead-out area on the outside. The binary code is stored on the disc as a series of pits (actually bumps) or non-pits. These pits are set in a continuous spiral starting at the inside of the disc and ending at the

394   THE AUDIO RECORDING HANDBOOK

outside. The pits are 0.5 micrometers in width, and range between 0.833 to 3.054 micrometers in length. Spacing between the spirals is 1.6 micrometers. Figure 14-22 shows a magnified section of a CD, allowing a visual inspection of how the data are stored. A small low-powered laser strikes the disc from the bottom; the light is either reflected back to the laser or scattered. The audio information is contained on the disc at a sampling rate of 44.1kHz with 16-bit quantization. A TOC (Table of Contents) is written in the lead-in area. There can be up to 99 tracks on a disc, each with up to 99 index points. Not all CD players read index points, however.

Figure 14.22   Illustration showing the pits and lands on the surface of a compact disc.

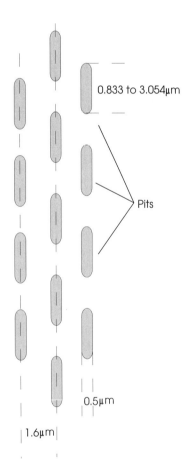

There are several types of compact discs, each of which has a standard that must be adhered to for compatibility between players. A set of these standards was compiled in a series of color-coded books by the Philips Corporation. To get a license to manufacture and produce compact discs, the manufacturer certifies that the compact disc being made meets one of these standards.

For example, the Audio Compact Disc standard is contained in the Philips "Red Book." The pickup arm, which contains the laser, reads from the inside of the disc to the outer edge. The disc rotates in a clockwise manner at a rate of speed that varies with the position of the laser. This permits a constant velocity output of the bit stream. The Audio Compact Disc Audio, or CD-A, disc holds 74 minutes of music but can be extended to about 82 minutes if necessary. Discs with capacities exceeding 74 minutes, however, may not play on all CD players.

Other Compact Disc Types

The Compact Disc Read Only Memory format, or CD-ROM, is principally used for computer data storage and distribution. It stores the equivalent of 950 1.44-megabyte high-density computer floppy discs. Since most CD-ROM drives include the ability to play audio-only CDs, units with both audio, instruction sets, and still- and full-motion video are found in the computer game market.

Compact Disc Video, or CD-V, contains digital audio and full-motion video. This should not be confused with the video CD, which is a subset of the CD-ROM standard. The CD-V is a combination of an audio CD and a video laser disc. These discs can store about 6 or 7 minutes of a video with audio plus another 20 minutes of audio only. They are thus ideally suited for the "music video" market as presented on MTV.

The Video CD, often incorrectly called CD-V, is a version of the CD-ROM standard. It uses a data-reduction system called MPEG (Motion Pictures Engineering Group), and can hold full-length motion pictures along with high-quality audio. It plays from a conventional CD-ROM drive and lets the user watch movies on the computer monitor. It was primarily developed for the Karaoke market.

Compact Disc Interactive (CD-I) is also a subset of the CD-ROM Standard. Like CD-ROM, it can combine text, video, graphics, and music. There are three modes for music: hi-fi, mid-fi, and speech mode, which roughly correspond in fidelity to a standard LP record, an FM radio broadcast, and an AM broadcast respectively. Interactive playback is made possible by using data-reduction schemes, which permit encoding of up to sixteen tracks of monaural music for a length of seventy-four minutes.

The Photo CD was designed to display high-quality still pictures. Images from conventional color negatives (print film) or color positives (slide film) are scanned digitally and stored on the Photo CD. A Photo CD can typically hold as many as 100 photographs. Discs can be played back on dedicated Photo CD players or on standard CD-ROM drives as long as the appropriate software is used.

Digital Versatile Disc (DVD) The newest product in the compact disc family is the Digital Versatile Disc, or DVD. It is also sometimes referred to as the Digital Video Disc, since at this time its primary purpose is for movies. The primary differences between the DVD and the standard CD are the pit size and track spacing. Such advances became possible as more economical shorter-wavelength lasers were developed. Although similar in appearance to the audio-only CD, the DVD may contain two layers of data on each side, calling upon the second layer when necessary. On the first pass, the laser focuses on the deeper layer; as the second layer is selected, the laser refocuses to read the nearer semi-transmissive layer. The DVD holds 4.7 Gb per layer (a dual-layer DVD holds 8.5 Gb on a single side), as compared with the 680 Mb capacity of the audio-only CD.

The Moving Picture Experts Group (MPEG) established a series of compression standards for video that is used for graphics and videos in many digital graphics formats. The codec analyzes the video picture for redundancy and discards the redundant information, which allows for more compact data. For audio encoding, three lossy compression schemes are in use; the Dolby AC-3 system is used in the USA and Japan, while the MPEG audio encoding system is used in Europe and Asia, although the AC-3 system is becoming more common in these countries as well. Additionally, a third type, the DTS system, is found on many discs. Because of these compression schemes, the DVD has space for 5 channels of digital sound. Movies can carry digital stereo sound or one of the multi-channel formats that provide discreet left, center, right, left-rear and right-rear sound. A common subwoofer channel, the "dot 1" in the 5.1 surround system, is included in all three systems. Commonly called the LFE (Low Frequency Effects) channel, this is usually used for low-frequency sound effects such as rumble and explosions, instead of music.

The DVD may also lend itself to the "Super CD" format, a format being looked at to expand the sampling rate and quantization rate of the existing audio-only CD format. This would result in expanded frequency response and dynamic range. At this time there are two systems available for consumer use, the SACD (Super Audio CD) and the DVD-A.

The DVD-A allows various combinations of channel arrangements, sampling rates, and word length. In addition to the typical 12-cm disc found in the movie format, there is the possibility of an 8-cm disc. Below is a chart showing the varieties available within the DVD-A specification.

Audio Contents	Channel Arrangements	Playback Times per Side			
		12-cm Single-layer	12-cm Dual-layer	8-cm Single-layer	8-cm Dual-layer
2-channel only	48kHz/24-bit	258 min	469 min	80 min	146 min
2-channel only	192kHz/24-bit	64 min	117 min	20 min	36 min
2-channel only	192kHz/24-bit	125 min	227 min	39 min	70 min
Multi-channel only (5.1)	96kHz/24-bit	86 min	156 min	27 min	48 min
2-channel and multi-channel	96kHz/24-bit 2ch and 96kHz/24-bit 3ch with 48kHz/24-bit 2ch	76 min	135 min	23 min	41 min

The DVD-A uses a non-lossy compression scheme called Meridian Lossless Packing (MLP) that is audibly very transparent yet allows storage of large quantities of digital data. The DVD-A may also contain an AC-3 or DTS stream to maintain compatibility with existing DVD-Video players. The specification also allows for a Red-Book layer; however, early releases have not contained this.

The SACD is a competing system using the DSD (Direct Stream Digital) encoding which is not compatible with the DVD-A and CD PCM technology. The SACD uses a sampling rate of 2.8224 MHz. Because of this high sampling rate and delta-sigma modulation, a 1-bit word length is possible. This then only requires a simple analog low-pass filter to be used for decoding. This allows a frequency response of 100kHz and a dynamic range of 120dB. Originally, the SACD was to be a hybrid and also contain a Red-Book 44.1kHz/16-bit CD layer as well as the DSD 2-channel and multi-channel data layers. However, it appears that with the introduction of DVD-A as a mass-market multi-channel carrier, the SACD will become an audiophile-only medium containing only the DSD data. Combination CD/SACD/DVD-V/DVD-A players are possible.

Compact Disc-Recordable (CD-R)

Compact Disc-Recordable is a write-once-read-many (WORM) system that allows a specifically designed recorder/player to encode digital audio information on a CD one time only. After the disc is completed, it can be played as many times as needed on a standard CD player. The disc is a pre-grooved blank enclosing a photochemical dye or a thin metal film usually made of tellurium. A digital bit stream modulates the power of a laser. The laser strikes the dye or metal film, changing its reflectivity relative to the bit stream. On playback, the pickup reads the changes in reflectivity as the pits or non-pits of a conventional CD. If the disk is written in the track-at-once (TAO) mode, it is only recordable and playable on the recorder until it is "fixed up." At that time a permanent TOC (Table of Contents) is written, and the disk can then be played back on an ordinary player. In the disc-at-once mode, the CD-R is written from an image file that contains the entire track and index points, as well as track timings and countdown information. Since the data are already formatted for the disc, the TOC is written first, followed by the data. This type of recorded disc, often referred to as a PMCD or Pre-Mastered CD, is acceptable for processing by the pressing plant. There it can be used to drive the LBR (Laser Beam Recorder) that makes the CD glass master. CD-Rs written using the track-at-once standard contain link blocks between the tracks and cannot be used as PMCDs.

CD-R systems are found as stand-alone devices containing the drive, the AD/DA converters, the line-level electronic circuitry, and the control systems. These stand-alone units usually have analog (+4dBm or -10dBm) and digital (AES/EBU and/or IEC-958 (SPDIF)) inputs and outputs that are accessible on the backs of the units. CD-Rs can also be found as internally- or externally-mounted peripherals for Digital Audio Workstations (DAWs) or computer-based systems for CD-ROM implementation. The internal models rely on the computer's power supply and are internally cabled to the host's SCSI adaptor, while the external models have their own power supply and case, and are cabled to the host by a SCSI cable. Another recent arrival is the DVD-R, a system similar in implementation to the CD-R, except that the wavelength of its laser is much shorter, and therefore the pits and lands are smaller.

CD-R and DVD-R have also been joined by the CD-RW (Read-Write), which allows the user to record on the disc and then erase it for future use. The CD-RW and DVD equivalents (DVD-RW and DVD-RAM) operate using the magneto-optical principal, described next. Here again, the DVD systems use shorter wavelength lasers to encode many more bytes of information in the same physical space.

Magneto-Optical Media

These recordable-erasable optical disks are gaining rapid acceptance and use in both the professional and consumer venues. The Sony PCM-9000 is said to be the logical successor to the PCM-1630 CD mastering system, and the MiniDisc is designed to replace the compact cassette in portable situations. The PCM-9000 uses a proprietary recording format, which essentially operates as follows:

As magnetic material is heated beyond a certain transitional temperature called the Curie Point (after Pierre Curie), its magnetization dissipates. As the material cools, it takes on the magnetic polarity of any magnetic field applied to it. The applied magnetic field then orients the spot on the disc in a north or south polarity, which corresponds to a one or zero of the applied digital bitstream. Once the spot falls below the Curie Point in temperature, that magnetic polarity is permanent until heated once again. The polarity of the material is then read by a laser using the Kerr principle, where the polarization plane of light is rotated slightly by a magnetic field. The professional version of this system uses a 133-millimeter single-sided disk and can hold over 100 minutes of recorded data. Once remagnetized, discs are sealed against air to prevent corrosion of the magnetic material. This system is designed to be used with up to 24-bit quantization levels and supports sampling rates of 48kHz, 44.1kHz and 44.056kHz. Figure 14-23 shows such a system.

Additionally, there are smaller systems that can record eight tracks on a disc. Several of these systems are used professionally, and they are a logical choice for high-bit, high-sample-rate output storage from digital audio workstations (DAWs). Figure 14-24 shows an eight-channel recorder/player of this type.

■ **MiniDisc**

The MiniDisc, introduced by the Sony Corporation in 1990, is a miniature version of the magneto-optical disc system just discussed. It operates according to the same opto-thermal principle but is smaller in size. To achieve the same time capabilities as the compact disc, however, the MiniDisc uses a data-compression scheme called Adaptive Transform Acoustic Coding (ATRAC), which reduces the amount of data by a factor of five-to-one. The disc itself measures 64

Figure 14.23 A high-quality magneto-optical disc recorder (Sony PCM-9000, Sony photo).

Figure 14.24 An eight-channel magneto-optical disc recorder (Genex photo).

millimeters in diameter and is encased in a rigid protective case, much like a 3.5" computer floppy disk. ATRAC uses a psycho-acoustic algorithm that takes advantage of the masking effect and the threshold of hearing.

Certain problems can arise, however, with data-reduction systems depending on how they handle multiple generation copies. With DAT and other linear PCM encoding systems, data are literally cloned, and the copy sounds identical to the original. When reduced data are played back through the ATRAC system, however, the information is converted to a 16-bit data stream with the missing data interpolated. This means that if the interpolated bitstream is copied, it is further reduced by a five-to-one factor. The data loss of the compression algorithm is additive, with each generation having less data than the preceding one, which causes a loss of fidelity. The reduced data cannot be copied because the machines do not allow access to them.

During recording, the MiniDisc system uses a recording head (the magnetic overwrite head) and a laser to simultaneously erase and record. The laser heats the magnetic layer to the Curie Point, allowing it to be remagnetized by the recording head. For playback, the same laser strikes the disc at a lower intensity. When reflected, the laser light rotates, or polarizes, according to magnetic field orientation.

Since the recordable MiniDisc is always updating its TOC (Table of Contents) whenever it records, editing existing recordings is a simple matter, done by changing the table of contents. Audio may even be divided between two segments on the disc with two minutes at one point and six minutes at another, for instance. Because the MiniDisc player has a read-ahead buffer, the connection between the two segments is seamless as the selection is played. This buffer system also allows the user to program the TOC so that it is possible to insert, delete, re-sequence, split, and join sections of music without any audible problems. In use, this acts like a mini digital editor. The so-called "Porta-Studio" that was designed as a portable four-track recording system using the compact cassette medium has been modified to use the MiniDisc format. Four- and eight-channel MiniDisc Porta-Studios are now available that let the user record multi-channel digital audio onto this format.

Summary of Optical Storage Systems

The compact disc (CD) and the digital versatile disc (DVD) are similar to the LP record in that they are playback-only media. As such, it is not possible to record directly onto them. However, the manufacturing process is so streamlined and automated that these discs can be produced very economically in large numbers. Even smaller numbers of these discs are within the financial capabilities of small organizations. The DVD has brought inexpensive multi-channel sound into the home video playback system, and is bringing audio-only surround sound to the home environment as well.

On the other hand, the CD-R, CD-RW, MiniDisc and DVD-RW family offer the user the ability to record over old information, be it digital audio or a copy of *Funk and Wagnall's Encyclopedia*. Their relatively low cost and ease of use allows them to be used as compact computer back-up systems as well as digital audio storage media for compiling the listener's favorite songs from multiple sources. Furthermore, the compression schemes used for digital audio on the DVD will soon allow the consumer to record multi-channel audio through the computer.

SECTION VI

Synchronization Systems

■ INTRODUCTION

Since the early days of multi-tracking, engineers have always been looking for ways to synchronize machines together. Today with the strong marriage of audio and video technologies, it has become essential to have frame accurate lock-up of multiple audio and video machine systems.

Although there is only one chapter in this section, it contains a lot of information that is important to the recording studio engineer. It begins with a discussion of timing systems and then addresses the SMPTE Time Code, both longitudinal and vertical versions. Following that is a brief discussion of the uses for time code, including audio and video synchronization.

The chapter ends with a brief discussion of one of the newer studio phenomena: the Musical Instrument Digital Interface, or MIDI. Even as of this writing, the MIDI standard is undergoing change and improvement, and the enterprising engineering student should attempt to keep up with the changing standards.

FIFTEEN

Time Code

Since the earliest days of multi-track recording, studio engineers have felt the need for an efficient and reliable means of monitoring time-related data. The most obvious requirement, as well as the one easiest to implement, is to establish a means of notating the playing time of the finished recording. Obviously, if this were all that were needed, a conventional stop-watch would be more than sufficient. The requirements for more sophisticated measurement methods within the time domain may have started with the introduction of overdubbing and mixdown techniques. These techniques will be described in detail in Section VIII, "The Recording Process."

As these techniques developed, the engineer needed a way to return again and again to the same location on the tape. This could occur during the artist's rehearsals, during punch-ins, and throughout the automated mixdown process. While this point was usually musically apparent, a slight change in the phrasing could carry the punch-in or overdub too far into the existing music, particularly if the engineer was paying more attention to the new material than to the old. An unambiguous method of noting and returning to the required points was needed.

The earliest technique, still used on many consumer tape decks, made use of a counter which generally indicated the number of revolutions of the recorder's supply reel. Directly linked to the reel motor, the system could be quite accurate. However, practical considerations prevented this system from gaining widespread acceptance.

The first drawback of a revolutions counter is that it has little or no relationship to the real-time world. Neither artist nor engineer can be comfortable with a recording that begins, for instance, at 3473 and ends at 8136, since these numbers convey absolutely no information about starting time, ending time, or duration of the interval in between. Worse yet, documentation for future reference is of little use, since if the tape segment gets transferred to another reel, with more or less tape on it, both the numbers and their mathematical relationship will be changed. This makes intelligent record keeping nearly impossible.

A more satisfactory system, as seen in Figure 15-1, would provide a time-based readout, typically derived from tape travel across an idler wheel placed in the tape path. Although such a system is certainly easier to read and interpret by the operator, it may be inaccurate due to tape slippage, especially if the tape is pulled away from the idler from time to time as in editing.

In either case, the systems just described will be of little practical value when tapes are exchanged between machines, or even studios, and of absolutely no value whatever in the synchronization of two or more machines.

■ RECORDED TIME DATA

A far more accurate and reliable system would be to record the time information as a coded signal on one of the tracks of the multitrack tape, or on a special track designed for that purpose. Once

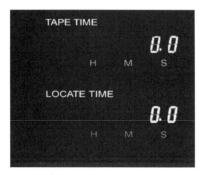

Figure 15.1   A typical tape timer displaying tape position as a function of time.

recorded, this time code would remain in perfect synchronization with the program material, and its future accuracy would not deteriorate due to various mechanical factors. For example, as the tape is shuttled back and forth between the reels, so is the time code. When the tape resumes playing, the time code readout resumes as well.

Like any other "stop-watch," the code does not usually tell the actual time of day. Instead, it indicates the running time of the recorded program. This may have been keyed to the actual time of day at which the recording was made, or to the elapsed time since the recording began. Thus, after 2 minutes and 37 seconds, on a 9:15 AM recording session that began 31 seconds late, the time code might read 09:18:08, or simply 2:37.

The earliest and most pressing need for a longitudinal time code came from video tape editors. Prior to 1956, when film was the only method for recording pictures, film editors used and counted sprocket holes to edit together different frames of the filmed material. However, in video recording, even though there were frames of video information on the magnetic tape, they were not directly visible as on film. Nor does the frame rate used in video match that used in film. Because of this, early editing was a hit or miss operation. Further advances in video systems, such as control tracks and timing pulses, were combined with telemetry technology from NASA (National Aeronautics and Space Administration) to develop a time code technology for video tape editing. However, different manufacturers developed different systems, none of which were compatible with each other.

■ SOCIETY OF MOTION PICTURE AND TELEVISION ENGINEERS (SMPTE) TIME CODE

In 1969 the Society of Motion Picture and Television Engineers (SMPTE) formed a committee to develop a time code standard that would provide videotape compatibility. Today, that time code is in universal use, and has also been adopted by The European Broadcasting Union (EBU). The complete standard is currently published as ANSI V98.12M - 1981: The American National Standard Time and Control Code for Video and Audio Tape for 525 Line/60 Field Television Systems. The standard is generally referred to as the SMPTE/

EBU Time Code, or, more simply, SMPTE Time Code. There are actually two versions of the SMPTE time code. One is longitudinal or serial code, and the other is vertical interval code. The latter is often called VITC (Vertical Interval Time Code) and is not particularly applicable to pure audio use. Longitudinal time code is usually recorded on an available track on the multi-channel tape, while VITC is recorded in the vertical interval between the recorded frames and fields on the videotape.

The SMPTE time code provides a readout in hours, minutes, seconds, and frames, to give the operator film-like frame editing accuracy. Thus, a readout of 23:59:32:21 indicates 23 hours, 59 minutes, 32 seconds, and 21 frames.

Frame Rates

There are several frame rates in use today, with different rates used in film, monochrome video, color video and European television. In film, a frame is simply one visible picture on a strip of celluloid. In order to understand the video frame, a short video primer is in order.

With NTSC (National Television Standards Committee) television, which is used in the United States, there are 525 lines across the television screen every frame. These are called raster lines. Five hundred twenty-four of these are full width, while one consists of two half-width scans at the top and bottom of the screen. To prevent any noticeable picture flicker, interlaced scanning is used. With this scanning technique, every other line of the picture is skipped during the first half of the frame, then filled in during the second half of the frame. The scans that occur during the first half are called field 1, and the remaining scans are called field 2. There are therefore two fields in every frame for monochrome video. Each of these fields is recorded on magnetic tape by one pass of the rotating video head(s).

At the end of each left-to-right scan of the raster, the electron beam returns to the left-hand side of the screen and drops down, skipping one line, to begin the next sweep. After sweeping a complete field, the beam returns to the top left corner of the screen to begin the first line of the next field. This return to the top of the screen is referred to as flyback. Just before and just after flyback, the electron beam is switched off, or blanked. This is called the vertical blanking interval and occupies a space equivalent to 21 raster lines. The black bar visible on your television screen when the picture is rolling is the vertical blanking interval, and it is normally unseen under proper operating conditions. It is here that VITC is recorded. This allows the time code to be read while the video machine is in pause (with the heads still rotating), a luxury that is not possible with longitudinal time code. However, since we are dealing primarily with audio, further discussions at this time will be about longitudinal time code. There will be a short addenda describing the differences between longitudinal and vertical interval time code at the end of the discussion on the make-up of longitudinal time code.

Figure 15-2 lists the frame rates in use throughout the world. It must be understood that color video speed is referenced to a different frequency than monochrome video speed, and that 30NDF (non-drop-frame) and 29.97NDF are exactly the same code but running at two different speeds. Conversely 30DF (drop-frame) and 29.97DF are also the same code but at differing speeds.

Note that the EBU and original NTSC standards correspond to half the line frequency used in the respective countries. A line frequency of 50Hz is found in EBU countries, while 60Hz is the US standard for AC power. Originally, NTSC video was broadcast on a carrier frequency of 6MHz, but with the addition of color in the 1960s, an

Figure 15.2 Frame rates in use throughout the world.

Frame rate (frames per second)	Application
24 fps	Motion picture film work;
25 fps	EBU television and film standard;
30NDF fps	NTSC monochrome video reference standard. Used for digital audio;
29.97NDF	NTSC color running at the monochrome video reference standard;
29.97DF fps	NTSC color video reference standard;
30DF fps	NTSC monochrome running at the color video reference standard. Used to match/correct speed.

additional carrier (for color) of 3.58MHz was used. This is referenced to a frequency of 59.94Hz instead of 60Hz. This difference caused the creation of drop-frame time code which will be explained later in the chapter. It is important to understand that the term "frame rate" defines the number of frames that pass a given point in one second, whereas the term "frame count" refers to the number of frames that increment before the frame field returns to zero.

Simply stated, the SMPTE time code is a longitudinally recorded signal, not unlike the digital data streams described in Chapter 13. The code provides an 80-bit digital word for each video frame. Unlike the coding schemes described earlier, a zero is defined whenever there is no transition within a bit cell, as shown in Figure 15-3a. In other words, there is only one transition per bit, and this occurs at the beginning of each bit. At 30 frames per second, 80 bits per frame, and 1 transition per bit, a continuous stream of digital zeroes will produce a square wave of $80 \times 30 \times 1 = 2400$ transitions per second. This is equivalent to a frequency of 1200Hz. A digital "one" is defined as a level transition occurring midway through the bit cell, as seen in Figure 15-3b. With 2 transitions per bit, a continuous stream of ones will produce a square wave of $80 \times 30 \times 2 = 4800$ transitions per second, or 2400Hz.

The EBU standard, with 25 frames per second, will provide a frequency of 1000Hz and 2000Hz respectively for continuous streams of zeroes or ones.

Figure 15.3 In the SMPTE time code, a zero is defined whenever there is no transition within a bit cell (a), while a transition within the bit cell (b) defines a one.

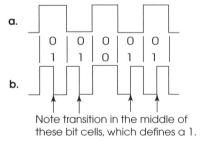

SMPTE Assigned Address Bits

Within the 80-bit digital word assigned to each frame, certain bit groups have been assigned to various time, sync, and user-defined functions as described below.

■ Time Code Address Bits

The SMPTE time code assigns specific bit groups to indicate hours, minutes, seconds, and frames. For example, Figure 15-4 illustrates bit groups 32–35 and 40–42, which indicate minutes (units) and minutes (tens) respectively. Using a binary coded decimal (BCD) system, bits 32–35 indicate 1, 2, 4 and 8 minutes, while bits 40–42 are 10, 20 and 40 minutes. Thus, 6 minutes is defined by a one in bits 33 and 34. If bits 32, 33, 34 and 41 are one, then the time is 1 + 2 + 4 + 20 = 27 minutes. Note that the units group is used to define minutes 0–9, and the tens group defines minutes 10–59. Therefore, 12 minutes is indicated by 2 + 10, and never as 4 + 8. Similarly, 60 minutes is indicated by a one in the hours (units) group, not illustrated here, and not as 20 + 40 or 2 + 8 + 10 + 40 within the bit groups presently under discussion.

The following is a list of the bit groups assigned to time:

Bit Group	Time Function
0–3	Frames (units)
8–9	Frames (tens)
16–19	Seconds (units)
24–26	Seconds (tens)
32–35	Minutes (units)
40–42	Minutes (tens)
48–51	Hours (units)
56–57	Hours (tens)

Time bits can sequentially count for a full 24-hour period. After 23:59:59:29 is reached, the next number is 00:00:00:00.

■ Drop-Frame Time Code Bit

As described so far, we have made the assumption that the black-and-white video frame rate of 30 frames per second is being used. However, as mentioned earlier, the NTSC frame rate for color is 29.97 frames per second. This means that a one-hour black-and-white program will contain 108 more frames than a one-hour color program. (In video-based digital systems, this corresponds to a sampling rate difference between 44.1kHz and 44.056kHz respectively.) Therefore, if a color program is clocked by a system calibrated to the black-and-white standard of 30 fps, it will take an additional 3.6

412   THE AUDIO RECORDING HANDBOOK

seconds until the readout indicates an elapsed time of one hour (108 frames / 30 frames per second = 3.6 seconds).

To compensate for this discrepancy, a drop-frame system can be used, in which certain frames are discarded, for a total of 108 frames during each hour. To accomplish this, the drop-frame time code omits the first two frame numbers (00 and 01) at the beginning of each minute, with the exception for the six minutes that start at 00, 10, 20, 30, 40 and 50 [(2 X 60) - (2 X 6) = 108 frames]. In other words, drop-frame time code omits the first two frames of every minute with the exception of the tens minutes. Thus the following sequence will be seen when in the drop-frame mode:

H: M: S: F
09:18:59:29 followed by
09:19:00:02 (not 09:19:00:00!)

-and-

09:19:59:29 followed by
09:20:00:00 (one of the six exceptions)

Figure 15.4 In each frame, bit groups 32, 35, and 40–42 define minutes (units + tens).

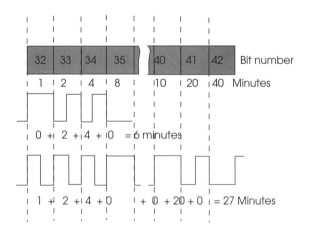

To implement the drop-frame code, bit 10 in each 80-bit word is defined as a drop-frame flag. When the drop-frame time code is enabled, a one is assigned to this bit. Otherwise, bit 10 remains a zero. 29.97DF (drop-frame) SMPTE code is used when the video time must match the wall clock time exactly. If color video (29.97) is run at a 30 fps frame count it is called 29.97NDF (non-drop-frame) code and runs 0.1% slower than real time. And conversely, if a digital audio tape at 30 fps, for instance, is run at the video speed of 29.97DF, the code would be 30DF code and would run 0.1% faster than real time. This is the reason for "pull-up" and "pull-down" that will be discussed later.

■ **Color-Frame Code Bit** Due to the nature of the color video signal, edits must be made between frame pairs, rather than between any two adjacent frames. (A color frame is four fields in NTSC: two for chroma and two for luminance.) The 1st and 3rd fields are defined as color frame A, and the 2nd and 4th fields are color frame B. To prevent a horizontal shift (or "H" shift) in the picture, edits must preserve the AB, AB, AB sequence. Bit 11 in each frame is the color-frame flag, and is encoded as a one to indicate color frame identification. This signifies that all even-number frames are "A" frames and odd-number frames are "B" frames for electronic editing systems. Otherwise, bit 11 is recorded as a zero. Many newer systems have a built-in color framer that is set at the beginning of the editing session or when a new roll of tape is mounted on the record video tape recorder. In this case, the color-frame bit should be set to zero.

■ **Sync-Word Bits** As with the digital pulse streams described earlier, the beginning and end of each digital word must be clearly defined. In the SMPTE time code, a permanently assigned sync-word occupies bits 64–79. The word consists of 2 zeroes, 12 ones, 1 zero and 1 one (0011 1111 1111 1101). This sequence, or its mirror image, cannot possibly be duplicated by any combination of bits elsewhere within the frame word, and so it is immediately recognized by the system. This series, which will appear on an oscilloscope as a burst of digital ones, defines the beginning and end of each word and tells the reader which direction the tape is moving.

■ **Plus-One Frame** Not all time code readers contain this feature, but it is an important aspect if accurate absolute timing is required. Note that the 16-bit sync-word just described appears at the end of the frame. When the sync-word is detected, the code will be updated to display the time data contained in the word that was just completed. This means that the time data are always displayed one frame late. The Plus-One frame function automatically adds one frame to the count to correct for this built-in error.

■ **User-Assigned Bits** Within each code word, 32 bits, in 8 groups of 4 bits each, have been reserved as "User Bits," to meet whatever unique requirements the user may have for encoding information. Some time code generators allow the operator to enter these data from a keypad attached to or on the unit. In multiple reel situations, these bits are often used to encode the reel number and date of recording. The user bits are bit groups 4–7, 12–15, 20–23, 28–31, 36–39, 44–47, 52–55 and 60–63. They are frequently referred to as binary spare bits, or binary groups.

■ **Unassigned Address Bits** Bits 43, 58 and 59 have been defined as permanent zeroes, until otherwise assigned by the Society of Motion Picture and Television Engineers. Formerly, bits 10 and 11 were also unassigned, until being defined as drop-frame and color-frame code bits as previously described. Bit 27 has recently been assigned as the Fieldmark Bit by SMPTE.

Bi-Phase Modulation In the practical application of the SMPTE time code, there will never be a continuous stream of either ones or zeroes. So, although the code may superficially resemble a square wave whose frequency is continually varying between 1200Hz and 2400Hz, it is correctly identified as a waveform with bi-phase modulation, square-wave-like in appearance but not in name.

To sum up the SMPTE time code format, we can say that the SMPTE time code is a bi-phase digitally-encoded data stream, in which each recorded frame is identified by an 80-bit word. The word contains 26 time-code address bits, a drop-frame bit, a color-frame bit, 16 sync-word bits, 32 user-assigned bits, and 4 unassigned bits. The code is longitudinally recorded, for most audio uses, on a specially defined or unused analog audio track. Figure 15-5 illustrates the complete time-code data structure.

Figure 15.5 The complete SMPTE time code structure. Each frame has a unique 80-bit digital word assigned to it, which identifies the frame number and the time (hours:minutes:seconds). Also included are eight binary groups (1–8) for user-assigned functions, as well as drop-frame and color-frame flags, and four unassigned bits (U). Each word is concluded with the 16-bit sync-word shown here.

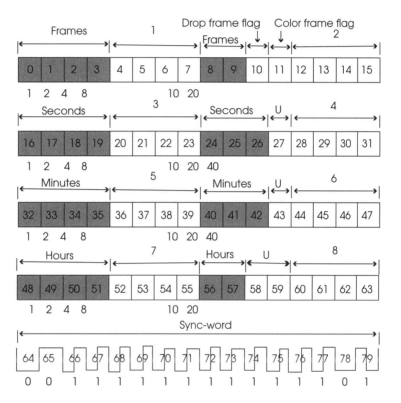

■ VERTICAL INTERVAL TIME CODE (VITC)

Vertical Interval Time Code, abbreviated VITC, is very similar in composition to longitudinal time code. It is used exclusively by video tape recorders, and it does contain a few more data bits than the standard SMPTE time code. Each of the nine data bit groups, frames/units, frames/tens, seconds/units, seconds/tens, minutes/units, minutes/tens, hours/units, hours/tens and the sync-word, are preceded by two additional sync bits. These sync bits replace the SMPTE time code sync-word.

In addition to this, a Cyclic Redundancy Check (CRC) 8-bit word is included. The CRC word provides a check on the accuracy of the word itself, similar to the error correction schemes used in digital audio recording. This gives us a total of 90 bits per frame for vertical interval time code.

VITC also uses one of the unassigned address bits as a field mark bit. This is bit number 27 in the longitudinal time code and bit number 35 in vertical interval time code. This makes VITC a field accurate code.

VITC is recorded in the vertical interval which, as mentioned earlier, uses 21 raster lines and is the space between fields where the electron gun is shut off during flyback. The major advantage of using VITC for video is that the code can be read during high-speed wind and rewind functions, as well as in slow-motion and freeze-frame modes. This is possible because even though the longitudinal motion of the videotape has stopped while the machine is in pause, the video heads continue to spin. This allows the information stored in the vertical interval to be read. Vertical Interval Time Code is recorded using a bit rate of 1.79MHz, which is one-half the color carrier frequency.

■ TIME CODE IMPLEMENTATION

In the typical operation involving time code, several pieces of hardware will be found. First is a time code generator which produces the time code signal to be recorded, and also gives a front panel readout in hours, minutes, seconds, and frames.

The time code reader is a device used in playback applications. It is used to read and display time code from an external source, such as from a previously recorded program.

Much of the hardware in use today combines the functions of a generator and reader in the same chassis. Even though they often share a common power supply, one part of the device may be generating code for a composite recording, while the reader may be switched to read the various machines providing source material.

The time code generator/reader may have an auxiliary video character generator output which is used to display the time code on the screen of a video monitor. For optimum visibility, the time code may be displayed within a luminous window, either at the top or bottom, on the screen. On more sophisticated units, there may be controls for adjusting the size and position of the display. For applications using generators or readers which do not have this function, a separate video character generator may be employed. Video loop-through connections are also frequently found, which allow video-tape copies of the program to be made with the time code display "burned in" or permanently recorded onto the tape copy.

Recording Time Code

Generally, SMPTE time code may be distributed and recorded as though it were a regular audio program signal. However, optimum performance suggests that certain precautions should be observed.

Time code is a fairly high-level signal, whose square-wave-like appearance is rich in harmonic content. As such, it is apt to cause cross-talk problems, especially if routed too closely to microphone-level lines. Time code lines and microphone lines should not be bundled together or run in the same multi-pair cable or snake if the best system signal-to-noise ratio is to be maintained.

Most if not all digital tape recorders and other digital storage devices such as the DAW have a dedicated system or track for storing SMPTE-EBU time code. However, many analog machines use one or more of the existing magnetic tracks, and it is generally recommended that the code be recorded on an outside or edge track of the multi-track analog audio tape recorder to keep inter-channel cross-talk at a minimum. If feasible, the adjacent audio track should be left vacant, or possibly used for some signal that will tolerate a moderate amount of high-frequency roll-off. This may be required to filter out audible high-frequency components of the time code that bleed over from the adjacent track. This is more critical on narrow formats, such as 16-tracks on one-inch tape, than on the larger two-inch tape versions. Some experimenting may need to be done to find a recording level that is low enough to keep cross-talk within reason, yet high enough to afford reliable playback of the time code. Figure 15-6 shows the generally recommended levels for recording time code on various format analog audio and video (non-VITC) tape recorders.

Some two-channel tape recorders have a special track, inserted between the two standard audio tracks, just for time code. These center-track time code machines may have special level requirements, and the manufacturer's recommendations should be followed.

The time code should be recorded for about 10 seconds before the musical program begins for most analog audio recordings. This allows ample margin for electronic editing or synchronization later on, in which it may be necessary to cue up to the beginning of the program, or to provide a pre-roll before the program begins. An exception to this is in the case of digital tapes prepared for Compact Disc mastering. Here, two full minutes of "digital black" on the video tracks (as discussed in Chapter 14) are required, along with continuous time code on audio track 1 of the ¾-inch videocassette. The time code must start at 00:00:00:00, and the program must start at 00:02:00:00. Two minutes of digital black with time code is also recommended at the end of the last selection that is to be transferred to CD.

In order to prevent code reader confusion, each time code address should only appear once on a reel of tape. For example, when recordings made on different sessions are later assembled onto a single reel of tape, it is quite likely that the same time code address will appear in more than one recorded selection. To electronically search into the middle of such a reel of tape, it will first be necessary to record a fresh time code along the entire length of the tape, as previously described.

Figure 15.6 Generally recommended recording levels for time code.

Machine type	Format	Location of TC	Level
Multi-track ATR	2 in. 16 tr.	Edge track	-3 to 0 VU
ATR—Audio Tape Recorder			
Multi-track ATR	2 in. 24 tr.	Edge track	-6 to -3 VU
Multi-track ATR	1 in. 16 tr.	Edge track	-10 to -6 VU
2-track ATR	Half-track	Channel 1	-3 to 0 VU
2-track ATR	Quarter-track	Channel 1	-10 to -6 VU
VTR	1 inch	Audio track 3	-5 to 0 VU
VTR—Video Tape Recorder			
VTR	3/4 inch	Audio track 1	-5 to 0 VU
VTR	Betacam	Time code track	-5 to 0 VU

Using a Synchronizer

As described so far, SMPTE time code generators and readers are valuable additions to the recording engineer's arsenal of production tools. If nothing else, they make all the time-keeping chores a lot easier to perform. However, the most valuable aspect of time code may be for those applications where it is necessary to lock two or more tape transports in synchronization. An obvious application is the recording of multi-track audio "in sync" with video. The audio must be recorded to coincide with the picture, and later on mixed down (again, in sync) and transferred to the videotape's audio channels. Even in audio-only album production work, it is not uncommon to find two or more multi-track machines synchronized together for what may be called "multi-multi-track" recording sessions.

In any case, SMPTE time code, or VITC, is recorded on each machine that is to be used. During subsequent playback, one machine is designated as the master, and the other(s), the slave(s). A synchronizer, such as the one seen in Figure 15-7, reads the incoming time code from each slave, and compares it with the master code. If the codes are not identical, it is an indication that the slave machine is not properly synchronized. A control signal is sent to the slave's transport system to bring it into synchronization with the master. The procedure repeats continuously so that the transport is kept in sync with the master.

In most operations, the slave will follow the master through all modes of operation (start, stop, fast forward, rewind). It is often possible, and may even be desirable, to enter an offset so that the

Figure 15.7 A typical time code synchronizer (TimeLine MicroLynx, TimeLine photo).

slave and the master will maintain a fixed time interval between them. In either case, once the master and the slave (or slaves) are locked together, the entire multi-machine system may be controlled as though it were a single machine. Figure 15-8 illustrates the interfacing requirements of a typical time-code-based multi-machine system.

When time code is locked to an external reference, such as a word clock, it is said to be resolved. When two or more machines are resolved, frame numbers are largely ignored as the phase relationship of the speed controlling PLLs (phase locked loops) are controlling the machines. If resolved to the same reference as when the recording was made, all machines are running in real time. If the reference is different, as will be mentioned later in reference to digital sample rate, this is often referred to as NTSC pull-down. This matches the audio, or film, speed to the video speed.

■ **Multi-Machine Applications** Multi-machine synchronization capability greatly enhances the versatility of multi-track recording work. For example, the basic rhythm tracks need not be played over and over again during a string overdub session. Instead of working directly with the master multi-track tape containing the rhythm tracks, a rough mix and the time code

Figure 15.8 Diagram of a multi-machine system under synchronizer control.

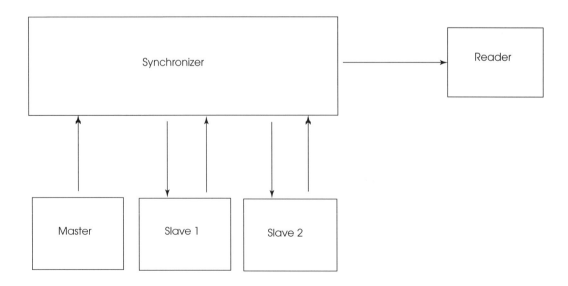

are transferred onto another machine—perhaps only an eight-track recorder. This may be played repeatedly until the string session is ready to be recorded on the remaining tracks.

Later on, the completed string tracks may be transferred "in sync" to the multi-track tape in one pass. In fact, the strings may have been recorded across town or even across the country, while the basic rhythm tracks remain safely stored away until needed for mixdown. For that matter, there may be no real need to transfer the strings to the multi-track tape at all, since the eight-track machine may be employed as a slave during the mixdown. Or, in another scenario, the strings may be transferred, and also played back from the slave during the mixdown, with an offset to create a doubling effect as discussed in Chapter 6.

Many times, even with a 24-track machine, the recording artists may feel that they need more tracks. One solution that will be discussed later in Section VIII, "The Recording Process," would be to combine and bounce tracks. This, however, means that the combined tracks will no longer be separately available for later balance adjustments. Another solution is to use two 24-track machines and a synchronizing system in each room, but that can get expensive quickly. As a more economical possibility, the studio need not equip each of its control rooms with two synced multi-tracks, but for those sessions that do require more track capacity, additional machines may be employed as slaves. Needless to say, the slave machines do not even have to be in the same control room since they become, in effect, extensions of the master system that controls them.

■ Single-Machine Applications

Time code may also be used to advantage on single-machine sessions. For example, with the appropriate software, the synchronizer may be programmed to automatically perform the punch-ins and punch-outs that were described at the beginning of this chapter. Once the appropriate punch-in and punch-out times are determined, this information may be entered into the synchronizer's memory. During recording, the synchronizer will perform the punch-in and punch-out chores without human intervention. If desired, the tape transport may also be instructed to play through the section to be recorded over and over again during the rehearsal. This would also allow the punch-in/out points to be trimmed and adjusted as necessary, prior to actually going into the record mode.

The modern synchronizer usually duplicates the tape recorder's transport controls, as seen in Figure 15-9. With the appropriate interfacing, these controls, in conjunction with the synchronizer's memory system, may be used to perform many of the functions traditionally associated with the transport's auto-locator.

Console Automation Console automation systems, which will be discussed in Chapter 18, rely on SMPTE time code as well. A track of the multi-track tape recorder is dedicated to time code, and the time code addresses are used to define the points where the various console adjustments need to be made. Hard-disk systems only require one dedicated track for code, while other types of systems may need more channels. The SMPTE time code generator and reader are usually built into the console automation system, and often the same code may be used for both automation and synchronization.

Jam Sync Jam Sync is an important feature of any time code system. It allows the repair or restriping of defective or broken code. The jam sync function is used to synchronize or "jam" the generator's output to an external time code signal. This can be from a recorded source or from another generator. Jam Sync can reconstruct defective code,

Figure 15.9 A synchronizer controller containing tape transport controls (TimeLine Lynx, TimeLine photo).

synchronize multiple generators to the same starting time, and generate new code to match old code. The jamming generator accepts the code from the external source and uses it to generate the new code.

In one-time jam sync applications, activating a front panel control sets the generator output in sync with the incoming signal. When the control is released, the generator output resumes independent operation, and any subsequent discontinuities from the incoming code are ignored.

In the continuous jam sync mode, the generator continues to read the input reference signal, and if there are any discontinuities in the incoming code, these are faithfully reproduced by the generator. This is often necessary in cases where the original time code addresses are needed.

Duplicating Time Code

In duplicating a tape, the time code should not be copied directly. This is because the bi-phase waveform often will not survive the process, and its rapid deterioration from one tape to the next will result in almost certain reader errors. For tape duplication work, the jam sync mode may be used as described above, to read the incoming time code signal from a previously recorded program. The generator's time code output is a regenerated replica of the incoming code, and it is this signal that is recorded on the tape copy. Assuming that the one-time jam sync mode is applied, anomalies in the incoming signal are ignored.

On the other hand, the master tape may contain time code signal discontinuities which were intentionally introduced during the editing process. In this case, subsequent copies should retain these discontinuities, and so the continuous jam sync mode should be used instead. In this case the generator output is a restored replica of the actual input code, and retains all the addresses that are present in the input signal from the master tape.

Many time code readers have a restored code output which may be used during copying operations. Note that the terminology introduces a possible source of confusion, since a restored code is of course a regenerated version of the original (continuous jam sync). However, the term "regenerated" is reserved for an entirely fresh code that simply used the original code once, as a reference point from which to begin (one-time jam sync).

Electronic Editing Applications

Most of the electronic editing techniques described in Chapter 14 require time code data for implementation. For example, when an edit point is located, the time code defining that point is stored in the electronic editor. The time code also keeps the recorder and player (master and slave) in sync, and when the edit point is reached, the master goes into the record mode, based on a comparison of the edit point's stored code and the instantaneous address of the master tape. Of course, these functions are not carried out under the control of the synchronizer described in this chapter. Instead, the electronic editing hardware discussed in Chapter 14 accommodates the necessary software for this work.

The time codes from various takes to be recorded onto the final master will rarely, if ever, be usable for future production work. This is due, in part, to the inevitable gross time shifts that occur at each edit point. Worse yet, if an edited program segment is followed by a segment from an earlier take, the code will jump backwards at the edit point, making future time code search operations virtually impossible. Therefore, it is common practice to record fresh time code along the entire length of the master tape prior to the electronic editing session. This is often called pre-striping. Now, as each new program segment is transferred to the master tape, its original time code is discarded in favor of the new code recorded earlier.

Time Code and Digital Audio with Film and Video

As stated earlier, non-drop SMPTE time code is based on a rate of 30 fps. It was this rate that was used in the early days of digital audio to derive what has become the standard sample rate of 44.1kHz for the compact disc. Since some of the earliest professional digital recorders used the ¾-inch U-matic format, this was chosen as the preferred recorder for storing digital audio that was to be converted to compact disc. As discussed in Chapter 14, the digital audio was modulated to a monochrome NTSC video signal and stored as RF video data on the tape. In order for the system to be compatible world-wide, a sampling rate had to be found that would also work with the EBU standard reference of 50Hz and a frame rate of 25 frames per second. Taking the 525 lines of resolution used in the NTSC standard and subtracting the lines not used for picture we arrive at a rate of 245 lines per field. If two samples are stored per line, we get a sample rate of 29.4kHz, which is too low. However, by storing 3 samples per line we generate a sample rate of 44.1kHz based on the 30 fps standard. This works well also with the European system in which there are 625 raster lines and 37 in the vertical blanking interval. As shown below, the same sample rate can be used throughout the world.

$$3 \text{ samples} \times 60\text{Hz} \times \frac{525 - 35}{2} = 44.1\text{kHz}$$

$$3 \text{ samples} \times 50\text{Hz} \times \frac{625 - 37}{2} = 44.1\text{kHz}$$

However, since color video runs at a rate of 29.97 fps, if digital audio is to be synchronized to the picture, the audio rate has to be slowed slightly as mentioned earlier. Here the reference clock is different than the rate used during the digital recording, resulting in the NTSC "pull-down." A frame rate of 29.97 does not correspond to a sample rate of 44.1kHz, but to a rate of 44.056kHz. This change would make the audio playback below its real pitch. This requires the engineer to "pull up" the sample rate to 44.144kHz when recording the audio, so that when it is slowed down to a frame rate of 29.97NDF to synchronize to the video, the pitch will be correct.

■ MUSICAL INSTRUMENT DIGITAL INTERFACE (MIDI)

Another system for synchronizing electronic devices has recently emerged on the performance side of the control-room window. This system, called MIDI (Musical Instrument Digital Interface), is increasingly found in the modern recording studio. MIDI, a relatively new tool in the recording studio, was developed as a system for creating communications links between electronic instruments such as synthesizers, electronic keyboards, and various rhythm machines. As such, MIDI is primarily used by performers, but the prudent recording engineer should be aware of the system and how it works since it is becoming more and more common in the studio.

Background To understand MIDI, it is important to see how and why it evolved. Early synthesizers were often criticized for their lack of depth or character in the sound. The instruments were monophonic, and most were only capable of providing one type of sound at a time. Many musicians solved this problem by playing on two synthesizers at once—organ stops with the left hand and piano sounds with the right. Digital and analog delay units were often used as well to thicken and fill out the sound.

Further developments with these instruments provided control-voltage outputs that would allow the first synthesizer to control the second with regard to such things as note-on, note-off, pitch and timbre. This system was practical as long as both synthesizers used the same control signals and provided them at the same rate to generate or control sounds. However, manufacturers seldom agreed, and there was no standardization. MIDI was developed by an association of manufacturers as a standard, so that all types of electronic instruments and devices could communicate using the same control signals and protocol.

MIDI is a digital bit stream very similar to the SMPTE time code data stream in appearance. However, the information contained and the word length and transmission rates are quite different. It is a serial interface using ten-bit words at a transmission rate of 31.25 Kbaud. (There are 8 bits in a byte, and a baud is a transmission rate of 1 byte per second.) Each MIDI word contains a start-bit, eight data- or status-bits, and a stop-bit. The first bit in each data group determines the type of word, and the remaining 7 bits express the value. Information is sent in multi-byte groups, which are usually made up of one status byte followed by two data bytes. Status bytes always start with a 1, and data bytes start with a 0.

There are two main types of messages: channel messages and system messages. MIDI can provide up to 16 different channels of information in the bit stream, but the more channels that are in use at one time, the slower the data rate per channel. The first four bits of the channel status word assign the rest of the multi-byte group to a specific device or group of devices as determined by the user. The remaining four bits determine voice and mode selection or status.

System messages do not carry channel assignments, but are divided into three groups called Common, Real-Time, and Exclusive. Common messages are read by all units in the system. Real-Time groups are also intended for all units, but supersede any previous instructions which may be running. Exclusive message units carry a manufacturer's identification number, and thereby are intended for only devices of that particular ID group.

As mentioned earlier, the remaining two bytes are data words. The controlled unit waits until both data bytes are received before acting on the information. The data byte contains information that controls events such as note duration, velocity (touch), and pitch.

MIDI Equipment

Equipment designed for MIDI applications comes in many varieties. A MIDI system can include a master electronic keyboard, acoustic-to-MIDI converters, drum machines or electronic drum pad controllers, sequencers, and signal processors such as equalizers, reverberators and signal delay units.

The master electronic keyboard is usually the control center of the system. It is from here that the artist plays and controls the MIDI ensemble. An acoustic instrument, such as a guitar or acoustic piano, may be used as a controller as well. There are also MIDI woodwinds and brass available that control MIDI parameters using the instrument's keys or valves and a sensor for air intensity. Drum machines can be programmed to act as a master tempo controller, or drum control pads can be used as triggers for specific events or sounds. MIDI-controlled reverberators, limiter/compressors, noise gates, etc., are also commonly available and easily interface into the system.

The MIDI sequencer is probably the most important component of the MIDI system. It is here that the digital control data can be stored or recorded for later playback. This is a recording, not of music, but simply of the digital data. For playback, each MIDI channel must feed a MIDI instrument. In effect, the recorded performance is recreated each time the sequence is played. A musician plays a melodic line into a sequencer using a keyboard or other controller, and the sequencer remembers the information and assigns it to a sequencer track. Other tracks can be added (in sync) until all the available sequencer tracks are used. Some computer-based sequencers offer up to 64 tracks. Actually, the number of tracks and the length of the song are limited only by the amount of memory available in the computer. On playback, the sequencer becomes the master, and the MIDI out data are sent to the MIDI slave machines. If the performance is not to the artist's liking, any sequencer track can be edited or modified until the desired performance is achieved. As there are many excellent texts on MIDI composition and implementation available, we will concentrate here on how MIDI fits into the professional studio.

Interconnection

The connections between MIDI devices are relatively simple. Most MIDI-capable equipment will have three jacks labeled MIDI-In, MIDI-Out, and MIDI-Thru. The MIDI specification states that 5-pin DIN connectors are to be used, but at this time only three of the pins (2, 4, and 5) are currently utilized. Some manufacturers use the 5-pin DIN plugs while others use XLR connectors or quarter-inch tip/ring/sleeve phone jacks and plugs. This requires the MIDI user to have on hand various types of adapters for interfacing different equipment. Cables are specified to be no more than 50 feet long.

MIDI-Out provides the master control signal that is sent to the various devices to be controlled. Most equipment designed for use with MIDI can be either a master or a slave. MIDI-In is where the signal from the master is accepted, while MIDI-Thru is simply a loop-through that passes the same signal on to another device. Because of losses due to cable length, no more than three devices should be chained together by the MIDI-Thru connectors. For systems requiring more than four units (one master and three slaves), it is recommended that a MIDI distribution system be used. Figure 15-10 illustrates these two conditions.

Figure 15.10 Two different methods for MIDI interconnection: (a) The MIDI "daisy chain" using MIDI thru connectors, and (b) using a MIDI distribution system or "multi-thru" box.

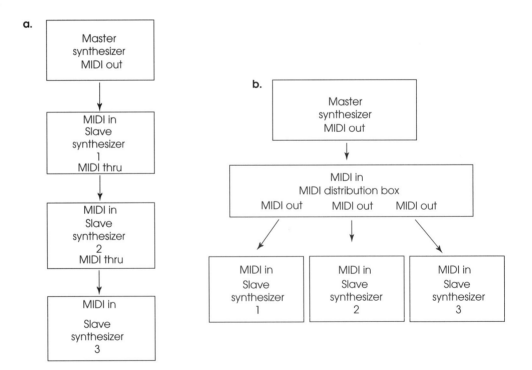

MIDI in the Recording Studio

The musician usually brings his or her MIDI system into the recording studio with the performance data already stored in the sequencer. Much studio time has been saved since the performance is essentially ready to go. Each instrument's audio output(s) are then sent to the studio console. Depending on the output levels and impedances of the instruments, direct boxes may be required. When all is ready and the console input levels have been set during a rehearsal, the recording can be made either direct to two-channel stereo (if it is a complete performance), or to multi-track tape so that more parts can be added later. Often, the MIDI system is interfaced to the recording console right in the control room. This allows better communications between the engineer and the artist, and no separation problems are created since there are no microphones in use.

During this process, the engineer will make use of whatever studio processing devices are necessary to provide the best possible recording. Reverberation may be added to a direct-to-two mix, while noise gates, equalizers, compressors, etc. may be required during the multi-track recording. It is essential that the engineer make sure that the MIDI interconnection cables are routed away from the audio signal cables as interference and cross-talk can occur.

■ THE SMPTE/MIDI INTERFACE

Using the SMPTE time code makes it easy to synchronize a MIDI sequencer to a multi-track tape. There are several SMPTE-to-MIDI synchronizers that allow the studio's SMPTE output to control the rate of the MIDI sequencer. This allows the virtual tracks recorded in the MIDI sequencer to be recorded or just run along in sync with the existing multi-track tape material. This can prevent tying up too many audio tracks on the multi-track tape recorder with synthesizer information.

In mixdown, the SMPTE output from the multi-track is fed to the MIDI sequencer, and the MIDI slave outputs are fed to unused channels on the mixing console along with the output of the multi-track tape. A typical MIDI system recording session synced to time code is illustrated in Figure 15-11.

Figure 15.11 A sophisticated studio setup using SMPTE time code and MIDI to produce a musical selection synchronized to an existing video.

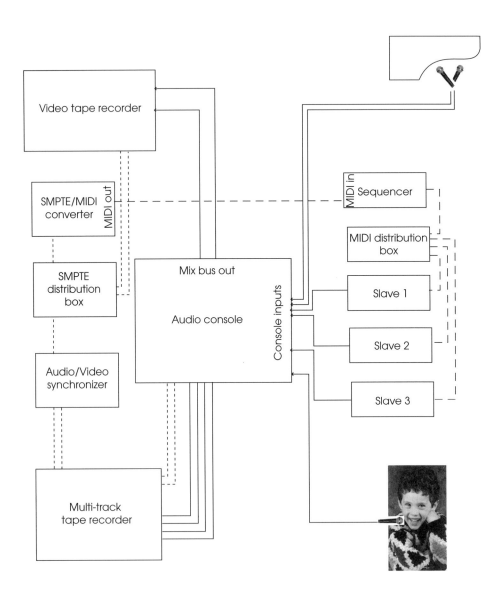

SECTION VII

The Recording Console

■ INTRODUCTION

At first glance, today's modern multi-track recording console may appear to be a hopeless maze of knobs and switches cluttering the path between the microphone and the tape recorder. Yet, despite its apparent complexity, the console is essentially a collection of similar control functions that are repeated over and over again. They give the engineer a greater command over each input from the studio and the signal output that is sent to the tape recorder. In fact, the console may be the single most important piece of equipment in the signal chain of modern recording, as all systems interface with the console. It is the heart of the composite system.

Chapter 16 of this section actually discusses the recording console several times. First the console is discussed briefly, to give the reader an understanding of its basic functions. Then each section is broken down into its component parts and described in detail. Finally, various secondary signal paths are introduced, and our examination of the complete console is concluded. Many of these concepts are common to both tracking and mixdown, but may be utilized differently in each scenario.

A more tape recorder tracks become available and used by the artist and engineer, the logistics of controlling these tracks manually in mixdown becomes a difficult job. Chapter 17 introduces console automation systems that are designed to make this task manageable. Mixdown will be discussed in depth in Chapter 20, but the basic concepts and functions of automation systems will be discussed here. Tape and hard disk based automation storage systems will be addressed.

SIXTEEN

Recording Studio Consoles

■ OVERVIEW

Under some circumstances, a completely satisfactory recording may be made by simply plugging a microphone into a microphone pre-amplifier and then sending that amplified signal to the input of a tape recorder. Particularly in the case of a single stereo microphone and a two-track digital tape recorder, there may be little point in inserting any type of intermediate control device into the signal path. The stereo microphone "hears" what the concert hall listener would hear, and this information is directly transferred to tape. Given a well-balanced musical ensemble, playing in an acoustically satisfactory environment, an excellent recording may be made in this manner. For a discussion of microphone techniques used in such circumstances, refer to Chapter 4.

For the majority of contemporary recording situations, however, somewhat more flexibility may be required and a recording console becomes a necessity. The console may be nothing more than a simple combining network where several microphone inputs are mixed together to provide one, or perhaps two, outputs to a tape recorder. At the other extreme, the console may be capable of mixing, in a seemingly endless number of combinations, the outputs of dozens of microphones, and may have perhaps 24 or 32 or more channels of outputs as well as various surround formats and stereo and monophonic outputs.

In either case, the console becomes a combining and routing point between microphones, line-level sources, processing equipment, and the storage devices and media. These devices or systems can be either in the digital domain or in the analog state. Regardless of the apparent complexity of a modern multi-track recording console, it may be analyzed as a combination of four major control sections. This does not change whether we are discussing a modern analog multi-channel console or one of the newer complex digital mixing systems. The four sections as shown in Figure 16-1 are:

1. The Input Section (inputs to the console from microphones, electronic instruments and recording system lines);

2. The Output Section (metered outputs from the console to the multi-track tape recorder, the mixdown recorders, or other storage systems such as DAWs);

3. The Master Module Section (master auxiliary sends, cue sends, reverberation returns, 2-mix or stereo outputs);

4. The Monitor Section (stereo metering, loudspeaker selection and sends, headphone monitoring, source/recorder switching).

Most consoles contain these four major sections, but as the complexity of popular music has increased, consoles have developed into two major types: the in-line or input/output module style, and the simpler split-configuration style. In the former, most of the functions for a particular input and output are merged onto one linear module. This module will also contain the monitoring functions for that I/O module, the stereo bus assignment and panning. The latter style of console maintains separate sections for the input modules and the output modules, often with the monitoring section located between the two sections. Although the basic functions and purposes of the consoles are similar, the radical difference in layout requires discussion.

Early multi-track consoles were larger versions of the split-configuration consoles developed for live recording and broadcasting. The change to the in-line style came about largely as a result of the proliferation of tracks on the multi-track tape recorder. The earlier console may have had 24 or more microphone inputs, but typically only four, or at the most eight, outputs were required. At the time, it made sense to treat (and to locate) these outputs separately from the inputs.

The eight outputs often represented eight complete mixing buses, although it would be a rare occasion when even half that number were required. For example, with 16 inputs, it would be impossible

Figure 16.1 The four major sections of a multi-track recording console.

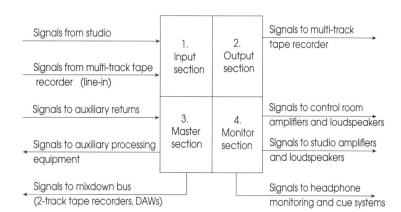

to use more than eight mixing buses, and that number could only be put to use if each bus was restricted to mixing no more than two inputs (a highly unlikely situation). In any case, for each additional microphone assigned to any mixing bus, the total number of buses that it is possible to use is diminished accordingly. Furthermore, as the number of recorded tracks increases, the requirement for mixing buses usually diminishes as well.

On today's console, there are apt to be as many outputs as there are inputs. The 24-track tape recorder or digital workstation is an industry standard, and it is not uncommon to link two such machines for even more output capability, or to find 32-track digital systems in use. Taken to its extreme, each microphone may feed its own signal to a separate track on the storage medium with little or no mixing taking place during the actual recording session, and with no need for mixing buses at all. Consequently, it is not unheard of to find a console with a 56-input by 48-output configuration; that is, 48 inputs for multi-track returns and eight additional for effects returns. Therefore, with more and more inputs taking up valuable space in front of the operator, it makes little ergonomic sense to allocate precious space for a complex mixing/monitoring section. The available space may be more efficiently put to use by "folding back" the output and monitor sections so that they appear in-line with the input section. Now, module 10 will contain the controls for microphone 10 and console output 10 as well. Thus we have the I/O module as the basic block of the in-line console. More recent analog and digital consoles have taken this a step further by simply adding a second input/output section with controls immediately above the first.

In the analog domain it is important to optimize signal-to-noise ratio, and the recording engineer will try to keep all recorded levels consistently at a maximum on the tape. With most of the balancing and mixing taking place later on, there is really little need for much more than the simplest sort of gain control device in the signal path between the microphone pre-amplifier and the tape recorder input. Furthermore, most equalization, reverberation, delay and other signal processing will be added later during the mixdown session. Certainly none of this represents a sudden departure from conventional multi-track recording techniques, but rather is the logical culmination of a practice that has been undergoing change since the earliest days of multi-track recording.

Today, the equalizer, and even the familiar linear fader, may be found and put to better use within the monitor system so that the engineer may hear their effects without these effects being recorded. That way, a change in the equalization or the mix of the component signals that seemed right during the recording session will not have to be undone later in the light of the mixdown session. Accordingly, the equalizer and the fader have been appropriately relocated to the monitor section. This monitor section not only provides the signals for the control room monitors, but is also the source for the 2-mix bus from which the final stereo mix is sent to the 2-channel tape recorder.

On the other hand, even with the proliferation of available channels, recording is still not limited to the exclusive assignment of one microphone to one tape channel. Although such a "one-on-one" technique certainly allows the engineer the greatest mixing flexibility later on, there are many occasions when it will be desirable to combine two or more signals during the recording session. At these times, the equalizer or linear fader or a duplicate input may be required within the microphone-to-tape signal path, and the appropriate signal path switching is usually provided for this purpose.

While the split-configuration console is still in use to some extent, it is more at home in the broadcast and sound reinforcement areas than in the modern multi-track recording studio. This type of console, found in both digital and analog versions, is infinitely more practical in live reinforcement situations where the house mix may feed many banks of amplifiers while the monitor mix is directed simultaneously to the stage. The uses and the architectures may be different, but the four basic sections are still the same. Our discussion will mostly deal with the I/O module style of the in-line multi-track console, but most of the functions and features are the same for the split-configuration mixer. Figures 16-2a though 16-2d show, respectively, an analog in-line console, a small split-configuration analog console, a small split-configuration digital console, and a digital multi-channel console.

A brief description of the various functions within each section is given here. This is followed by a sequential listing, along with detailed explanations, of the many components which make up the total signal path through the console, from microphone to tape recorder.

■ THE INPUT/OUTPUT (IO) MODULE

Input Section   The first stage in each signal path is the microphone pre-amplifier, where the low-level microphone signal is amplified to the console's internal line-level or the level required by the digital converter. Next, assuming that the module's equalizer and linear fader will not be used at this time, the signal passes through a simple gain control. This may be nothing more than an amplifier with a rotary potentiometer. In a digital console, the analog-to-digital converter is often found just after the microphone pre-amplifier, and is then followed by a gain adjustment that is accomplished in the digital domain.

Output Section   In electrical work, a bus is a length of copper or aluminum bar stock that is used as a central feeder line, as in a circuit breaker panel. In a recording console, a bus is a common signal path, to which the outputs of many input module signals have been routed.

After this elementary level control section just mentioned, the signal may be routed to one or more output buses. Although these can certainly be used for mixing purposes, they are often referred to as channel buses, simply to distinguish them from mixing buses, which will be described later. If more than one channel bus is selected, the signal will travel through a pan-pot. Instead of using the channel buses, the signal may be routed to the tape recorder channel associated with each I/O module (for example, input 3 to output 3, etc.). This direct mode is the one that would be used most often for so-called one-on-one recording. This may be preferred because it is the shortest route from the microphone to the tape.

Figure 16.2 a. A large modern analog in-line console (API Legacy, API photo); b. A modern digital in-line style console (Soundtracs DPCII, Soundtracs photo); c. A digital split-configuration console (Yamaha Photo); d. A small analog split-configuration console (Mackie Photo).

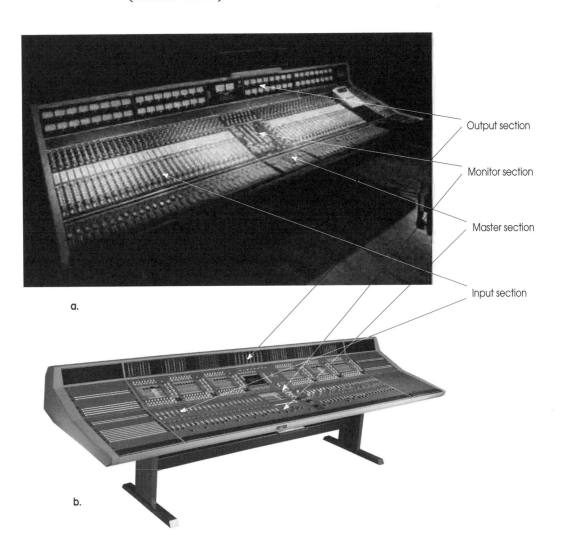

Figure 16.2 (continued)

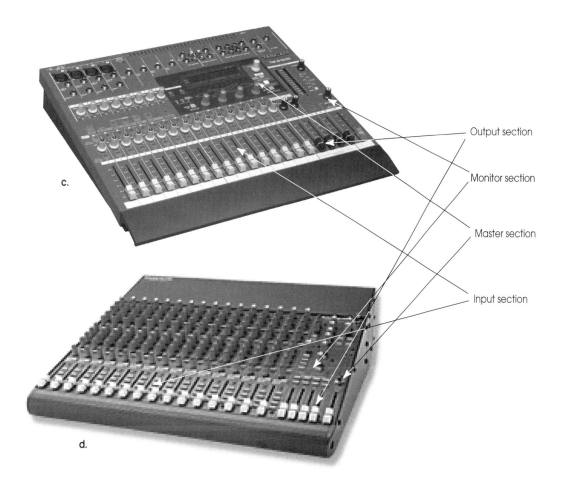

At this point the signal is sent to the multi-track tape recorder either as a digital signal with its own level adjustment, or as an analog line-level signal through a buffer amplifier which may have its own rotary potentiometer or "trim" control. The channel path metering signal is usually derived at this point and sent to the I/O channel meter. The channel line-out signal is also routed to the I/O module's monitor input section.

The Monitor Section Here, the signal passes through the module's equalization section and linear fader. After equalization and level adjustments, the signal passes to a monitor pan-pot from which it will be routed to the left and right mixing buses. At this point in our description, these mixing buses are being used solely for monitoring purposes. Later in the remix process, they will be used to feed the two-channel recorders used for the mixdown session.

■ THE MASTER MODULE

Additional space must be provided for various master controls for the mixing bus outputs, cue lines, auxiliary send and effects return lines. These are commonly found on a master module located to the right of the I/O modules.

As will be noted below, the engineer/producer may wish to lower the listening level of previously recorded tracks, while concentrating on whatever is being recorded at the moment. On the other hand, the studio musicians must easily hear what was recorded before, if they are to play along in accompaniment. Accordingly, the master module section usually provides a separate set of sends for the headphone monitoring in the studio. These controls are independent of both the recording levels and the listening level in the control room. These lines are often called foldback or cue send lines, and large consoles may have several separate systems.

The auxiliary buses are used to send and return signals from outboard processing equipment such as reverberation systems. The returns allow the signals from these devices to be mixed into the monitor path and consequently the 2-mix bus. Since most processing of this type is done during mixdown, these returns are usually not sent to the channel outputs. The auxiliary buses often may be used for both cue and processing sends.

■ THE MONITOR MODULE

The recording engineer will frequently need to listen to various signals out-of-context with the level at which they are to be recorded. For example, he may want to listen to the input or output signals one at a time to verify his signal output arrangement. In the case of a tape containing some previously recorded tracks, the producer/engineer may wish to concentrate on the recording of the new material. However, during playback he will want to hear the entire program in the proper balance.

The monitor functions of the console will provide the engineer with the necessary controls to adjust the relative listening levels without affecting the channel recording levels. Of course, if the engineer adjusts his recording levels, these adjustments will be heard through the monitor bus as well. However, adjustments intended for monitoring purposes only will not find their way onto the tape. The engineer may also wish to compare the sound of the channel outputs from the console with the actual recorded tape. This comparison, or similar comparison of one program with another, is popularly known as an a/b test, and reference may be made to "a/b-ing" the tape (or other program).

Since monitor level and panning for each channel has already been established within the I/O module, the monitor module of the console may contain little more than a sophisticated switching system which is used to select the mix bus outputs, or the outputs of one of several 2-channel tape recorders that are used during mixdown. With an analog console, most of the channel/tape a/b monitor switching will be found on the I/O module, but additional switches in the monitor section allow the engineer to monitor any of the auxiliary send lines which will be described later in this chapter. When using a digital system, the channel/tape monitor switching may be accomplished at the I/O module or in the Master section of the console as well. Master controls for the solo bus and the PFL (pre-fader listen) bus are found here also. There will be separate switching systems for both studio and control room monitor sends, as well as the associated gain controls, speaker switching and talkback systems. Stereophonic, monophonic, surround and bus metering, and any phase coherency monitoring and metering is also included in this section. It is possible that the metering section may also include a third-octave spectrum analyzer.

One of the largest differences between analog and digital consoles is the number of controls per channel that may be available. On an analog console, each module typically has the same types of knobs or controls. For instance, you will find the channel-out pan-pot on every I/O module, whereas on a digital console there may be one control for the pan-pot that is shared by all channels or a bank of channels. Often, in a digital system, one control or knob may serve the same function as various single controls in the analog domain. This control could be virtually anything at various times—a panpot in one mode and an equalization or dynamics control in another. Such a control is thus often referred to as a "virtual control."

Figures 16-2a-d identify the console sections just described for both an in-line console and a split-type console, while Figures 16-3 a-c picture a typical analog in-line console I/O module and master module, an I/O module from a digital board, and an input module and an output module for a split-configuration analog design. Note that in Figure 16-3b there is a bank of eight I/O modules on a screen, with one set of controls to the right for equalization of all eight channels, a set of controls below the screen for auxiliary sends or pan-pots, and a set above for microphone and/or line-level trim and phase switching.

■ COMPONENTS OF THE IN-LINE RECORDING CONSOLE

The analog console to be described here is of an in-line design, with an equal number of inputs and outputs, and a two-channel monitor system. The digital console has a dual-input-style design with both an upper and lower input that are identical. Only one of the inputs is shown. Figures 16-4a and b are simplified block diagrams of the typical signal path through the I/O modules in these consoles. Refer to these illustrations as each new point on the signal path is introduced. Each important point has been assigned a letter (A-Z-CC), and a detailed description of these points follows. Additional modules will be introduced as they make their first appearance within the signal path. While most functions are common to both the analog and digital consoles, some are specific to one or the other and this will be noted as we go along.

Due to the nature of the in-line design, all controls and components within a certain category are not necessarily found within a single module. For example, on the analog console the monitor panpot is located on the I/O module, while the monitor (mix) bus ACN (Active Combining Network) amplifier is located in the master module. The control-room monitor level itself is found on the monitor module.

The digital console uses so-called soft-assign, where a number of controls serve many functions. By touching the desired module on the screen, these controls become active for that I/O channel.

RECORDING STUDIO CONSOLES   443

Figure 16.3 a. An input/output module and a master module from an in-line recording studio console (Sony MXP-3036VF; Sony figure); b. An input/output section for an in-line digital console. All faders can be soft assigned to serve either function (Soundtracs DPCII; Soundtracs illustration); c. An input and output module for a split-configuration console (Sony MXP-2000; Sony figure).

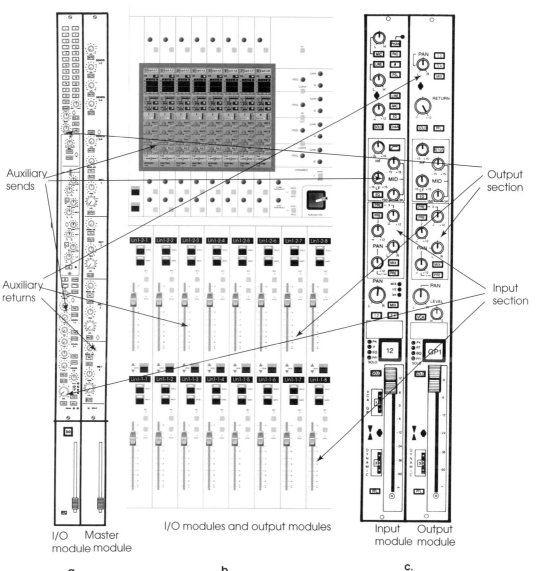

Figure 16.4 a. A simplified signal flow diagram through a typical I/O module for an analog console. The heavy line indicates the channel path, and the dashed line is the monitor path; b. A simplified signal flow diagram through a typical channel of an in-line digital console.

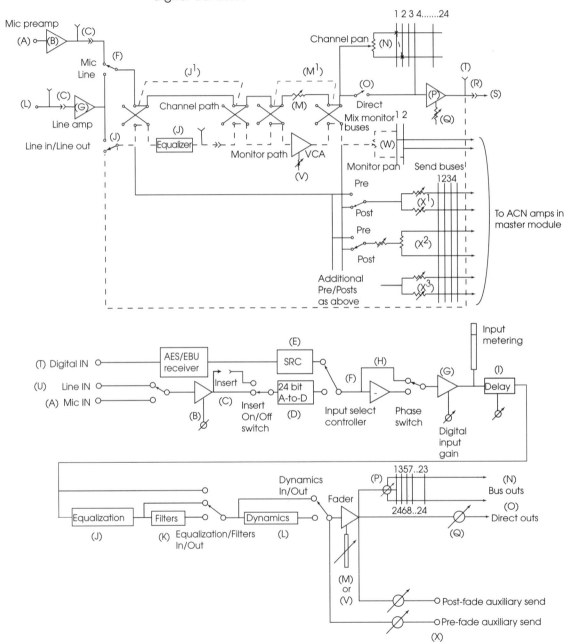

Instead of the traditional channel/monitor path configuration of the analog console, the digital system simply places another I/O module above the first one. This way, the more familiar channel/monitor path can be emulated, or a different style of console, such as a split configuration, can be easily implemented.

A) Microphone Input Plug

At the console, each arriving input signal line is generally assigned to a specific input, more or less permanently. Each microphone input on the studio console is numbered for identification (Figure 16-5). When the console is installed, the microphone lines may be soldered in place, making later changes difficult. Another approach is to solder the microphone lines to various multi-pin connectors such as "Tuchel" or "CPC" connectors. In other consoles, designed for quick changes or portability, the connections may be made through XLR plugs (sometimes called Canon plugs).

B) Microphone Pre-Amplifier

As discussed in Chapter 3, microphone output levels are quite low, compared to tape recorder or line levels, and therefore require amplification before any signal processing can take place. A microphone pre-amplifier just after the input connection boosts the signal to the console's internal line level. This is usually between -2dBm and +4dBm.

Just prior to the input of the microphone pre-amplifier is where the phantom power bus is applied. A switch on the I/O module is

Figure 16.5 Microphone input connections

used to turn 48 volts DC off or on as required by the microphone. The precision 6.8K ohm dropping resistors are tied to pins two and three of the input connection and then terminated to an appropriate grounding point.

C) Patch Points On many consoles, the microphone pre-amplifier is immediately followed by a set of patch points which are physically located in the console's jack or patch bay. These insert points may also show up on an outboard patch panel, that is, one not in the console, and there may be a switch on the console to implement the insert point. These points allow the engineer to insert external (outboard) processing devices into the signal path, or to reroute signals for special applications.

D) and E) A/D Converter and SRC The next two sections are not found on the analog console. The first (D) is an analog-to-digital converter. It is here that the signal, now at line level, is converted into the digital domain. From here on the audio signal stays in digital form. The SRC (E), or sample rate converter, takes a digital signal from any source and reclocks it to the console's internal data or sample rate if required. It is advisable to switch this out if the incoming sample rate is the same as the console's master rate as additional processing is unnecessary.

F) Input Select Controller or Mic/Line Switch This switch or selector allows the engineer to select what type of input is sent to the I/O module. Only one input can be sent through an I/O module at any one time. In the traditional analog console, the choice is usually between a microphone-level signal or a line-level signal. In some cases, the switch may select between two lines and/or two microphones. The digital console often offers a third choice, that of a digital input signal. Since it is already in digital form, it does not need to pass through the analog-to-digital converter, but may need to be routed through the SRC. There may be a listed set of digital input sockets that can be selected as the source for the selected I/O module.

G) Line-Level Gain Control Here, the gain of the post mic/line switch can be adjusted to compensate for various types of sources. Adjustments can be made to bring consumer-level equipment, electric musical instrument levels or broadcast levels to the same amplitude. In the digital console this is usually done in the digital domain after the analog-to-digital converter. There may also be a phase reversal switch associated with this control, but it is more commonly found nearer the equalization section.

H) Phase Reverse Switch

This is a section where the digital signal can have the phase reversed prior to any further processing or signal manipulation.

I) Delay Section

This short delay line is only found in the digital console. One of the issues the engineer needs to be aware of when using a digital console is what is known as propagation delay. This occurs when a signal that is already in the digital domain needs to be converted back to analog and then again to digital form. For example, the engineer may decide to use an outboard analog processor because of its unique sound. Here, the digital signal is converted to analog as it leaves the digital console and goes to the external analog processor and then is converted back to digital when returned to the console to be mixed in with the original. The additional set of converters may delay the processed signal enough so that when combined with the original, unintended comb filtering occurs. By delaying the original signal a small amount (typically in the 6ms to 10ms range) prior to combining it with the processed version, this distortion can be avoided.

J) Console Equalizers and Channel/Monitor Equalization Switch

The equalizers seen in Figure 16-3 are typical of the types found and built into each individual I/O module. For a detailed discussion of equalizers, refer to Chapter 7.

Note that in the simplified analog block diagram, Figure 16-4, a pair of ganged double throw switches (J^1) on either side of the equalizer is shown. This permits the signal to bypass the equalizer. Typically, switching is controlled by a single pushbutton, labeled CHANNEL/MONITOR. This pushbutton switches the equalizer either into the channel signal path or into the monitor signal path depending on the preference of the engineer and the requirements at the time. Here again, since the digital console usually has a stacked input arrangement, the lower control can be used as the monitor fader and the top one used as the channel fader, each having its own equalization section.

K) Filters Switch Many consoles have, just prior to or just after the equalization block, a set of high-pass and low-pass filters. This is shown as (K) on the block diagram of the digital console (Figure 16-4b). They will also have a channel/monitor pushbutton to allow them to be switched, independent of the equalization, into either the monitor path or the channel path, or will have a separate set of switches for each path.

The phase reversal switch is also usually found here, and follows the channel/monitor switching of the equalizer. This electrically reverses the phase of the signal causing a 180° change.

L) Dynamics Section Both analog and digital consoles may have a dedicated dynamics section in each I/O module. This will usually include compressors and expander/gates, and almost always follows the equalization section. This way, changes in equalization that result in level changes can be caught by subsequent dynamic processing.

M) Input Level Control and Fader Exchange Switch In the typical multi-track recording session, fine adjustments of the channel fader are not required. Usually, the channel level is set so that the playing reaches or just slightly exceeds 0VU on the analog multi-track tape recorder or approaches 0dBfs on a digital device. This produces maximum signal-to-noise ratios on all tracks, thereby minimizing the cumulative effects of tape noise or quantization noise in mixdown. Whichever storage system is used, after the initial level is set, there is little need for change. Consequently, another pair of double-throw switches (M^1) routes the input signal through a simple gain-adjust device. As noted earlier, in the analog console this may be a simple rotary potentiometer, or perhaps a linear fader with a shorter throw than the customary fader which will be described later on. As with the equalizer section, a single control performs the necessary switching, allowing either the primary fader or the rotary or shorter fader to be placed in the channel path while the other is placed in the monitor path. The primary fader can be used as both the channel path and the monitor path fader in many digital consoles simply by toggling one function or the other, while in others there may be two full-length controls.

N) Channel Assignment Switches (Output Bus Selector Switches)

After each input signal has been suitably processed, it may be routed to one or more outputs by depressing the appropriate bus selector switches, or several input signals may be combined and sent to one output. For example, the engineer may wish to use a microphone and a direct box on an electric guitar, mixing their outputs together to form one composite signal which can be routed, for instance, to track number 2 on the tape recorder. By depressing the channel 2 selector switch on both input modules, the two signals are sent to channel 2's ACN where the signals are summed.

As a space-saving consideration, channel assignment numbers are usually laid out in two vertical rows. Depending on the particular design, there may be a single button for each odd-even channel pair, or a separate button for each channel. The virtual digital console may have an assignment section where various inputs or sockets are configured with the name of their respective tape recorders or storage devices during installation. Opening a screen or configuration file lets the engineer assign the signal to the proper output. Various examples of channel assignment switches are shown in Figure 16-6. In each case, a nearby pan-pot usually allows for panning the signal between the selected channels.

Often, in a digital console, there will be what is called an output routing pool which is made up from a number of hardwired input sockets that are soft assignable to the output of an I/O module. This will be discussed further in the section on console architecture.

O) Direct Assignment Switch

In the typical one-on-one recording setup, there is little need for the traditional mixing bus since each sound source or microphone input will be directly routed to a separate channel on the multi-track tape recorder. Therefore, rather than going through the exercise of assigning input 3 to channel bus 3, etc., a single direct-assignment switch on each module accomplishes the same thing. The direct switch usually bypasses the channel assignment switches and their respective ACN amplifiers. This reduces the length and complexity of the signal path and thereby reduces the channel noise level. One of the advantages of a digital console is that the signal passing through the I/O module can be routed directly to many different output channels simultaneously with no increase in channel noise. So, for instance, a microphone can easily be assigned to track 5 on the multi-track recorder, tracks 2 and 3 of the DAW, and simultaneously to a backup archival system. A digital direct out assignment panel is shown in Figure 16-6b.

Figure 16.6 Channel assignment switches are usually found at the top of the input/output module. Typical arrangements may be: a. a single row of switches for odd/even pairs of channels; b. each channel has its own individual switch, and c. a soft assign section for both buses and direct outs.

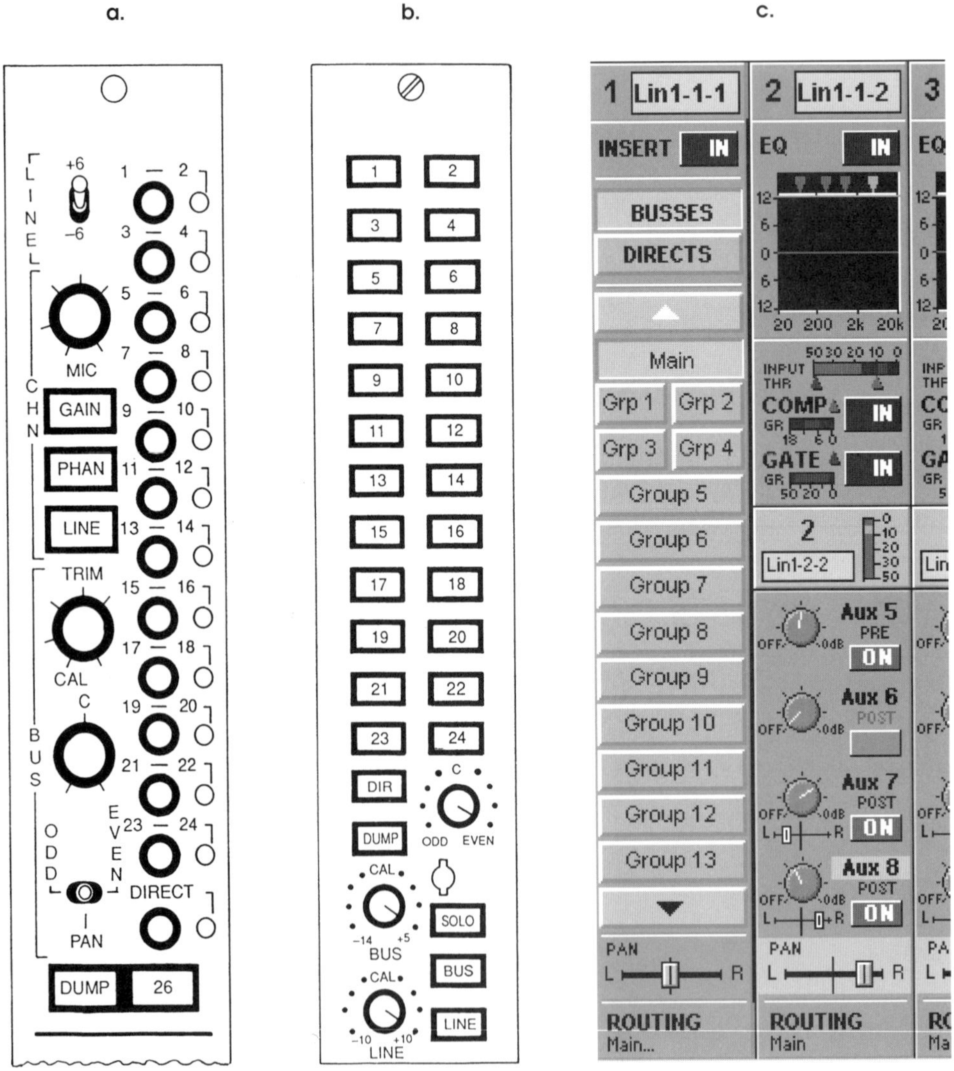

P) Channel Active Combining Network Amplifier

At this amplifier, all the input signals that have been assigned to the channel output bus are combined into one composite output signal. There is a similar amplifier for each output channel on the console.

Q) Channel Trim (Output Bus Level Control)

It used to be customary (and still is in split-type designed consoles) for each output channel to have its own linear fader for over-all gain riding of the signals which had previously been combined. Given the nature of present multi-track recording practice, that is, keeping all tracks separate until mixdown, a full-scale fader at this point in the signal path may now be quite unnecessary. Therefore, a simple rotary channel trim potentiometer may be commonly found instead. This is certainly adequate for making coarse adjustments in channel gain, and may even be used for a certain amount of simple gain riding during the recording session. However, when more demanding gain riding is required, a grouping function may be used, and this will be discussed later in the chapter. On many modern consoles, the direct assignment switch bypasses the channel trim control as well.

R) Channel Out (Output Bus) Patch Points

Prior to the output plug that is sent to the multi-track tape recorder, there may be a set of patch points or insert points similar to those described at C. An external signal processing device inserted here will affect all of the individual signals that were assigned to that bus. In a digital console, it is important for the engineer to know whether these insert points are before or after the digital-to-analog converters.

S) Channel Output Plugs

At this point, the signal path leaves the console and goes to a specific track on the multi-track tape recorder. Once again, a variety of output plug or assignment configurations are available.

T) Line Input Plug

Here, the tape recorder output is returned to the console. Although this signal will actually be routed to several points within the console, at this time only the monitoring path is shown as the dotted line from (T) to (J) in Figure 16-4. In a digital stacked input console the tape recorder output is usually assigned to the second input fader in the I/O strip, then assigned to the main mixdown bus.

- **Master Line-In/Line-Out Monitor Switch**

Many consoles, both analog and digital, may have a switch similar to this, particularly where there are no dual I/O modules. Depending on the position of this switch, the engineer may monitor either the console line-out (channel out), or the line-in signal from the tape recorder. This single control will act as a master line-in/line-out switch so that the entire console monitor system can be switched between the two. There are also usually line-in/line-out switches on the individual I/O modules, but the master switch always takes precedence. This switch is sometimes referred to as the tracking/mixdown monitor switch.

U) Master Line-In/Line-Out Switch

The line-in/line-out switch just described is a master control typically found on the monitor module, since it selects the mode which is to be monitored. When activated, it switches all I/O modules simultaneously.

However, it is often desirable to be able to have some modules in one mode and to switch one or more individual modules to the opposite mode. That is, when the console is switched to monitor the multi-track tape recorder outputs, a few selected modules may need to monitor the channel outputs, and vice versa. A line-in/line-out reverse switch on the I/O module is usually provided for this purpose.

On the analog block diagram, note that when the switch is in the normal position (as shown), the module mode agrees with the master line-in/line-out status that was selected on the monitor module. However, when the switch is in the reverse position, the individual module mode is always the opposite of the master status. This function may be a convenience when overdubbing in the sel-sync mode on a multi-track tape recorder that is not set up for automatic input/output (monitor) switching.

With the console monitoring the tape recorder outputs, those I/O modules that are being used as microphone lines for the overdub parts are put into the reverse mode so that the operator can hear the microphone outputs as well as the previously recorded material.

On the digital console being described, there is no master line-in/line-out switch. It is simply a matter of selecting either the recorder returns (often the upper row of faders) or the mixdown monitor path, which is most often the lower row.

V) Monitor VCA Fader

As already noted, the familiar linear fader is now more often found within the monitor lines, or identical faders are found in both places, as indicated in Figure 16-4. Furthermore, the actual gain-riding action is frequently regulated by a VCA (Voltage Controlled Amplifier) within the audio signal path. As before, the gain is determined by the fader position, but now the signal passing through the fader is a direct-current control voltage whose amplitude determines the gain of the amplifier. A motorized fader may be used in place of the VCA, and now the DC voltage controls the fader position and the audio passes through the fader if analog, or, if in the digital domain, the fader controls the gain of the bit stream. This VCA control or motorized fader system allows great flexibility in mixing and is the basis for automated mixdown, which will be discussed in detail in Chapter 17.

As described earlier, the function of this linear fader can often be interchanged with that of the rotary or shorter channel fader by means of a fader reverse switch.

W) Monitor Pan-Pot

In the earliest days of multi-track recording, it was not uncommon to find four monitor loudspeakers at the front of the control room. This was a more-or-less reasonable culmination of a practice that began still earlier as mono gave way to stereo. As the number of available recorder channels doubled, so did the number of speakers. As three-, and very soon thereafter, four-track recording was introduced, the number of loudspeakers kept pace. Fortunately, common sense prevailed as the industry progressed from four to eight and then to 24 or more channels, and the number of speakers was held at four. Many recent control room designs favor the two-channel monitor system, which is of course a more accurate representation of what the consumer will eventually hear at home. But, with the advent of surround sound in film, video, and audio-only media, a return to multi-loudspeaker monitoring is already underway.

Accordingly, on an in-line console, a monitor pan-pot or joystick will be found that sends the monitor signal from its respective I/O module to either the left or right monitor bus, or to a surround bus to include any proportion in between. Surround can be found in three basic varieties at this time. The most common is the 5.1 system which contains 5 loudspeakers and a .1 (dot-one) or subwoofer channel. The subwoofer channel is not really intended for music. It is specifically for effects, and as such is often called the LFE (low frequency effects) channel. The word pan-pot actually stands for panoramic potentiometer, which is simply a control made up of two or more concentric rotary potentiometers, with opposite tapers and multiple outputs, mounted on one shaft. As the level is raised on one, it is lowered by the same amount on the other. When the control is placed in the center, equal signal is sent to both the left and right buses or equally to all 5 surround channels. Since the monitor bus is providing the signal for the mixdown as well as for the monitor system, the monitor pan-pot assigns the signal to the respective channels of the mixdown tape recorder as well.

When in the surround mode, the console may also have several controls to supplement the panning. These controls are the Sub control and the Divergence control. The Sub control determines the amount of signal that is sent from the I/O module to the LFE channel of a surround bus or output. The Divergence control limits the width effect of the Pan control when sending to surround buses. Divergence of 0 allows full right/left, front/back panning while a setting of maximum prevents any panning away from the center of the panned image.

- **Left and Right ACN or Bus Summing Amplifiers** It is at this point that all signals to be routed to the monitor buses are combined. As mentioned previously, the ACN amplifier outputs feed the various mixdown tape recorders as well as the monitor system.

- **Speaker Level Control** This potentiometer functions as a master gain control for the combined signal fed to the loudspeaker systems. As discussed before, there will be separate controls for control room and studio loudspeaker systems.

- **Master Audio Fader** The Master Audio Fader at the outputs of the mix bus ACN or summing amplifiers regulates the signal level that is fed to all mixdown recorders, as well as to the speaker level control just described.

- **Monitor Output Plugs** At this point, the monitor signal leaves the console and goes to the power amplifier(s), and then to the monitor loudspeaker systems.

X) Auxiliary Send Controls These controls (Figure 16-4b), usually rotary potentiometers on an analog console, are found in the I/O module or on the virtual control screen assignment section of a digital console. They allow the engineer to feed some portion of any input signal to an auxiliary bus that is summed at the master section. These buses are then sent to various outboard devices such as reverberators or other processing equipment. The auxiliary bus outputs are also often used to feed the headphone cue system. As in earlier designs, where these buses may have been called Echo Sends, these lines may be fed from before (pre) or after (post) either the channel fader or the fader used for monitoring. These auxiliary buses are often used for multiple purposes. For example, the same send line can be used with the cue system during recording and as a reverberation send during mixdown, or it may be used at any time to feed some external device such as a digital delay line.

In Figure 16-4a, the *pre* position picks up the signal before the equalizer in the monitor path. Several alternative send systems are also shown on the analog console. In the first (X_1), separate send-level controls are permanently assigned to send buses 1 and 2. Alternatively, there may be a master send level plus a pan-pot (X_2). As another method (X_3), there may be separate send levels plus pushbutton (or toggle) switches to select any of the available send buses. Depending on the specific console design, between four and eight of these systems may be found. Within the digital console, the number and type of auxiliary buses are often determined in the console set-up process. Many of these consoles can have as many as 20 auxiliary buses, depending on available resources.

A detailed drawing of a typical auxiliary send system for an in-line console is shown in Figure 16-7. Note that the signal can be picked up from a point before or after the channel rotary fader (E) or before or after the monitor linear fader (N) depending on the positions of the pre/post switch and the channel/monitor switch. As shown, the signal to be sent to the auxiliary bus must first pass through the channel fader. Consequently, the fader's position influences the auxiliary send signal level. On the other hand, if the selector switch is in the pre-fader position, the signal level to the auxiliary bus becomes independent of the position of the channel fader.

There are at least two applications for a pre-fader send position. If the channel fader is brought way down to attenuate a signal going to the output bus ACN or summing amplifiers and assignment switches, there may not be sufficient level after the fader to provide the send level that is required. In the pre-fader position, a full-level signal is available at the auxiliary send line, regardless of the input fader.

As a second application, when the auxiliary bus is being used as a reverberation send, a satisfactory blend of direct and artificial reverberant sound may be established with the monitor fader at some average level position. Now, as the fader is raised, the direct sound level increases, while the reverberant sound changes, and as

Figure 16.7 A simple block diagram for an auxiliary send system.

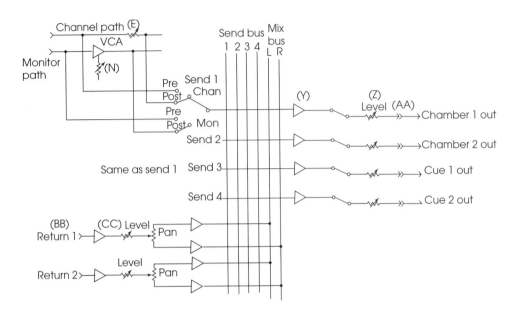

the direct sound becomes louder, it seems to move closer to the listener. On the other hand, as the fader is brought down, the direct signal fades out, leaving only the reverberation, so the signal seems to recede into the distance.

Often, one or more pairs of auxiliary buses will be normalled to the reverberation devices(s) while others are reserved for alternate uses. On the other hand, within the digital system, the pool of input and output sockets can be soft assigned to the auxiliary bus.

Y) Auxiliary Send ACN or Summing Bus Amplifier This amplifier combines all the signals that are routed to that particular auxiliary send bus. Provisions are also often made to send the monitor mix bus to the auxiliary send ACN summing amplifier that is normally used to feed the headphone cue system. This allows the engineer to quickly send what is being heard in the control room to the musicians.

Z) Master Auxiliary Send Level

A rotary potentiometer that regulates the level of the combined auxiliary send bus may usually be found next. Any auxiliary bus metering is usually derived directly after this control. In the digital system this again may be a rotary potentiometer or fader that can be used for other purposes, such as grouping, depending on which function is needed at the time.

AA) Auxiliary Send Plug

Here the auxiliary send signal leaves the console, going by way of the patch bay or output assignment pool to the input of the selected reverberator or other device.

BB) Auxiliary Return Plug

This is a dedicated series of auxiliary inputs, designed to accept the return signal from the external reverberation system or other processing equipment and then routed to the summing amplifier of the mixdown bus. When, as in many digital systems, there are no dedicated aux returns, one of the many additional input channels can be used, and assigned back to the mixdown bus.

CC) Auxiliary Return Level

Since the external system is not necessarily a unity gain device, this potentiometer and associated pan-pot give the engineer control over the output level and placement of the signal returned to the console.

As will be discussed in Chapter 17, it is generally best to record the basic tracks "dry," that is, without any artificial reverberation. Reverberation will be added later on during the mixdown. Consequently, the auxiliary return buses are fed to the monitor/mixdown bus only. It is possible to return effects or reverberation to the channel outputs that go to the multi-track tape recorder, if necessary, by patching the effects return to a spare I/O module and assigning that module to the desired channel ACN bus. However, precautions must be taken to prevent a feedback loop when recording reverberation on the same channel with the direct signal that is feeding the reverberation device.

It is often possible and desirable to return reverberation to the auxiliary sends that are being used to feed the headphone cue system. This way, the musicians in the studio can hear the headphone mix with or without reverberation as needed. Often, reverberation is required so that the musicians can hear the proper blending of instruments that they are accompanying.

■ SUMMARY OF SIGNAL FLOW PATH—ANALOG CONSOLE

The microphone input signal is boosted to line level, after which a pair of patch points permit the insertion of external signal processing devices. The signal follows the channel path through the console and directly out to the associated channel on the multi-track tape recorder. Alternatively, the signal may be sent to one or more of the channel buses for routing/mixing to any channel out.

The signal is then returned to the front end of the console where it is routed along the monitor path through the equalizer and linear fader and then to the monitor loudspeakers and the 2-mix bus. Pre (equalizer) and post (fader) lines route the signal to one or more of the auxiliary send buses, as desired.

A) Microphone Inputs, L) Line Inputs, F) Mic/Line Selector Switch

So far, the recording console has been discussed as a routing and processing point between microphones and a tape recorder. A secondary, yet no less important, function is as a control point between one tape recorder and another. For example, after a multi-track tape has been completely recorded, it must be mixed down to two tracks if a stereo master tape is needed, as is usually the case.

So, instead of the console ACN channel outputs, the multi-track tape recorder outputs are routed to the input of the console monitor section at point (J) in Figure 16-4. The multi-track tape recorder is normalled to these (L) line input plugs, that may be found near or adjacent to the microphone input plugs.

The line-in/line-out switch determines the input to the console monitor section, and will be discussed in greater detail later on. The mic/line switch (F) on the I/O module determines whether a line-level source or the output of the microphone pre-amplifier is applied to the channel input. When in mixdown, the mic/line switch also allows the channel paths to be used as additional auxiliary sends.

C) Patch Points

A detailed drawing of the signal path through some of the patch points in the recording console patch bay is shown in Figure 16-8. Note that the act of inserting a patch cord into patch point (1) allows this signal to be routed elsewhere without interrupting the normal signal flow. However, insertion of a jack at patch point (2) does interrupt the circuit, allowing the path through some external device to take the place of the normal signal flow. This wiring convention, although by no means standard practice, allows the engineer considerable flexibility in changing signal routing paths to meet the needs of the recording session.

Foldback (Cue) System

When the studio musicians are acoustically isolated from each other, the engineer must be able to feed a well-balanced program into a headphone system so that each musician will be able to hear what the others are doing. Previously recorded tracks must also be sent to the headphone lines so that the musicians may play along in

Figure 16.8 a. Patch points not in use—no interruption of regular signal path; b. Patch points in use—the regular signal path is interrupted at patch point 2. Note that a meter could be inserted at patch point 1 without affecting the normal signal path; c. Detail of the flow path through the patch bay when an external device is inserted in a signal path.

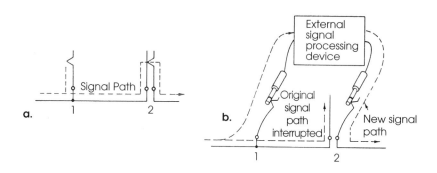

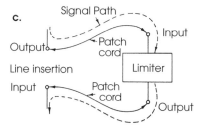

accompaniment. The auxiliary send buses are used for this purpose, with usually one particular pair normalled to the cue system. Often the dedicated cue send buses mentioned earlier can access the monitor mix as well.

Mute (Channel Cut Switch) The mute or channel cut switch is a simple on/off switch that allows the engineer to remove the signal flow entirely without disturbing any of the level or auxiliary controls. This facility is particularly useful for briefly shutting off a microphone or tape track that must later be turned back on at the same relative setting. When the channel is grouped (discussed later), this switch may mute all signals assigned to that group.

Solo Function At times, it may be desirable to listen briefly to the output of one microphone only, in order to check its performance or placement. The solo function is one of the methods available that allows the engineer to do this without affecting the recording in any way. From each I/O module, a solo send line is permanently routed through the solo switch to the solo combining amplifiers. When the switch is depressed, the input signals reach the combining amplifiers. Another set of contacts on the same switch energize a relay which routes the output of the solo combining amplifiers to one of the control room amplifiers in place of the signal regularly assigned to it. At the same time, as shown in Figure 16-9, the signal path to any other control room speakers is interrupted so that the solo signal alone is heard. Two or more solo buttons may be depressed at once, and the appropriate inputs will be combined and routed to the solo circuit.

Most of today's modern in-line consoles furnish the engineer with a stereo non-destructive solo bus. The solo signal is derived post-fader, post-pan but pre-mute switch. This allows the engineer to monitor the signal at the selected level with proper panning and with equalization and effects if selected. Earlier consoles made use of a mono solo bus only with the solo signal going equally to both control room monitor speakers. There is usually a solo switch for both the channel path and the monitor path on each I/O module. This allows the engineer to solo the channel ACN bus out, or to just hear the individual channel signal.

Before stereo non-destructive solo became commonplace, an earlier version used the VCA controls on the monitor circuit to achieve this effect. Depressing the solo-in-place switch for any channel would mute the VCAs for the remaining channels, allowing the operator to solo the input signal with proper panning and equalization, along with its reverberation return, if any. (Other sends to the reverberation lines were muted along with the channels themselves.) Note that the solo-in-place function could not be used during recording since it performs its function by muting all other channels. For solo during recording the regular solo button on each I/O module would be used. This solo-in-place function is still retained in many consoles, and its implementation is relegated to the automation software, which will be discussed in Chapter 17.

Pre-Fade Listen Another way for the engineer to listen briefly to an incoming signal is with the pre-fade listen circuit. It is similar in operation to the solo circuit, except that the signal is derived just after the mic/line switch. This allows the engineer to hear the quality of the incoming signal prior to any level control, equalization, or processing. The pre-fade listen circuit is strictly a monophonic bus, and often the signal is sent to a small monitor speaker built into the console meter bridge or to just the left control room monitor while the other speakers are muted.

Figure 16.9 The solo function. When any solo button is depressed, a relay interrupts the normal signal path to the monitor switches, and the solo signal only is sent to one path of the control room.

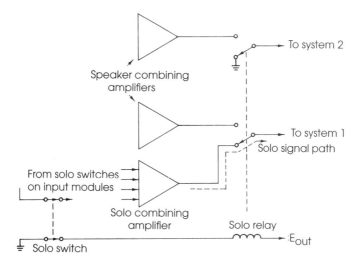

Output Bus Selector Switches and Pan-Pots

As described earlier, the bus selector switches allow the engineer to route an input signal to one or more of the channel output buses. A somewhat more involved switching system will permit the input signal to be routed to two or more buses in unequal proportions. A pan-pot (panoramic potentiometer) is used, and is shown in Figure 16-10a. Note that there are now two sets of bus selector switches, and that bus number 1 has been selected in one set and bus number 4 in the other. The pan-pot wiper arm is closer to the lower set of switches, so although the signal is routed to both bus 1 and bus 4, the power distribution is unequal, with bus number 4 being favored somewhat. Assuming that these buses are monitored on separate speakers, for example, number 1 on the left and number 4 on the right, the position of the pan-pot will determine the apparent location of the signal. As shown, the signal would appear to be slightly right of center.

As mentioned earlier, the pan-pot is really two concentric but reverse gain controls. It is designed so that at its midway position, the signal to both sides is attenuated by 3dB, as compared to the level if the signal is routed to one side only. Since a 3dB drop represents a halving of power, both buses add up to the same amount of power as when the signal is routed to one side only, and therefore up 3dB. Thus, if the signal is gradually panned from one bus to another, there is no change in apparent level as the sound source moves from one loudspeaker to another. Without the 3dB drop feature, a signal would get louder as it moved to the center.

When only one output bus is selected, the pan-pot is usually bypassed in the circuit regardless of its position. Only when two or more buses are selected does it come into play.

Direct Outs

As just described, the bus selector switching system with pan-pots offers great flexibility in mixing input signals to one or more channel ACN amplifiers. However, when a single microphone is to be routed to one track on a tape recorder, there is little point in going through such a switching system if a more direct route to the tape recorder is available.

In perhaps the majority of recording sessions, very few mixing buses are required at one time. For example, although the drums, string section, and chorus may be recorded using several microphones each, perhaps everything else is done with just one microphone per recorded track. Therefore, a complete mixing bus for each track is not necessary; in fact, the extra circuitry lowers the signal-to-noise ratio of that channel.

As shown in Figure 16-10b, there may be a direct out plug after every channel fader. By way of these plugs, input 1 is routed directly to tape recorder track 1, and so on. A single microphone, to be routed to track number 7, for example, is simply plugged into console input number 7 at the microphone panel in the studio. In this case, the mixing bus assignments are not used. The buses are only used when several microphones are to be combined onto one track. Then, a convenient output bus is selected and the output of its

Figure 16.10 a. Detail of a flexible bus selector/pan-pot switching system, showing additional direct output facility; b. Graph of relative power delivered to buses 1 and 2 by pan-pot. At any position, $P_1 + P_2 = 0$ dB.

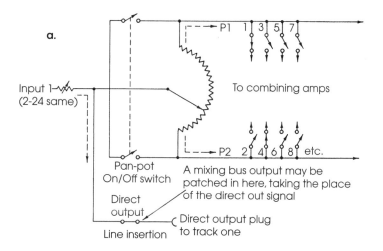

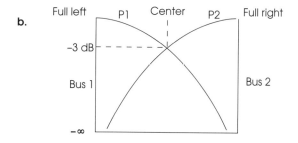

combining amplifier is routed to the appropriate track on the tape recorder. When the direct switch on an I/O module is selected, that channel's ACN is disconnected from the output plug and the direct out is substituted.

Combining Amplifiers

As illustrated in Figure 16-11, the combining amplifier has many inputs, one for each I/O module in the console, and each auxiliary, solo, PFL and monitor channel requires one as well. These multiple inputs are usually in the form of a large number of resistors, as shown in Figure 16-11a. The resistors prevent a group of inputs that are routed to one output from shorting each other out and, in the case of inputs routed to more than one output, keep the outputs isolated from each other.

In many consoles, these resistors are found in the input section of the various I/O modules, as seen in Figure 16-11b. The resistors are wired to the output bus lines, and each bus is routed to a separate combining amplifier, which in this application may be called a line amplifier.

Broadcast Mode

From time to time, it may be necessary to record a multi-channel program and a stereo version simultaneously. This can certainly be done without any modification to the console as presently described. Of course, gain adjustments made in the channel path will alter the stereo mix as well, as is desirable during the typical multi-channel-only recording.

However, in the case of a simulcast stereo broadcast/multi-track recording, for example, the integrity of the stereo mixdown (now being broadcast as well) must not be compromised by multi-track recording gain adjustments made while the broadcast is in progress.

The Broadcast switch seen in figure 16-12 permits the operation just described. Note that when the switch is in the Broadcast position, both the channel and the monitor paths are directly fed from a common point (the mic/line switch), regardless of the position of the channel/monitor switch (D). However, the channel and monitor paths are now completely independent of each other so that adjustment to the multi-channel buses will have no effect on the stereo mix. Note that while in the Broadcast mode it is not possible to monitor the channel outputs.

The Broadcast mode is also extremely useful when recording direct to two. This is when any number of microphones are mixed directly to a two-channel tape recorder. This is usually the way that classical music is recorded due to the nature of the music. The

Figure 16.11 a. A typical combining amplifier; b. In many consoles, each input module is connected to each output bus as shown here.

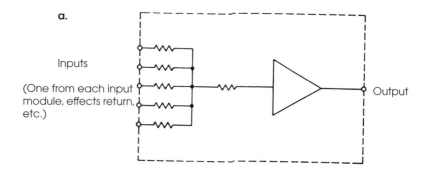

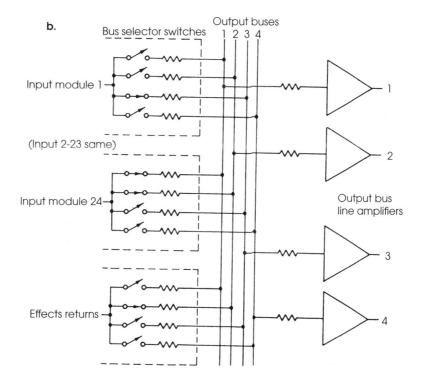

ensemble required for this music is best maintained by recording all parts simultaneously, with subtle balance changes made by the engineer as the music is recorded directly onto the two-channel tape recorder.

The Broadcast mode permits the engineer the full use of the filters, equalization, auxiliary buses, etc. while mixing the signals from the microphones onto the monitor bus. The channel buses are not used during this process.

Grouping When signal levels are under VCA control, it is relatively easy to use a single fader to control the levels of a group of several signals. Simply stated, the direct-current control voltage from the fader is used

Figure 16.12 Detail drawing, showing Channel Reverse and Broadcast Mode switches. In practice, the Line, Channel, and Reverse modes are usually accomplished by FET switching. (New circuit details are indicated by heavy lines.)

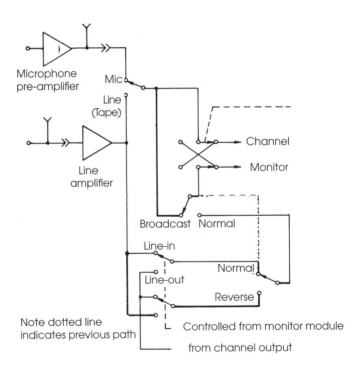

to control all the VCAs within a selected group. With the exception of this group control of levels, the signals remain completely isolated from each other. Thus, the signals passing through modules 3, 5, and 23, for example, may be controlled with (if you like) fader 24. If it happens that these signals are to be combined onto a single channel, then fader 24 may be considered a temporary group master channel control, obviating the need for a permanently assigned channel level control to perform this function. In other words, for simple level adjustments to the channel, the regular channel trim pot may be used. When more extensive control is required, the grouping mode may be used.

Of course, the modules just mentioned may just as easily feed separate channels, with fader 24 continuing to serve as a master level control for the group. The aforementioned situation actually parallels the function of the channel submaster on a split-type console. However, when the inputs are assigned to varying channels, VCA grouping is required.

Group-Select Switch

In a typical implementation of grouping, a multi-position thumbwheel group-select switch or software assignable grouping control will be located near each channel fader, as seen in Figure 16-13. To assign several faders to Group 3, for instance, simply select Group 3 on the appropriate thumbwheel or pushbutton switch. Next, determine which of these faders is to be the group master by depressing the master switch usually found quite near the group assignment switch. The fader so designated will function as a group level control for all faders assigned to Group 3. Each of the faders within the group may also be independently adjusted. In many applications, the group master may be any fader that is conveniently located and otherwise unassigned, or it may actually be one of the faders within the group, thereby controlling its own signal and the group's as well.

Some consoles have a separate set of faders (typically three to eight) permanently assigned as group masters. In this case, there may be no master buttons near each channel's group select switch. In the case of a digital console, there may be a panel or menu that can be called up on the screen, and where the individual channel path gain controls can be assigned to a master controller.

A third option places the grouping function under the control of the console's automation software, thereby doing away with all switches, thumbwheels, separate group masters and the like. This function will be discussed in detail in Chapter 17 on console automation.

Figure 16.13 Monitor fader modules showing (a) rotary switch and (b) software assignable grouping. Also seen are some automation controls and indicators; c. a control group assign panel with touch screen control. When the groups are selected, the faders found below the screen are the group controllers and recall their previous settings. When de-selected, the same faders become auxiliary masters, bus masters, or whatever else is selected at the time, with the respective settings previously used.

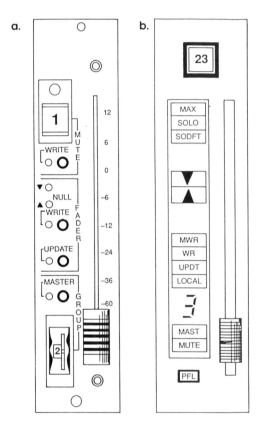

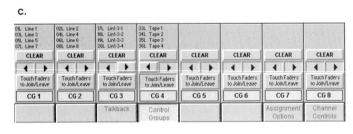

DD) Monitor Pan to Channel Buses (Dump)

From time to time, the engineer may want to record a microphone group as an equalized stereo mix on two of the multi-channel buses, while simultaneously recording each microphone separately on other individual channels. For example, consider a five-microphone drum pickup. The five microphones are to be recorded on tracks 7, 8, 9, 10 and 11, with a stereo mix going to channels 3 and 4. In this case, the microphones must be plugged into inputs, and routed directly (O) to the corresponding channel outputs, thus freeing the channel buses (for the moment).

Each microphone is monitored in the usual manner, as described earlier. In addition, the monitor pan-to-channel or Dump switch (DD) on each I/O module allows this monitor feed to also be routed to the channel bus selector, as shown by the drawing in Figure 16-14. Now, by selecting channel buses 3 and 4 on each of the five I/O modules, a stereo mix of the drum microphones is recorded on these two channels. Of course, the stereo mix-only could be recorded simply by not assigning the five inputs directly (O) to any channel outputs, or the five previously recorded tracks could be combined or "bounced" to channels 3 and 4 by using the dump switch.

EE) Monitor Pan to Auxiliary Send Buses

A similar switching system, also seen in Figure 16-14, allows each I/O module's monitor feed to be sent to the (cue) send buses as well. This may prove to be convenient during particularly complicated recording sessions, since it obviates the need to take the time to establish a separate cue mix.

FF) Monitor Mix (2-Mix) to Cue

A similar function may be available just after the master fader. This allows the entire stereo mix to be routed into the cue system. In this case, the cue system might also contain more (but never less) of a certain signal by continuing to feed the appropriate I/O module to the cue lines (as described earlier), in addition to the stereo mix cue feed.

Summary of the Analog In-Line Console

The console described in this chapter is representative of a typical in-line multi-track recording console. It is important to remember that the specific nomenclature, physical appearance and relative position of each function will vary from one manufacturer to another, and even among different models from the same manufacturer.

Figure 16.14 Detail drawing showing monitor pan-to-channel buses and monitor pan-to-auxiliary send buses.

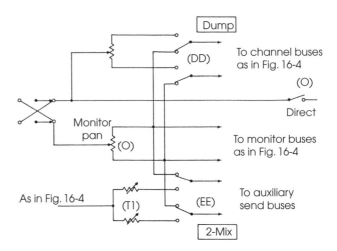

Furthermore, not every function described here will appear on every console.

For the less-experienced engineer who may have learned on the simpler split-configuration type of console, the in-line design will take at least a little "getting used to." The primary difficulty may be in becoming familiar with the monitor system. For example, although I/O module 3 always represents microphone 3 in the studio and tape track 3 in the control room, the two may not otherwise be related. If the channel bus selectors are in use, perhaps microphone 3 is being routed to track 7, for instance, while microphone 18 feeds track 3. In this case, I/O module 3 is used to monitor microphone 18, while microphone 3 may be heard over I/O module 7. Even when one remembers that the monitor linear fader is controlling its respective channel signal and not necessarily the microphone plugged into that I/O module, the situation can be challenging.

Such potentially confusing situations may be minimized, if not eliminated completely, by a judicious mixture of pre-session planning of microphone assignments and some re-routing at the patch bay, as required. Figure 16-15 shows the main block diagram or flow chart for a professional in-line analog console. Compare it to the simplified block in Figure 16-4.

Figure 16.15 Block diagram for typical I/O, master, and monitor modules.

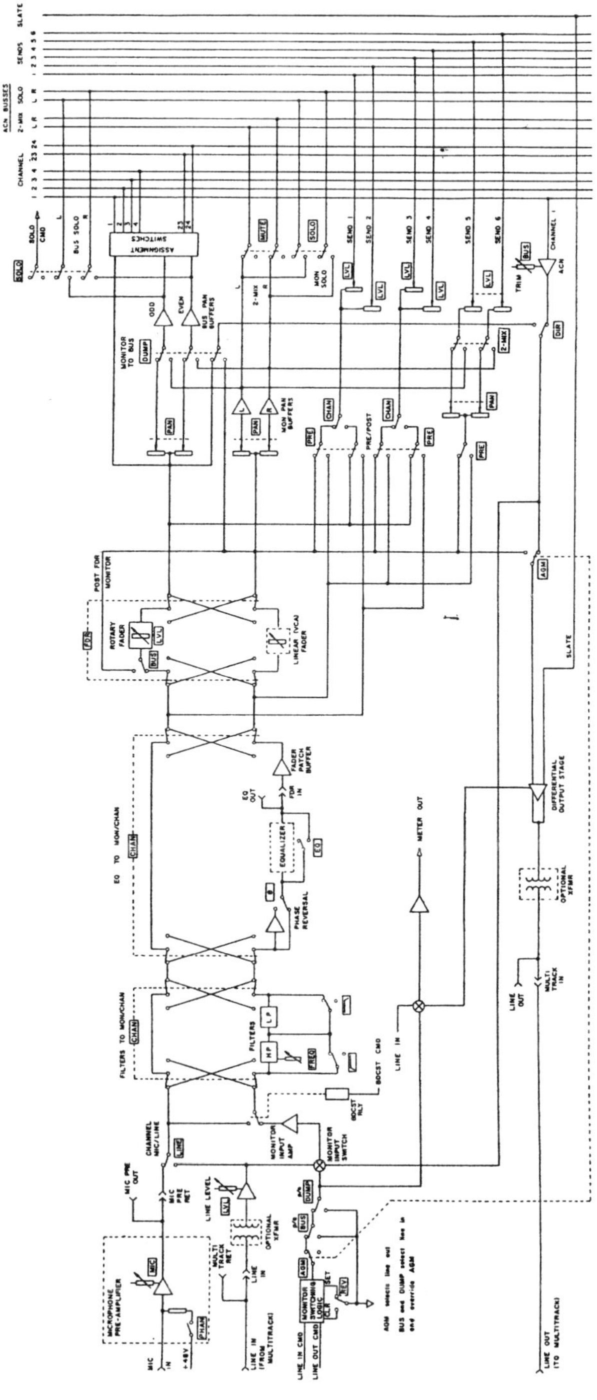

Compare with the simplified diagram in Figure 16-4. The console is a modern professional in-line multi-track recording console with VCA automation, vacuum fluorescent metering and wild faders.

■ THE DIGITAL DIFFERENCE

In the first section of this chapter, we looked at the simplified block diagram of a console. There were many similarities between the analog in-line console and the digital mixing system. And although an engineer with experience on an analog console should have no trouble migrating to a digital product, there are some differences which will be discussed in the following sections.

Signal Flow Paths in the Digital Console: Additional Details

While most analog consoles have their electronics contained within the console frame, or at most have the AC-to-DC power supplies some distance away, most large digital consoles do not have their electronics in the same location as the control surface. In some cases it is because of cooling fan noise, while in others it is simply a matter of space. Digital electronics can exhibit higher levels of generated RFI (radio frequency interference) than analog versions, so it is often advisable to keep the signals as separate as possible. Additionally, large digital console electronics can either generate enough heat to make the studio space uncomfortable, or require air handling units that can raise the noise floor of the studio space and make the room less than ideal for monitoring.

The control surface of the digital console often houses only the controls used to manipulate the signal. Even the DSP (digital signal processing) chips are located remotely. In other cases, although the microphone, analog line, and digital I/Os are remotely located, the processors are within the console control surface frame.

■ CONTROL SURFACE TO ELECTRONIC RACK INTERFACES

Since the control surface is separated from the I/O electronics, a reliable method for transporting data between these two locations is required. One of the most used methods is the MADI protocol. If you recall from Chapter 13, the Multi-channel Audio Digital Interface (MADI) protocol is very much like the AES/EBU code in format, but carries 56 channels of data multiplexed together. The console we will examine here has 4 electronic racks. Each electronics rack can contain 56 I/Os and is connected to the control surface with a MADI cable. The console can have a total of 4 racks for a total of 224 I/Os. In addition to the MADI interfaces, the rack-to-console path usually contains lines carrying SMPTE time code and video sync pulses if desired. The control surface of this console is shown in Figure 16-16. The figure shows a 64-fader version of the console although there are a total of 160 I/O channels (80 for the top row of faders and 80 for the bottom row). By simply "paging" in groups of eight over to the other channels, the virtual controls then become assigned to the higher numbered I/O modules. In fact, a console with only two rows of faders, that is, 16 physical I/O modules, will still have 160 input channels available. However, there will only be 16 physical, visible controllers that can be manipulated at one time.

Figure 16.16 A detailed illustration of the control surface of a modern large in-line digital console, showing two of a possible five banks of I/O sections and the center master/monitor section (Soundtracs DPCII, Soundtracs illustration).

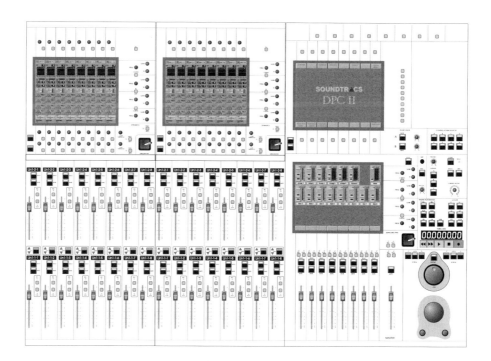

Each rack in the system houses a series of interfaces usually on plug-in cards that connect the console to the rest of the studio. A typical installation may have two racks of digital I/Os and two racks of analog I/Os. The digital racks will be a combination of AES/EBU interfaces, T-DIF (Tascam Digital Interface) cards for DTRS recorders, optical interfaces for ADAT type recorders, and possibly a separate MADI interface for carrying 24 or 48 tracks to and from a digital multi-track tape recorder. The racks are connected to the work surface with MADI cables.

■ SYSTEM CONFIGURATION FILES

In the system configuration files, all the input and output sockets that show up in the input and output assignment pools are named and assigned to physical devices. When the console arrives, these sockets are named something like socket 1-2-3, which would be rack number 1, row 2, socket 3. Remembering or even setting up an assignment chart for all these meaningless numbers would be difficult, so each socket can have a meaningful name assigned to it that will show up in the selection pool. Now socket 3 in row 2 of rack 1 can simply be labeled PCM-3324 track 3 input. Now, when the engineer opens the input assignment panel, a list describing the actual sources available for selection is shown.

■ CLOCKING THE DIGITAL CONSOLE

One of the most important issues that needs to be addressed when using a full digital studio is the clock source. When digital signal is sent from one device to another, it is imperative that they have their clocks synchronized, that is, the sample rate pulses are timed from the same master source. In mostly analog studios where an analog console is feeding two or three DAT recorders, and occasionally one DAT may be sending signal to another, this may not be an issue since the source digital machine is providing word clock along with the digital audio signal to the receiving device. However, in a large all-digital system, problems will arise if a common clock is not used. In some cases, the console itself, as the main component in the system, may be the clock source. And even if a separate word clock connection is not made between devices, the console can provide a master clock through the AES/EBU signal bus.

A better way is to use an external dedicated highly accurate word clock that is distributed to all digital devices in the studio, including the console. This generally guarantees that all devices are synched together to a word clock. The word clock selection panel for a digital console is shown in Figure 16-17.

■ THE INPUT POOL

Determining which source goes to a specific input is done by opening the input selection pool screen and assigning a device to the upper or lower input fader of each I/O strip. This is where the names that were assigned to sockets in the configuration file appear. Selecting a general category, such as microphones, opens another panel where the available microphone sources are listed. When the desired microphone is selected, an option for phantom powering appears and may also be selected. When a line-level group or source is selected, this option is not available. Here the signal can be labeled, using the console keyboard, and the label will appear at the top of the module and may even appear in a window just above the

Figure 16.17 An audio sample clock and synchronization panel for a digital console (Sountracs DPCII, Soundtracs illustration).

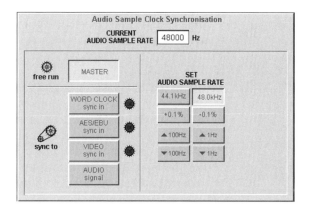

fader if the console is so equipped. Here you will also find the input gain adjustment and phase switch which often resides near the mic/line switch on an analog console. However, instead of having a dedicated control for this adjustment, one of the rotary controls below the screen is used to virtually adjust the gain. By selecting a socket that is connected to a microphone or line level source, the type of input is automatically determined, and a conventional mic/line switch is unnecessary. A typical input assignment panel is shown in Figure 16-18.

■ PATCH (INSERT) POINTS

Most digital consoles have two types of insert or patch points. There is usually one just after the analog microphone pre-amplifier and before the A/D converter. Additionally there is usually another within the digital signal chain. The former can be used to insert an analog processor in the signal chain prior to digitization, and since this is after the microphone pre-amplifier, a standard line-level (+4 dBm) will suffice. The digital insert point is used if a processor such as a digital reverberation unit is desired in the signal path. However, the signal would have to be routed to one of the digital output sockets, sent to the input socket of the device, then routed from the device back to another digital input socket. In all, to insert a digital device in the path, 4 digital sockets will be used. Although it is easy for the engineer to assume that with 244 sockets available this will not cause a shortage, it is wise to plan efficiently so as not to deplete the available socket pool. Typically, a set number of sockets are dedicated as digital insert points, and configured as such at the time of installation.

■ AUXILIARY SEND BUSES

Like its analog cousin, the digital console has auxiliary buses as well. At the beginning of each session, a console is configured, and the engineer can specify how many and what type of auxiliary buses are desired. It is important to remember that each auxiliary bus requires a certain amount of dynamic signal processing (DSP), and

Figure 16.18 The input channel selection panel screen for a soft assign digital console (Soundtracs DPCII, Soundtracs illustration).

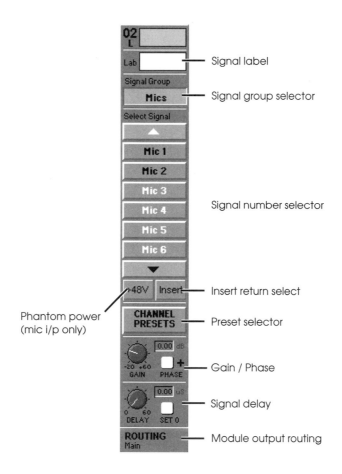

that if more buses are specified than are actually needed, digital processing that may be needed later may not be available. The console we are discussing has a total of 40 available internal buses which can be divided among the Main, Group, Auxiliary and Solo systems. For instance, if too many auxiliaries are selected when configuring a surround console, there may not be enough DSP power to fully configure a surround Solo or PFL bus.

The auxiliary buses serve both as effects sends and as cue sends. When used as the former, the effects returns are simply routed back to any unused input, and then assigned to the mix bus. When used as cue sends, the buses are sent to the headphone cue system amplifiers.

■ THE OUTPUT POOL: BUSES AND DIRECT OUTS

Like the input assignment section, the output assignment section of each virtual I/O strip opens a window listing the available outputs from the dedicated socket pool. Unlike the input pool, however, there are two types of outputs, the bus outs and the direct outs. The bus out section is where the assignment to the mono, stereo or surround mix bus is found. Here, on some consoles, the auxiliary buses can be assigned to either outboard devices or cue sends. On others, there may be a separate section for auxiliaries and cues. Additionally, there are assignment switches for the mono, stereo and surround group buses as well.

The direct outs are usually the multi-track outputs, as in the analog console. The direct outs can usually be configured as pre- or post-fader sends. In the pre- mode, the input fader simply controls what is sent to the mix bus and does not adjust the signal level to the track on the multi-channel tape recorder. This level is set by using the microphone trim control or the output bus trim control. In the post- mode, any changes made at the input fader affect what goes to the multi-track and the mix bus. In this way, the console can be configured as a conventional in-line I/O console with one row of faders (upper or lower) assigned to the direct outs and controlling the channel path, while the other row is assigned to the mix buses and used to adjust signals that are sent to the monitor path and 2-mix bus. Alternatively, each channel can simply be an input with its output controlled by the fader and assigned to the mix bus, with the signal also assigned to a direct out that is trimmed at the beginning of the session for the best level on tape for that instrument. The latter frees all the available physical faders for use as mix controllers. Figure 16-19 shows an output assign section.

As mentioned earlier, there may be a separate section for auxiliary and/or group outputs. The main output assign may be in this section also. Here the main 2-mix out is sent to the mixdown tape recorders or DAWs. Figure 16-20 shows the output screen for part of the master section of a digital console. Below this screen is a group of eight faders which can be assigned to either the upper row or lower row of output buses shown in Figure 16-20a. Figure 16-20b

Figure 16.19 The output bus and direct out selection panel for a digital console (Soundtracs DPCII, Soudtracs illustration).

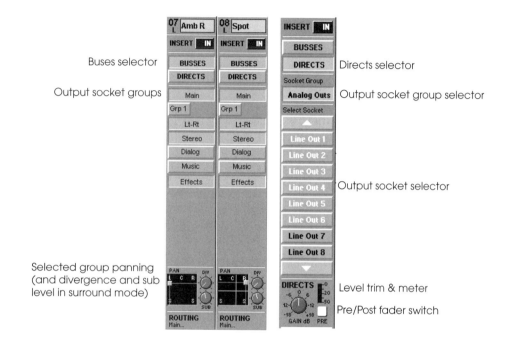

shows the trim control, the dynamics section, the metering and the routing that is available for each of the outputs shown in part A. Note that in part A some buses are mono, while others are stereo or surround.

The digital console described in this section is typical of the newer style of sophisticated digital production consoles. Although these consoles are very sophisticated and flexible, we are at the beginning of the development curve for these kinds of tools, and change is rapid. At this time, one of the main advantages of the digital desk over the analog version is simply the flexibility and myriad routing and processing variations that are available to the engineer. The ability to literally configure the console for the project at hand expands the possible scope of work that can be efficiently accomplished in that studio or remote truck. We will leave the discussion as to which type of console sounds better for others to debate.

Figure 16.20 a. The software configurable auxiliary master sends, group sends, and stem control sends assignable to the eight faders below the central master touch screen on a modern digital console; b. When touched, each auxiliary master send, group bus out, or stem send opens to show an output control section with metering, level trim, limiting and bus routing (Soundtracs DPCII, Soundtracs illustration).

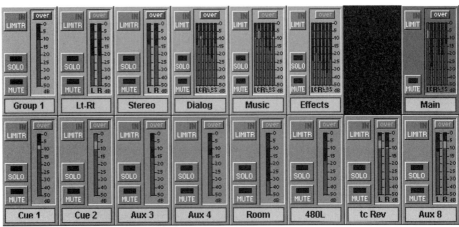

SEVENTEEN

Console Automation Systems

From our previous chapter on audio recording consoles, it may be noted that an engineer working alone will have a difficult job manipulating all of the controls required for a complex mixdown of 24 or more tracks. As is often the case, a "second" engineer or assistant may be required to help mix large and complicated orchestrations and arrangements. As will be explained in Chapter 19 (on the mixdown procedure), controlling fader levels, channel mutes, reverberation and effects returns, equalization and other outboard or inboard processing equipment can be a difficult process indeed. These problems can be compounded by the design of the modern in-line console. What makes great ergonomic sense for recording basic tracks and for unassisted mixdown operations doesn't necessarily allow a lot of elbow room for two engineers working together on a closely spaced in-line design. In fact, in some of the larger film and video studios, the audio console is physically split into three in-line sections, with a section for music, one for dialogue, and one for sound effects. This type of design allows three engineers to work as a team without getting in each other's way.

It is also often desirable to set the levels on a group of faders, such as the drum tracks returning from the multi-track, and then to be able to control that group with a single fader without affecting the group's internal balance. Split-configuration consoles allow this, in a way, by using the sub-group masters. However, all inputs assigned to a particular sub-master will be summed in mono. To maintain the left-to-right orientation of the drum kit, the multi-track returns of the drum tracks would have to be assigned between two different sub-masters. The two sub-masters could then be moved together to vary the overall stereo level of the entire drum set. This approach can work very well, but if there are many such sub-groups required, such as a drum mix, a back-up vocal mix, a keyboard mix, etc., the available number of sub-masters will soon be exceeded.

Recording console automation systems solve these problems and many more. Automating certain functions within the console frees the engineer's hands for other functions during the actual transfer of the multi-track signals to the two-channel mixdown tape recorder. The functions that are most often automated are the individual mixdown fader levels and the respective channel "mute," or off and on status. Some analog consoles and most digital consoles allow for automating other functions as well, such as equalization, panning, and effects returns.

The data that are generated by and for these automated functions can be stored for later retrieval. This allows the engineer to work on small sections of the mix one part at a time. It is possible, for instance, to establish the rhythm mix for the first verse alone, and to then store that information and hear the results while adjusting the lead solo levels and equalization. This complete mix of the first verse can then be stored while work proceeds on the second verse, bridge, and chorus. In fact, since the stored data can be retrieved at will, it is possible to create two different balances of the same section (such as the rhythm instruments) and to then compare them within the context of the remaining tracks.

When the entire song has been mixed to everyone's satisfaction, the multi-track tape recorder is rewound to the beginning of the selection and played. The console automation system takes control and operates the various automated functions as the tape is played, and the resulting mix is transferred to the two-track mixdown machine.

■ THE VOLTAGE-CONTROLLED AMPLIFIER

These functions would not be possible without a way to encode the various fader positions. One way this is accomplished is by using a fader as a simple voltage attenuator. A known voltage level is sent through the fader, and its output, when measured (relative to the reference voltage), will represent the fader position. This changing voltage is used to control the gain of a voltage controlled amplifier, more commonly referred to as a "VCA."

In earlier chapters, it was stated that many amplifier circuits were passive; i.e., a fixed-gain amplifier was preceded by a variable attenuator or "fader." An exception to this was in the case of an active microphone pre-amplifier where the actual gain of the amplifier was adjusted by the microphone trim control. These active circuits allow greater flexibility and lower noise than the equivalent passive design. The gain of the amplifier is adjusted by turning a potentiometer that varies a DC voltage that controls the bias and negative feedback on a transistor or integrated circuit. The greater the applied DC voltage, the greater the attenuation of the amplifier. This technique gives us a gain range of 100dB or greater, a very wide range indeed.

In a typical VCA-equipped console, the monitor path faders vary a DC control voltage that changes the gain of the VCA proportional to the movement of the fader. The audio signal is routed through the voltage controlled amplifier and not through the fader itself. This method of level change is transparent to the engineer, since the signal behaves exactly as if the audio were going through the fader. This control voltage is often then digitized and stored in binary form for later retrieval. It is possible, with this technology, to automate any function involving an amplifier. This includes channel mutes, grouping, group mutes, and solo functions as well as fader levels. It is even possible to automate functions such as compression and gating, if desired.

Another form of fader automation is possible by using similar DC control voltages. Instead of controlling the gain of a VCA, the resultant voltage drives a servo-motor. These small servo-motors are attached to the actual audio taper fader, and the control voltages (when regenerated) energize the motor and move the fader. In these so-called "moving fader" systems, the audio signal does go through the actual fader. This was often preferred by many engineers since they felt that the VCA lacked the quality of a passive fader. Today, however, it is the preference of the engineer that determines which system is selected. Figure 17-1 shows a simplified signal path for a split-configuration console submix bus, a VCA-controlled group, and a moving fader system.

■ AUTOMATION DATA STORAGE

For the automation system to be of use, the representative data must be stored and then retrieved in synchronization with the

Figure 17.1 a. Simplified signal path for a split-configuration console submix bus; b. Simplified automation signal path for a VCA-controlled console group; c. Simplified automation signal path for a moving fader console group.

a.
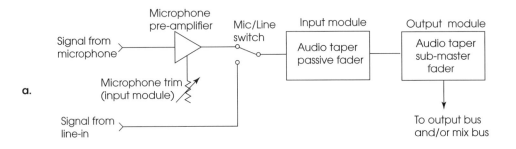

b.
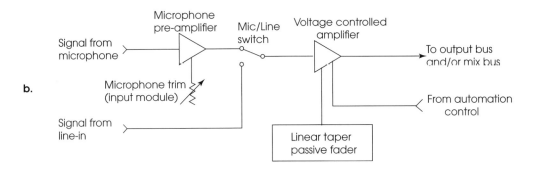

c.
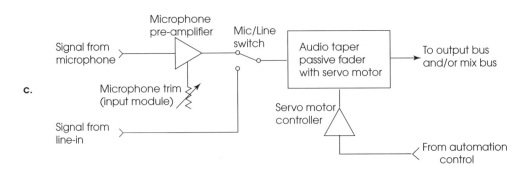

music material. One way of doing this is to record the data on a track of the analog tape along with the music. This vertical alignment will ensure that the automation data are synchronized with the program material. To prevent distortions of the recorded control voltages, distortions that can be caused by analog tape recording systems (see Chapter 10), the control voltage for each VCA is periodically sampled and quantized. (For a further discussion on digitizing a signal, refer to Chapter 13, "Digital Audio.") The automated controls are sampled and quantized in a pre-specified sequence called scanning. The scanning rates of different automation systems vary, but they are generally in the 10Hz to 40Hz range. This means that the amplitude of the control voltage for every VCA is measured every one-tenth to one-fortieth of a second depending on the system. Between samples, the value is held in a buffer memory similar in nature to the sample-and-hold circuit of digital recording systems. This stream of binary code, generated by the automation computer or digitizer circuit, is then sent to the storage medium. Of course, there can be a delay between the time of the last scan and the movement of a fader or control, particularly if the fader continues to move between scanning samples. Obviously, the higher the scanning rate, the less noticeable this effect is. In fact, when in the automated mode, the engineer hears exactly what the automation system has recognized as valid data since the fader sends its signal to the automation computer, which in turn controls the VCA. If the scan rate isn't fast enough, what is heard is often referred to as the "Zipper Effect" as the gain jumps from one level to another. When the automation computer is turned off, control of the VCA is then accomplished by the voltage change generated by the fader itself. Most systems have a smoothing filter at the output of the D/A converter to remove as much of the step functions as possible from the control voltage.

Some systems have a bypass mode that allows the engineer to switch the audio signal to a second conductive element within the fader pack, thereby bypassing the VCA altogether for non-automated requirements. This allows the linear fader to be used for basic tracks or direct-to-two recording without the VCA in the signal path, yet gives the engineer the flexibility of automation for the mixdown of multi-track masters. A separate element is needed since audio faders use a logarithmic taper, while DC controllers use a linear one. Moving fader systems use a similar principal in that there is a logarithmic taper for the audio signal, while a linear track is used for voltage control of the servo motor.

Some older analog systems use two tracks on the multi-track tape recorder for storage of the automation information. The signal is recorded and played back "in-sync" with the existing material in a manner similar to the way that an overdub is accomplished. In this case, however, the musical material is monitored from the playback head while the automation data are recorded on a separate track by the record head. This records the automation data non-vertically aligned, that is, offset by the spacing between the record head and the play head on the multi-track tape. To ensure that the intended level changes of the keyboards, for instance, are accomplished in time with the music, the automation data are played "in-sync" from the record head to maintain the proper time relationships. Changes to these stored levels are accomplished by generating new data based on the original data plus any further changes, or by generating new data and recording this information on a second track. By alternating between these two data tracks, the mix can be continually revised until the desired mix is attained.

A disadvantage of this type of storage system is that as more and more revisions are made, the automation data lag further and further behind the musical material. During the process of getting the perfect mix, the data are continuously read from one track of the tape recorder, revised and re-recorded on the other data track and vice versa. There is an approximate delay of .5 ms to 1 ms between the two events, depending on the speed of the computer in use. This delay is cumulative, and can become a problem with multiple revisions.

This problem and others related to tape-based data storage, such as drop-outs and edge damage, have been solved by the use of computer-based non-tape storage systems. These automation systems store their data on a hard disk like those found in current personal computers. A 2-gigabyte hard disk can hold many hours of mix data, and if it becomes full, the desired mixes can be downloaded onto a conventional Zip drive or other removable storage medium. This information can then be stored with the multi-track master tape for future reference or updates. Synchronization between the multi-track tape and the hard disk is accomplished with SMPTE time code (refer to Chapter 15). The code is provided by the built-in time code generator and reader found in the automation system. Many machines have a track dedicated and optimized for time code, with a separate input and output on the back of the machine. On analog machines lacking a dedicated time code track, the code is usually recorded on track number 24 of the multi-track tape prior to the beginning of the mixdown session. When the tape is running, the multi-track tape recorder becomes the master and the hard disk is locked in sync by the time code comparison circuit.

This ensures that the automation data are synchronous with the program material. These systems can be frame accurate or better, and there is no cumulative lag time with multiple revisions since each revision is referenced to the original time code track. Most systems will support standard and drop-frame time code as well as film and EBU standards.

Hard disk storage systems can be used with either VCA or moving fader systems. A bonus provided by this type of system is that the output of the multi-track time code channel can be sent to synchronizers and SMPTE/MIDI converters as well as to the automation system. This can provide multi-machine lockup, audio/video synchronization and other post-production techniques that require SMPTE time code.

Automation can be used in two fashions. If the engineer manipulates the faders and other controls during the actual playback of the tape, this is called "on-line" or real time automation; that is, all systems are on-line and functioning at play speed during the mixing. However, after an initial pass has been made, it is possible for the engineer to open the automation file and edit it either as a text file using a keyboard, and/or by using a mouse or pointing device to manipulate a graphic representation of the data. This is called "off-line" automation. The terms are taken from the video post-production industry where off-line editing is the manipulation of EDL (Edit Decision List) data in non-real time. In either case, the basic functions and modes of the systems are similar.

■ PRIMARY AUTOMATION FUNCTIONS

There are five primary modes of operation for automation control in most systems. These are the Isolate mode, the Write mode, the Mute-Write mode, the Read mode, and the Update mode. These functions may often be combined, such as Write and Mute-Write, and all functions can be used independently on any or all channels. Some of the moving fader systems do not specifically delineate these functions, but the system function modes are similar. Transitions from one mode to another are initiated whenever a fader is moved or a mute button is pressed, whereas in the VCA systems, the engineer must manually select the mode for the function desired. Figure 17-2 shows several examples of automated fader packs, both VCA and moving fader types.

Figure 17.2 At (a), an input automated moving fader on an automated digital console (Soundtracs DPCII, Soundtracs illustration); and at (b), a moving fader input section for an automated analog console (Martinsound Flying Fader system, Martinsound picture). The system at (b) was formerly marketed as the Neve "Flying Fader" system.

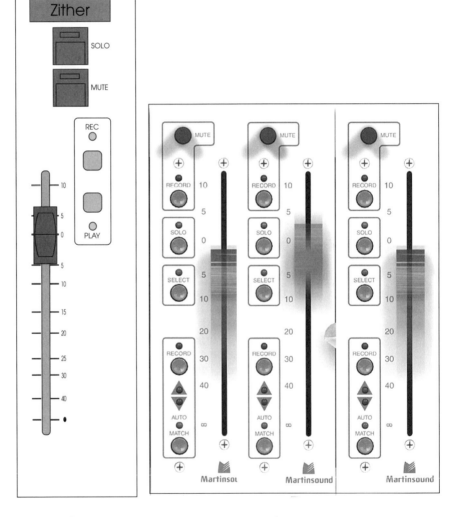

Isolate Mode This is, in effect, the "automation off" mode as applied to individual channel faders. In this mode, the VCA or servo motor fader is active, and changes to the fader affect what happens in the mix. However, none of the data are stored for later retrieval. The engineer may use this mode to set up a basic mix before recording the first pass. The same result can occur simply by turning the entire automation system off, although on some consoles grouping is not available unless the automation system is active.

Write (Automation Record) Mode The Write function is the initial mode that is selected at the beginning of the automated mixdown procedure. This is sometimes called the automation record mode. In this mode, the scanned voltage levels are digitized and stored in synchronization with the program material. All revisions will be based on these initial write levels. A "master" or "global" write mode is used to place the entire console in Write and Mute-Write simultaneously for this initial pass. The Write or automation record mode is the only function where information is generated for each fader without regard to previous data. This mode stores data from all automatable parameters as determined by the console manufacturer or the system configuration file.

Mute-Write Mode In some cases, the Mute-Write information is written as a separate signal from the fader level data. It is important to note that, even though a channel is muted, its fader level information is still written to the storage medium. This mode, usually used on the initial automation pass in conjunction with the Write mode, is used to program whether the channel is off or on at any given time. When the channel is muted, its signal is removed from the monitor path or two-mix bus, and therefore from the mix that goes to the two-track mixdown tape recorder. If there are auxiliary sends derived from the channel that is muted, they will also be muted unless the send is originated pre-fader. Mute-Write is often a separate function, even though a global Write command will usually cause the system to enter Mute-Write also. As such, it can be used on individual channels or all channels alone or in conjunction with other modes (except Read).

Update Mode   The Update mode is used to revise existing fader level data. Update uses the existing automation data and adds to or subtracts from those levels. Some systems with non-servo faders automatically "null" the fader when the Update mode is entered, and this is often called the relative trim mode. That is, the current physical position of the fader is the position that becomes the starting point for additions or subtractions of level. It is important not to enter the Update mode when relative trim is selected with the fader in either the fully open or fully closed position, since movement in one or the other direction would be impossible. The engineer should always make sure that there is enough fader movement available for the planned change prior to entering the Update mode.

Consoles with VCA automation systems have arrows or plus and minus LEDs (light emitting diodes) to indicate the null point relative to the fader. This indicates in which direction the fader needs to be moved to reach the null point. This allows the engineer to return the fader to the null after making an update, so that the transition from the new data back to the old is as smooth as possible. If the fader is not nulled prior to leaving the Update mode, there will be a jump in the fader's level and the sounds passing through it.

Servo-motor-based automation systems usually physically null the fader when one of several conditions occurs. If the system automatically goes into Update when a fader is moved, it usually auto nulls when the fader is released. This may also happen if the automation record or play button is pushed, depending on the system. The length of time that it takes for the fader to return to either its original position (relative trim) or to the current setting (update) based on the music is usually measured in SMPTE time code frames, and is often adjustable in the configuration file or setup menu of the system.

Many engineers prefer to record a baseline mix on the initial pass of the automation system and to then make all further revisions in Update. Other engineers would rather write fader data for a few channels at a time and create mix revisions by "re-Writing" the remaining faders singly or in small groups. Both methods have their proponents, and both methods will achieve similar results.

Read Mode or Play Mode   When the Read mode is selected, operation of the console VCA levels, mute status, and whatever else has been assigned or configured in the automation system is under the total control of the automation computer. That is, operation of the mute functions, automation solos, fader movements, etc. will have no effect on the signals being sent to the mixdown bus. However, the engineer can make changes in the non-automated functions of the console, such as equalization, panning or filters. If a console has a separate post-fader solo or solo-in-place system, individual channel equalization, panning, etc. can often be adjusted while in the Read mode. Many so-called Active Systems remain ready to drop into Record, Trim or Update mode as soon as a control is touched, and revert back to the Read or Play mode when the control is released.

There are buffer memories in the Read circuit that hold the current level until new data are received. These buffers make sure that there are no unwanted level changes between scans, or level changes caused by tape drop-outs on the data track or data discontinuities from the hard disk system. In actuality, the Read mode is probably the most frequently used function in the system since, when updating a mix, only the channels that are to be changed are switched into Update while the majority remain in Read.

Group Functions

VCA grouping is an important and useful function of the automation system. Although automation is not necessary for VCA grouping, they are usually found together.

As discussed in the opening paragraphs of this chapter, it is often desirable to control the balance of an entire group of faders with one single master control. Most consoles that contain VCA grouping provide for up to eight or nine different groups. Group selection can be assigned by a mechanical thumbwheel or by software controls through the automation system. Either way, the faders that are to be assigned to a specific group are selected to a group number. Any number of faders can be assigned to the selected group, and a group master is selected. This is usually a fader that is a member of the group, but may be a separate control. The relative balances among all the members of the group should be determined prior to selecting a group master. Once the group master is selected, moving that fader up or down will cause the entire group to mimic this action while maintaining their relative relationship with each other. With moving fader systems, the actual faders in the group will move as the master is adjusted, while with VCA-equipped systems the gain of the VCAs in the group will follow the gain changes of the master VCA. The group's balance should be set prior to selecting a group master since it is often cumbersome to change the individual VCA or fader level on a channel that has been selected as a group master without affecting the other members of the group. Figure 17-3 shows a system with separate faders for each assigned group. Some systems may use the term "ganging" to define traditional grouping where the master is a member of the group.

For instance, let us say that we wish to group a number of channels, such as 1 through 8. These 8 channels represent 8 drum tracks that were recorded from microphones placed on the kick or bass drum, the overhead cymbals (left and right), the high-hat, the snare, and the three tom-toms. We can pan the drums, in the stereo monitor mix, between 9 o'clock and 3 o'clock and assign all 8 of these channels to a group, such as group 5. The relative levels between the 8 channels are now adjusted to achieve the desired drum mix within the stereo image. We can now assign one fader, say, fader number 1 (the kick drum track), to be the master for the entire group. Later in the mix, when it is desired to raise or lower

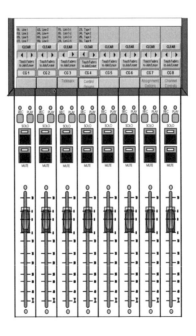

Figure 17.2 Automation controls (software assignable faders) in the master section of a large digital console (Soundtracs DPCII, Soundtracs illustration).

the level of the entire drum mix, this can be accomplished by moving fader number 1. The levels of the 8 drum tracks will follow the gain adjustments made to fader number 1 without losing their internal balance or panning. And, as long as all members of the group are in either Update or Write mode together, the automation system will store the changes for the entire group.

Any number of groups, such as vocals, keyboards, or entire string section tracks, can be assigned, thus allowing the engineer to change the relative balances between large instrumental and vocal ensembles as a block while maintaining the internal ensemble balances.

The Solo and Mute Systems

As mentioned earlier, channel mutes can be written separately from the fader VCA data. Once again, this means that even though a channel is muted, its VCA level is being recorded and stored whenever the channel is in Write or Update. This fact may be useful if a previously muted channel needs to be retrieved, such as when the producer changes his mind or a new version of the song is required.

The fact that mutes are part of the automation system allows several other helpful functions to be used. They are the group mutes, solo-in-place with effects, and group solos.

Group mutes are simply the system's ability to mute any number of channels simultaneously with the push of one button. This is accomplished by using the previously mentioned group assignment system. Any member of a group may be muted independently but, if the assigned group master fader is muted, the entire group will be muted also. For instance, if we desired to mute the tom-tom tracks from our previous example, the mutes on channels 6, 7, and 8 would be depressed individually. However, if it was desired to mute the entire drum mix (channels 1 through 8), the engineer would depress the mute button on the group master (fader number 1) to mute the entire group.

Another function of the automation mute system is the so-called "automation solo-in-place." When solo is enabled on the automation system, any channel that is selected will cause all of the remaining channels to mute. This condition may or may not be written to the automation storage system depending on the console's Mute-Write status. However, care must be taken since all channels, except those that are soloed, will be muted to the monitor path and therefore to the two-track mixdown bus. This is often called a "destructive solo." The advantage of this type of solo is that, by muting all channels except the ones selected, these channels can be heard by themselves in the stereo image along with any effects returns. For instance, if the engineer desires to hear the solo guitar with its associated reverberation (derived from the auxiliary sends and returned to the mix bus), the only way to do so is by muting all the other channels. This is accomplished by depressing the automation solo button for that monitor path channel. The channel's track solo button (on the input module) will place only the guitar in one or both monitor loudspeakers, while the monitor path's solo button will only place the guitar in the monitor loudspeakers in stereo (in place based on the monitor pan-pot). Only the automation solo will, by muting all the other inputs, allow the engineer to hear the individual track with its derived effects returns.

Any number of channels can be automation soloed at once, and by storing this information in the automation system, a quick change from, for instance, a large string section of 8 or 10 tracks to a string quartet (the section principals) of only 4 tracks, can be accomplished on cue. Group soloing is accomplished the same way, and is an inverse function of group muting. By soloing the group master, the entire group is soloed and all other channels are muted.

The Automation Master

The automation master allows the engineer to "fade-out" the entire console if required and to have this fade recalled for later playback. In effect, the automation master is a group master for all of the console VCAs. Lowering or raising the level on the automation master causes the gain of all the channel VCAs to vary accordingly, whether they are grouped or not. Only those channels that have been completely removed from automation control (local only mode) will be unaffected by the automation master. The automation master differs from the two-track or mixdown master in that the mixdown fader only attenuates the output of the mix bus and does not affect the gain of any VCAs.

The functions and devices just discussed are what are referred to as dynamic automation events. They are synchronized to occur in time with the program material, often changing slowly, such as in a fade-out.

Snapshot Automation

Another type of automation is referred to as snapshot automation. Here, a processor records a setting of all controls at a specific instant in time. This is particularly handy if, for instance, the first piece on a program requires different gain and trim settings as well as panning from the second piece. Formerly, at the rehearsal, the assistant engineer would laboriously write down on a setup sheet all the settings for each piece. Then, during the event, he or she would quickly change settings at the end of the first piece and hope to be finished before the second piece started. Often, in a multi-track session with many settings, a Polaroid camera was used to record the console settings, and the photo would be stored in the box with the tape for future use if required.

Now it is possible on many consoles to store that information as an electronic snapshot, and to have the console revert to these settings when the file is called up and executed. A snapshot usually is used to implement instant changes to the whole console.

Merges It is possible, with hard-disk storage systems, to combine sections from various versions of the automated mix. These so-called "merges" allow the engineer to take a section of revision 1, for example, and place it within the confines of revision 3. Likewise, if it is desired to have the same mix each time the chorus section of a song is repeated, the automation levels for that section can be merged into another revision of the mix whenever the chorus appears. This means that the engineer only has to mix the chorus once, and then that data can be plugged in or merged into the final mix whenever the chorus is repeated. The location or beginning and end of a merge are defined by the SMPTE time code values on the master tape. Automatic merges are created by the system each time an update is performed on just a part of the mix. For instance, if an update is made of a small section of the second verse, when the tape is stopped, data from the previous revision are filled onto either end of the revised section to form a complete version of the song. It is therefore extremely important to null the faders that have been updated before stopping the tape. If this is not done, level jumps between the new data and the old may occur.

Cues Often, various cue points within a song can be stored for later recall. These can be the points required for merges, or simply defined points for roll-back to the beginning of a particular section. These points are defined in terms of their SMPTE time code values, which avoids confusion since no two points can be the same. These points can often be defined "on the fly" (while the tape is running) with the push of a button. When the tape is stopped, the points can then be labeled or listed for later use.

Rehearsal Modes There is usually some type of rehearsal or practice mode available on most systems, or, if there is sufficient storage space, a subfile for each automation pass is written to disk. This allows the engineer to rehearse or practice fader moves without actually storing the automation data in the rehease mode, or without destroying the original mix data in the later system.

Off-line Automation With the addition of hard-disk storage to automation systems, it became possible to call up the file and edit it as you would a word processor document, that is, in non-real time. Originally, the engineer was limited to taking a section of the mix, identified by time code, and, using cut and paste techniques, moving that control information to another part of the song, or to a completely different take or version of the song. Now, however, with the addition of the GUI (Graphical User Interface) to all computer platforms, editing in the drag and drop fashion with a mouse or pen is possible. In fact, with a pen and tablet system, the engineer can draw his own custom fades and gain changes. Figure 17-4 shows an automation graphical editing screen, while Figure 17-5 shows a section of the graphical off-line editor containing 2 channels where there are several events.

Figure 17.4 The off-line editing screen for a console automation system (Soundtracs DPCII, Soundtracs illustration).

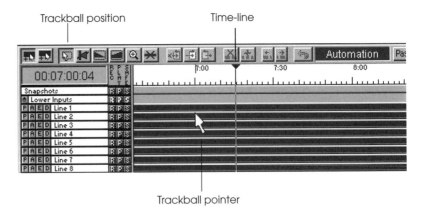

Figure 17.5 Closeup of an off-line automation editing system showing fader level and mute status (Soundtracs DPCII, Soundtracs illustration).

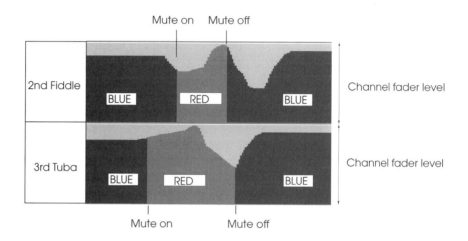

There are generally two types of events that can be automated: level events and switch events. Level events refer to changes that occur over time, such as a gain change, a panning sweep, or an auxiliary send level change. Switch events are instantaneous on/off events such as mutes, dynamics switching in/out, or an auxiliary send switching on/off. It is also possible to dynamically fire snapshots of the console, and to edit the timing of the event off-line. The snapshot itself cannot be edited except by replacing it with another. Figure 17-6 shows an off-line edit screen where two channels are shown and the second channel has an equalization section open for editing. There are both dynamic and switch events shown. Figure 17-7 gives a list of most functions found on a modern multi-channel digital console showing which functions can be dynamically automated and which can be included in a snapshot.

■ AN AUTOMATION SESSION

In summary, we can list the steps involved in an automated mixdown session as follows. The reader is also advised to refer to Chapter 19 on mixdown to supplement this list.

1. The tape is striped with SMPTE time code, or the two data tracks on the multi-track tape recorder that are to be used are defined and patched into the proper locations in the automation system. If code is used, the type of time code must be determined by the intended product use.

2. In a hard-disk system, the tape is played for a 30-second period to calibrate the time code reader to the type of code in use (standard, drop-frame, EBU, etc.).

3. The faders are positioned, either in the rehearse mode or to some approximate mix position, and the system is placed in the Write (and Mute-Write) or automation record mode. Groupings can be selected at this time after the rough mix is ready to be stored.

4. The music is played from beginning to end to define the beginning and ending points in the code for the selection. The automation system "writes" the fader levels and other data in synchronization with the program material.

5. The system is put in the Read or Play mode and the tape is played, while the engineer (in a non-total automation system) determines what panning, equalization and effects are desired. These decisions may be partially made prior to the initial Write mode as well.

6. The entire system or individual channels are put into the Update mode, and the balances for the song are revised until everyone is satisfied. Mutes and other functions may be updated independent of fader levels by selecting Mute-Write if applicable. If a particular channel or number of channels has so many wrong level changes that an update is impossible, those channels can be placed in Write (Re-Write) mode and new data can be written.

7. The system is placed in the Read mode and the resulting mix is recorded on the two-track mastering tape recorder.

8. With a hard-disk system, the information is downloaded to a floppy disk or other storage medium, and stored with the multi-track master tape.

CONSOLE AUTOMATION SYSTEMS 499

Figure 17.6 An off-line automation screen showing fader and switch events for the equalization section of one channel on a digital console.

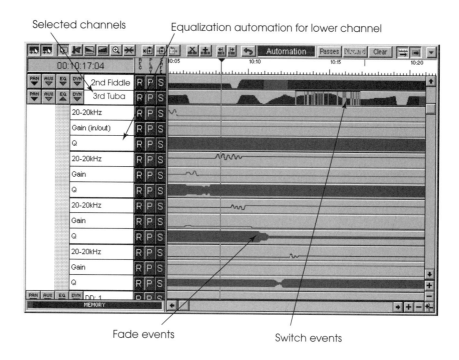

Figure 17.7 List of typical dynamically-automated and snapshot-automated functions for a digital console automation system.

Control	Included in Snapshot	Included in Dynamic Automation	Control	Included in Snapshot	Included in Dynamic Automation
Input source	Yes	No	Auxiliary pre/post	No	No
Input gain	Yes	No	Auxiliary pan	Yes	Yes
Input phase	Yes	No	Channel pan L/R	Yes	Yes
Input delay	Yes	No	Channel pan F/B	Yes	Yes
Analog insert on/off	Yes	No	Sub trim	No	Yes
EQ in/out	Yes	Yes	Sub mode	Yes	Yes
EQ frequency	Yes	Yes	Divergence	Yes	No
EQ gain	Yes	Yes	Channel mute	Yes	Yes
EQ "Q" control	Yes	Yes	Channel fader	Yes	Yes
EQ curve	Yes	No	Direct routing	Yes	Yes
HPF/LPF in/out	Yes	Yes	Direct level	Yes	No

Figure 17.7 (continued)

Control	Included in Snapshot	Included in Dynamic Automation	Control	Included in Snapshot	Included in Dynamic Automation
HPF/LPF frequency	Yes	Yes	Direct pre/post	No	No
Compressor in/out	Yes	Yes	Bus routing	Yes	Yes
Compressor threshold	Yes	Yes	Insert on/off	No	No
Compressor ratio	Yes	No	Insert sockets	No	No
Compressor attack	Yes	No	Bus output level	Yes	Yes
Compressor decay	Yes	No	Bus trim levels	No	No
Compressor gain	Yes	No	Bus limiter on/off	No	No
Gate in/out	Yes	Yes	Bus limiter threshold	No	No
Gate threshold	Yes	Yes	Bus limiter release	No	No
Gate attack	Yes	No	Bus output mute	Yes	Yes
Gate hold	Yes	No	Bus output routing	No	?
Gate decay	Yes	No	Talkback routing	No	No
Gate range	Yes	No	Talkback levels	No	No
Gate key source	Yes	No	Studio feed input source	No	No
Dynamics configuration	Yes	No	Studio feed level	No	No
Dynamics stereo link	Yes	No	Studio feed mute	No	No
Auxiliary on/off	Yes	Yes	Studio feed L/R	No	No
Auxiliary send level	Yes	Yes	Studio feed output routing	No	No

SECTION VIII

The Recording Process

■ INTRODUCTION

In this final section of the book, various aspects of both the recording and mixdown sessions are described.

In Chapter 18, overdubbing and the selective synchronization or "Sel-Sync" sessions are discussed, along with the techniques of "bouncing tracks," "punching-in," and the remote control of the tape recorder. Later, pre-session preparation of the control room and the studio is covered, along with a discussion of the use of signal processing devices while recording.

Chapter 19 concludes the book with a discussion of the final step in the recording process: the mixdown session. The mixdown session is actually a form of recording. However, in place of microphones, a previously recorded multi-track tape is routed back through the console, mixed down to a stereo program, and recorded onto another tape recorder. Both automated and non-automated functions will be discussed.

EIGHTEEN

The Recording Session

Before the introduction of the multi-track tape recorder, recording studio procedures were reasonably standardized. A song or symphony would be recorded in its entirety in one sitting. All the musicians would be present, and they would play—and replay—the music to be recorded until the engineer and producer were satisfied with the balance, the performance, the room acoustics, the soloist's interpretation, and so on.

Longer works might be recorded in sections, which would later be spliced together to create the complete performance. Chances were the musicians would make several recordings or "takes" of each section, and often the best segments from several takes would be edited together to form the ideal composite recorded performance. The editing process will be discussed in greater detail in the next chapter.

Beyond the editing process, little could be done to modify the recorded music. Nevertheless, the luxury of tape editing represented a major advance over earlier recordings made directly to disc. Here, the performance was permanently cut into the groove at the moment of recording, and there was no practical way of making even a simple edit later on.

■ OVERDUBBING

Once magnetic tape became the standard studio recording medium, it was only a matter of time before musicians began adding accompaniments to their recordings by playing along with a previously recorded tape. Both the new and the previously taped performance would be mixed together and recorded onto a second tape recorder. This technique became known as overdubbing. At about the time that it came into wide use, studio tape recorders with three or four separate tracks were pretty much the industry standard. Today, digital multi-channel tape recorders with 48 discrete tracks, and DAWs with as many as a hundred virtual tracks are found in everyday use.

During the initial session, all the tracks would be used. Then, the tape would be rewound and played while the musicians added other musical lines or parts, listening on earphones to the first machine and playing along in accompaniment. The engineer would mix the original recording from the first tape recorder with the new material from the microphones in the studio to the record head of the second machine. A typical example of the overdub process is illustrated in Figure 18-1a. Earlier, the orchestra was recorded on a four-track tape recorder, and the soloist now listens to the four-track recording and sings along with it. The engineer mixes the four tracks plus soloist down to a mono or stereo master tape. If, on playback, the balance is judged unsatisfactory, the soloist will have to be recorded again, while the engineer makes the necessary adjustments.

In many cases, the orchestra might be originally recorded on only two or three tracks, which would be directly transferred to the second tape recorder while the soloist was recorded on the third and/or fourth tracks during the overdub session. Later, these would be "mixed down" to produce the final mono or stereo master tape, as seen in Figure 18-1b. Although this development allowed more flexibility in arriving at an ideal balance later on, it added another generation of tape noise to the final product.

The overdub process brought a measure of efficiency and economy to the recording session. As the orchestral background was being recorded, complete attention could be given to the instrumental balance. Later, the solo could be added without the time and expense of having the orchestra make repeated takes while the vocalist searched for the perfect interpretation, and the ideal instrumental accompaniment could be assembled, by editing from several takes, before the solo was added. In either case, since the accompaniment was recorded first, the soloist became, in effect, the accompanist, as he or she would be forced to follow the tempi and phrasing of the previously recorded material. However, at the cost of some spontaneity, a technically superior recording could be made, since unsatisfactory balances of soloist to accompaniment could be redone at only the loss of the engineer's time. In the vocabulary of

THE RECORDING SESSION 505

Figure 18.1 In an overdub session, new material is mixed with the previously recorded program to produce the master tape.

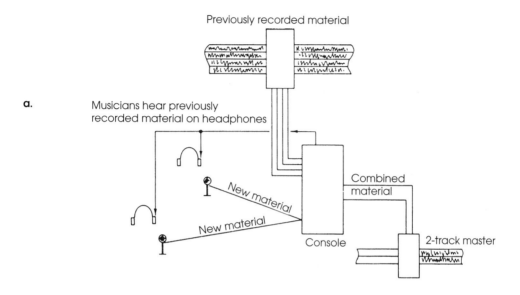

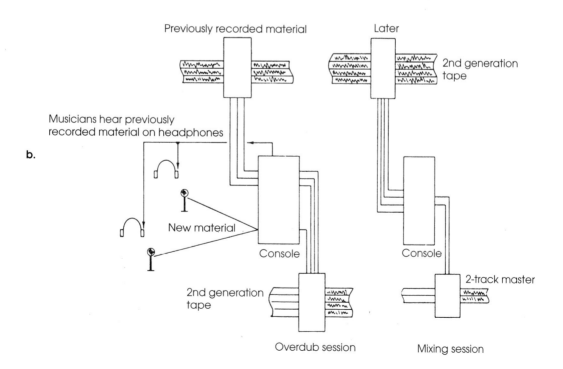

the recording studio, the instrumental background became known as the basic tracks or simply the tracks.

Although the overdub process expanded the capabilities of the recording medium, each successive overdub required an additional generation of tape. This presented no real problem on a single overdub; however, multiple overdubs could be troublesome since each would add another generation of tape noise. In the case of old material being mixed with new, the only possible balancing that could be done was between the part being recorded at the moment and everything that had gone before. In the case of a recording consisting of many tracks, recorded sequentially, it was difficult or impossible to predict the ideal balance until the recording was complete. On a tape that was a product of multiple overdubs, there would be no way to correct the balance of the third overdub, for example, without scrapping everything that was recorded subsequently and beginning again with a new version of the third overdub.

The Sel-Sync Process (Selective Synchronization)

The Sel-Sync process overcomes this very serious limitation of the overdub process. Sel-Sync, a term trademarked by The Ampex Corporation, really became popular with the introduction of 8-track tape recorders. The recording begins in the usual manner, using as many tracks as are required, but leaving at least one track—and usually more—unused, or open. After recording the basic tracks, the tape is rewound, and the new material is recorded on the open tracks while the musicians listen to what was previously recorded. In many cases, the rhythm section (drums, bass, rhythm guitars, keyboards, etc.) is recorded first, and these instruments comprise the basic tracks. Later on, perhaps strings and/or back-up vocals or keyboards may be added. These additional sessions are popularly called sweetening sessions. Last, but not least, the solo lines can be added.

Of course, if the basic tracks are monitored in the usual manner, from the playback head, the new material will be recorded on the tape out-of-sync with the old. This is illustrated in Figure 18-2. Imagine a simple two-track tape, with a basic track already recorded on track 1. If the musicians listen to the tape by way of the playback head of track 1, the new material recorded on track 2 will be out-of-sync later on playback, since it is being recorded about two inches behind the original material. The actual distance is the spacing between the record and playback heads, and the amount of signal delay depends on that distance and the speed of the tape.

To prevent the out-of-sync effect, the previously recorded track (or tracks as the case may be) can be monitored from the record

Figure 18.2 When a previously recorded program is monitored from the playback head, new material is recorded out-of-sync.

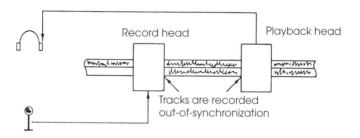

head while the same record head is adding the new material to the tape. Of course, no individual track within the record head is performing both the record and the playback function at the same time. Rather, as in this particular example, track 1 of the record head is acting as a temporary playback head, while track 2 is functioning in the normal record mode. Since the musicians are now listening to the tape at the precise point at which the new material is being recorded, new and old tracks will be in perfect sync. Later on, after the recording is completed, the tape will be monitored from the regular playback head in the normal manner. The Sel-Sync process is illustrated in Figure 18-3.

On a machine equipped for Sel-Sync, the finished recording could be monitored in its entirety from the record head. This is, after all, the way an inexpensive home machine functions, with one head performing the dual function of record and, later on, playback. However, since the design parameters for playback and record heads are not identical, the optimal record head leaves something to be desired as a playback head, and the playback head cannot make an optimal recording.

Although on recent machines the record head performs quite well as a playback device, such was not always the case. On the earliest machines, playback from the record head was conspicuously inferior, particularly at high frequencies. At the time, this was not considered to be a significant limitation given the obvious advantages of Sel-Sync over previous overdubbing techniques. Record head monitoring was merely a production convenience. Critical listening and mixdown would come later on, after the recording work was completed. At that time, the regular playback head would be

Figure 18.3 The Sel-Sync® Process. Previously recorded material is monitored from the record head while new material is being recorded, using the same record head. (The term Sel-Sync is a trademark of The Ampex Corporation. Therefore, other tape recorder manufacturers use a slightly different term to describe the same process, such as sync, self-sync, etc.)

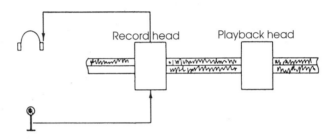

used, and so the poor playback response of the record head was of no consequence. Many current machines now have separate equalization circuits to optimize the frequency response and signal-to-noise ratio of the record head when in the Sel-Sync mode.

Transferring, or "Bouncing" Tracks

Of course, there inevitably comes a time when the number of available tracks is not enough, no matter what that number is. In a typical situation, with 23 out of 24 tracks recorded, it may become desirable to have three more tracks available.

Theoretically, this presents no problem. Instead of recording new material on track 24, three of the previously recorded tracks (5, 6, and 7, for example) can be monitored from the record head, mixed together, and re-recorded onto track 24, as shown in Figure 18-4. Now, track 24 contains a mono mix of tracks 5, 6, and 7, and these three tracks may be erased and reused for new material.

Any imperfections in the record head's behavior as a playback device will show up in the mixdown of tracks 5, 6, and 7. So, when bouncing tracks becomes necessary, the playback response of the record head can no longer be ignored. The frequency response and level must be as close to that of the regular playback head as is possible. This requirement becomes even more important when a noise reduction system is being used. Refer to Chapter 9 on noise reduction for a discussion of tracking error due to level differences.

Figure 18.4 Transferring, or "bouncing" tracks. At the console, tracks 5, 6, and 7 are mixed together and routed to track 24.

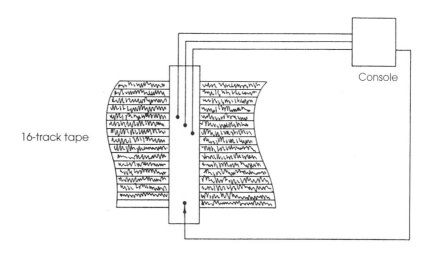

As just mentioned, most state-of-the-art tape recorders have separate circuitry for use with the record head. Thus, during alignment, the record head playback output and frequency response should be adjusted using the standard alignment tape.

Transferring Onto Adjacent Tracks

In planning a session that may require transferring tracks, it is important to remember that material from any track may not be bounced to an adjacent track. For example, although tracks 5, 6, and 7 may be bounced to track 24, as just described, they may not be bounced to either track 4 or 8, since track 4 is adjacent to 5 and track 8 is adjacent to 7.

The reason for this restriction is that the inter-channel crosstalk and separation within the record head is by no means infinite. Consequently, while recording onto track 8, the track 7 section of the record head "hears" some slight portion of what is being recorded (as does track 9). This has no particular significance unless track 8 is being fed some mixture that includes track 7 (or 9). Now, if track 7 is being fed to track 8, and track 7 is also picking up some of track 8 internally through the head stack, the feedback squeal thus created will be recorded on track 8, preventing the use of the track. This condition is illustrated in Figure 18-5.

"Punching In" In conventional recording, a series of takes are made until a satisfactory recording is achieved, or, by editing, the definitive performance may be assembled by splicing together sections from two or more takes. However, when recording tracks on a tape that contains previously recorded material, the manner of doing retakes is quite different.

To illustrate, assume that the instrumental background has been recorded and a chorus is now being added to the master tape. The chorus sings the first verse properly, but a mistake is made within the second verse. A new take of the second verse cannot be started on a fresh piece of tape, since the chorus must, of course, fit in with the previously recorded material on the master tape. So the master tape is rewound to a point before the beginning of the second verse.

Figure 18.5 Recorded material cannot be transferred to an adjacent track.

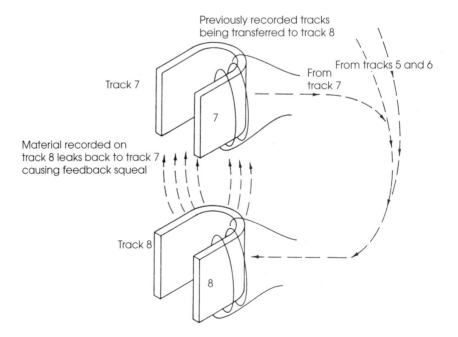

The tape is played, and just before the point at which the second verse begins, the chorus track is placed in the record mode, while the chorus sings the second verse again. The new performance takes the place of the old, while the previously satisfactory first verse remains as is. The process may be repeated again and again until a complete performance has been recorded, perhaps even phrase by phrase.

The practice of inserting the new material in this manner is popularly known as "punching in," since the engineer spends so much of his time punching the record button.

While awaiting the punch-in point, the musicians may wish to listen to the earlier performance, and perhaps play along with it so that the new material may closely match the previously recorded parts that are to be saved. If the playback response of the sync output has been properly aligned, there should be no distracting differences in sound quality as the musicians hear first the performance which is being saved, and then the new performance of the sections being redone.

Remote Control of Record/Playback Mode

On any session involving the Sel-Sync process, it is important that the record/safe and input/sync/playback mode of each track be independently controllable. For example, it may be necessary to record on tracks 9 through 12, while the material previously recorded on tracks 1 through 7 is monitored in the sync mode. The remaining tracks are reserved for future use. Obviously, a single record button that puts all tracks into the record mode at one time would be useless. On the other hand, 24 or more separate record buttons would be very difficult to operate and would make it very easy to accidentally erase a wanted track.

Figure 18-6 shows a typical multi-track remote control unit, providing individual control over the mode of each track. When the single record button is depressed, the machine will record only on those tracks whose safe/ready switches were previously put in the ready position. These switches are, in effect, standby switches, readying the appropriate tracks for recording while protecting the other tracks from accidental erasure. Previously recorded tracks may be monitored from the record or the playback head, depending on the position of the input/Sel-Sync/reproduce (i.e., record head/ playback head) switch, also seen in figure 18-6.

Most multi-track tape machines also provide automatic switching so that when the record circuit is energized, the channel in ready

512 THE AUDIO RECORDING HANDBOOK

Figure 18.6 A remote control unit allowing individual control over each track along with monitor switching, locate memories and transport control. (Sony PCM-3324)

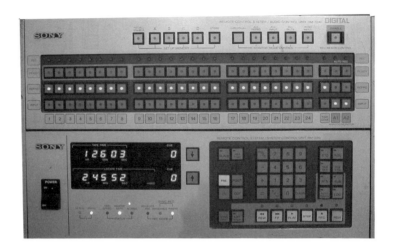

goes into the record mode and the input/Sel-Sync/reproduce switch changes from the Sel-Sync mode (record head playback) to input monitoring on the selected channels. This allows the musicians to hear from the record head on the selected track until record is selected, and when record is depressed, the signal in the headphones switches from the previously recorded material to the new material being played and recorded.

The Console in the Sel-Sync Mode

As described earlier, in Chapter 16, the multi-track recording console will contain monitoring facilities which enable the musicians in the studio to hear previously recorded tracks. In Figure 16-10, the tape recorder outputs, returned to the console through the multi-track tape returns or line inputs, are routed to the cue system as shown. These previously recorded tracks are also routed to the monitor path and control room monitor system so that the engineer/producer may likewise hear them. In the detailed drawing of the console monitor section (Figure 16-4), the line-in/line-out switch (M) would be in the line-in position for each previously recorded track.

If the multi-track machine in use has a standby monitor setting, the entire console can be put in the line-in mode. Positioning the standby monitor to the input monitoring position on the input/sync/reproduce switch of the tape recorder and setting the master monitor switch to Sel-Sync will cause the tape recorder to enter the input monitoring mode when stopped. The machine will automatically be switched to sync mode when the record mode is entered, with the exception of any channel that has had "ready" selected. These selected channels will remain in input mode as previously discussed.

Figure 18-7 is a simplified description of the signal routing through the console in the Sel-Sync mode. Note that the engineer must provide the musicians with a balance of previously recorded and new program material, while at the same time the new material is being added to the tape.

The controls for the control room monitor system and the auxiliary send headphone monitor system are usually independent of each other so that the engineer may route a suitable headphone mix to the musicians, while in the control room he may listen to, perhaps, just

Figure 18.7 Simplified signal flow paths during a Sel-Sync® recording session.

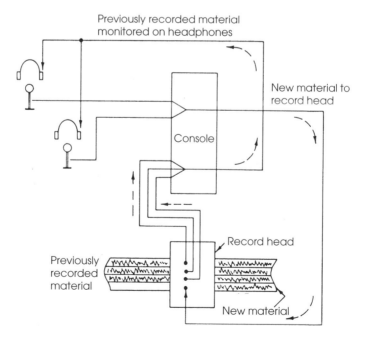

the new material to make sure that it is being recorded and added to the tape properly.

■ HEADPHONE MONITORING

Depending on the acoustic isolation between the instruments in the studio, the engineer will probably have to set up a well-balanced headphone cue mix of the program being recorded so that the musicians may hear each other in addition to what was previously recorded. In fact, even when no material has been previously recorded, headphones may be required, as the isolation between instruments prevents the musicians from adequately hearing each other without them.

Often, several different headphone balances will be required to suit the needs of various musicians. The drummer, for instance, may not need to hear himself in the headphones, yet the guitarist in an isolation booth will certainly need to hear the drums in his headphones if he is to play along in time. In short, a flexible headphone monitoring system is a very important part of any modern recording studio console, and the engineer must be prepared to establish quickly one or more headphone balances in addition to the basic recording balance heard through the console's monitor section.

Selection of Headphones

Headphones with open cell foam ear pieces should be avoided in most studio applications, despite their wearing comfort and frequently excellent reproduction capabilities. Although the headphones allow the musicians to hear each other directly, these types of phones also permit a considerable amount of the headphone monitor mix to "bleed" or be heard in the studio. Consequently, the microphones in the studio pick up a lot of this headphone leakage which then gets recorded onto the tape again, along with the new material. In severe cases, a feedback squeal may be produced if the microphone picks up too much of a headphone feed that already contains a portion of the microphone's output. For the same reason, inexperienced musicians should be cautioned against hanging their headphones across an active microphone or microphone stand between takes.

When a large number of headphones is in use, the engineer should periodically listen to just the program being recorded at the moment over the control room monitor system. Very often, some musicians will remove their headphones without unplugging them, and the headphone program is then heard in the studio and picked up by the microphones again. If a musician decides not to wear headphones, they should be unplugged before any recording is done. A periodic check of the material being recorded will verify that headphone leakage is being kept to a minimum. A very quick check can usually be made by depressing one or more of the solo buttons and listening for headphone leakage during a pause in the part being recorded.

Loudspeaker Monitoring of Cue System During Rehearsals

When strings and/or brass are to be added to a basic track, the conductor may wish to rehearse without headphones in order to better communicate with the musicians. At this time, the musicians may listen to the basic tracks over loudspeakers as they rehearse their parts. Although this is certainly no technical problem, the producer should understand that the engineer will probably need some additional rehearsal time, once the loudspeakers are shut off and the musicians are wearing headphones again. As long as the loudspeaker is active, all the microphones in the studio will hear the cue feed, making it difficult or impossible to establish a meaningful balance of the program about to be recorded. In addition, if the microphones themselves are being routed to the cue system, their output will have to be kept at a minimum so long as the loudspeakers are on, to prevent feedback.

■ TRACK ASSIGNMENT AND DISK ALLOCATION

Beyond operator convenience, there are a few precautions to be observed in assigning instruments to the various tracks of the modern multi-track tape recorder. Whenever possible, it is a good idea to assign the same instrument to the same track on each successive recording session so that later recording and mixdown work may proceed with a minimum of switching changes. For example, if strings and brass overdubs are to be added to several songs, a great deal of time will be saved if the same tracks are available on each song. In addition, once a basic cue mix has been achieved, it will probably suffice for much of the session, provided that the track assignment has remained constant during the earlier sessions.

The mixdown session will be discussed in detail in the next chapter, but here it should be noted that mixdown work will proceed a lot more efficiently if the tracks are arranged in some logical order prior to recording. For example, rhythm instruments may be recorded consecutively from track 1, while sweetening tracks begin at track 24 and work backwards towards the center of the tape (unless tracks 23 and 24 are reserved for time code and/or automation data). This leaves the center tracks open until last, so that additional instruments may be recorded towards the rhythm or sweetening sides as appropriate. The lead vocal can then be recorded in the middle.

In the early days of multi-track recording, the outside tracks were often reserved for the least important instruments, since there was apt to be some problem with tape-to-head contact. However, this should no longer be a problem, provided that the analog tape recorder is in proper mechanical alignment. With modern digital multi-track recorders, this is not an issue.

Another important consideration when recording direct to a DAW or hard-disk recorder is disk allocation. Some systems have a problem with playing or recording too many tracks at a time from one disk. Due to the large storage space required for high resolution digital multi-channel recordings, a number of separate disks are usually used. Between 8- and 16-track recording/playback is normally possible, depending on the system, the disk rotational speed and access time, and the resolution of the signal. Therefore, it is necessary to assign which tracks are to be recorded on which disk. The prudent engineer will not only make sure that there is enough disk space for all the tracks, but will also plan so that all of the bass track, for example, for a particular tune is recorded on one disk and does not cross over to another when space becomes unavailable.

In the case of the modular digital multi-track where only eight channels can be put on a tape, it is wise to group instruments together. For instance, if three DTRS machines are to be used, it would be best to keep all the rhythm tracks together on the first two machines. This allows the engineer to record the solos and leads on the third machine, and if the artists feel the results of the first takes are unsatisfactory, simply placing a new tape in the third machine makes eight new tracks available, while allowing the first set to be retained in case they are needed later.

■ PREPARING FOR THE MULTI-TRACK SESSION

Seating Plan In preparing the studio for the recording session, a little pre-planning will go a long way towards improving the efficiency with which the recording work progresses. The first step is to work out a seating arrangement that will be most comfortable for the musicians. Often, they are spaced as far apart as possible under the mistaken impression that this will keep leakage at a minimum. In a very dead studio this may in fact be true, but in most cases the wide spacing creates more problems than it solves. In the average studio, there is not that much attenuation as the distance between instruments is increased. The further away one musician moves from another, the more reflections each one hears, and the resultant signal delay makes it much more difficult for the musicians to play together. The performance probably suffers far out of proportion to any improvement in the recorded sound.

If leakage is that much of a problem, a far greater improvement might be made by moving the microphone a few inches closer, rather than moving the other musicians a few more feet away. For example, if two musicians, A and B, are one and three feet away from a microphone, this ratio can be doubled by moving musician A six inches closer to the microphone, or by moving musician B back an additional three feet. The first alternative will usually be the most satisfactory.

Microphone Setup Once the seating plan has been worked out and set up, the microphones may be placed in position. When running microphone cables across the studio, a great deal of confusion may be avoided if the cable is plugged into the microphone input panel first and then brought to the microphone, rather than the other way around. The cable slack is laid near the base of the microphone stand so that if the microphone has to be moved during the session, it will be possible to do so without too much difficulty. This also prevents the slack from every cable in the studio from accumulating in front of the microphone input panel, which would make it difficult to make quick changes or trace problems later on.

Console Preparation

It is a good idea to neutralize or "normal" all console controls before beginning any session. That is, make sure that all signal routing switches are off, and that the equalizers are in their zero position. With in-line console designs, it is also important to make sure that all switches relating to the monitor and channel paths are in their monitor path position. By following this precaution, signals will not be routed to the wrong place due to a depressed switch that went unnoticed, and unwanted equalization will not be applied to any channel path.

When an automated console is in use, it is sometimes a good idea to ready the automation system for a write pass sometime before the end of the overdub session. Since with the in-line console design the VCA faders are feeding the monitor system, a basic mix that can be used as the starting point for the mixdown session can be stored during the last overdub.

■ RECORDING

Slating

Prior to each take, the tape is usually slated. The term is taken from the film industry, where a slate board with the take number is held in front of the camera before each shooting. In the recording studio, the slate consists of a spoken announcement, such as "take 3," usually made from the control room. Some consoles feed a low frequency tone to the tape recorder, along with the voice, during the slate announcement. Later on, when the tape is rewound at high speed, the tones are heard as high pitched beeps which identify the beginning of each take. By counting these beeps, the engineer may quickly locate the beginning of a desired take.

Takes are numbered consecutively, and sometimes each new song begins with "take 1." However, it is often a good idea to continue the count for the duration of the session or for all the songs on the CD or other media. Thus, if one song ends after take 10, the next one will begin with take 11. This is a great help when recording unfamiliar music, or for tapes that will be sent to other studios for additional recording, editing, or mixing. If there is only one "take 17" on the entire collection of master tapes, there can be no question as to the identity of the take.

Count-Offs When it is known that additional material will be added to the tape, beginning at the first beat, a lot of trial-and-error time may be saved by recording a spoken count-off during the first session. On subsequent overdub sessions, the musicians will hear the count-off and be able to enter on the proper beat without resorting to guess work or complicated visual cues from the control room.

When there are to be long pauses during the recording of the basic tracks, during which additional instruments will enter later on, it is a little more difficult to provide satisfactory entrance cues. If someone in the studio counts time, it will be heard over most of the microphones, especially since there will be no music masking it. This will make it very difficult to remove the count later on. In such cases, it is better to have someone in the control room or in an isolation booth count time onto just one track. Later on this track may be erased with little difficulty before the mixing session. This track may often be called the cue track. Another possibility is for the drummer to click his sticks together in rhythm near one of the drum microphones such as the hi-hat pickup. This sound is soft enough, particularly since the drummer is usually slightly separate from the group, that the sound does not show up in other microphones. Later, that cue track can simply be muted in the mixdown process.

Tuning If any sort of tuned instruments are to be added later on, it is a good idea to record a reference tone (e.g. A-440) during the recording of the basic tracks. This will enable the musicians on later sessions to tune to the basic track in case there is a slight pitch variation between one session and/or tape recorder and another. Or, if the tape recorder in use has variable speed, the machine's speed may be tuned as required.

Recorded Levels In multi-track recording, it is a common practice to record each track at as high a level as possible. The desired control room monitor balance is achieved by adjusting the monitor fader of each track. In this way, each individual signal is recorded as high above the noise floor as possible. As mentioned earlier (Chapter 1), the engineer should realize that VU meters are not an accurate indication of the program's peak levels, and meters with PPM ballistics should be used if available, especially on percussive instruments.

Similarly, when a digital multi-track is in use, the prudent engineer will make sure that the recorded level does not exceed the maximum quantization levels as discussed in Chapter 13. In general, digital recording levels should be more conservative since signal-to-noise ratio is not dependent on recording level.

Using Signal Processing Devices

Signal processing should be used with some caution during the recording session. Equalization, phasing, or compression effects that may sound just right during the recording session often turn out to be unsuitable later on, after all the succeeding tracks have been recorded. Most of these effects can and should be added during the mixdown session. It is difficult and often impossible to remove the effects of signal processing devices that were applied during the recording session. An exception to this rule is noise gating, which is often used to help keep the individual tracks as noise free as possible during recording.

Of course, as before, when several inputs are combined during the recording session, it is necessary to process each one as desired before they are mixed together.

Artificial Reverberation

Generally, artificial reverberation is best applied after the recording session, during the mixdown session. As noted earlier, stereo reverberation requires two outputs, which are fed left and right. If stereo reverberation were added during the recording session, two tracks would have to be used. This of course wastes tracks needlessly and adds tape noise. Besides, a more suitable reverberant field can probably be created during the later mixdown session when the individual tracks are heard and processed in context.

Of course, if it is necessary to add reverberation to one or more signals which are being combined during the recording session with other unprocessed signals, this must be done as recording is taking place. For example, the rhythm guitar may be being taken direct and through a microphone with the combined signal sent to one track. A slight amount of reverberation added to the dry direct signal may enhance the sound, yet no reverberation may be desired on the microphone signal. So, during recording, the direct sound feeds the reverberation device while the microphone does not. In this case, if the reverberation was added later, during the mixdown session, it would affect the total guitar sound rather than just the sound that came from the direct box.

Often musicians will want to hear some reverberation in their headphone auxiliary mix. The producer may also wish to hear the control room mix with reverberation. It is important to make sure that the required reverberation only goes to the monitor and auxiliary paths, and not to the channel paths. Reverberation added to the tracks of a multi-track recorder becomes additive in mixdown. This may result in a muddy or distorted sound. For this reason, any reverberation added to the multi-channel tape during the recording of basic tracks and overdubs should be done with extreme caution.

End of Recording Many contemporary popular recordings eventually fade out rather than coming to a definite musical ending. The fade-out is usually made during the mixdown session as described in the next chapter. During the recording, the musicians usually play the final phrase over and over again, giving the engineer sufficient material to use during the fade later on. At the recording session, it is often advisable to simulate the eventual fade-out by bringing down the monitor path master fader or control room monitor gain, thus verifying that the music continues long enough to make a suitable fade during the mixdown session.

For those songs that do not end in a fade-out, it is important to maintain a few seconds of silence at the end of the recording so that the reverberation (natural or artificial) will have a chance to die away.

■ END OF SESSION

It is as important to return the studio and control room to their original states upon the completion of recording as it is to properly set up before the recording session begins. The console should be returned to its normal state (all switches in their proper positions), microphones should be put away and cables returned to their normal storage place in readiness for the next recording session.

NINETEEN

The Mixdown Session

During the mixdown session, the completed multi-track tape is played back through the console. Each track takes the place of a microphone input, and is routed, processed and combined with the other tracks to form a stereo program which is then recorded onto another analog or digital tape recorder.

■ EDITING

After the recording session, the takes that are to be used for the mixdown session are usually removed from their various reels and placed on one or more separate reels. The "out takes" may then be filed or discarded so that the engineer does not have to be encumbered with them while mixing. Often, this type of editing work is done just after the basic tracks are recorded, and before the overdub or "sweetening" sessions.

It is important to remember not to remove any of the count-offs at the beginning of each master take, and to leave enough extra tape at the end for any string, brass or effect parts recorded as overdubs that may last longer than the basic tracks. If the tape has been "striped" with SMPTE time code during the basic sessions, the editing engineer should remember to leave at least 10 or 15 seconds of uninterrupted code prior to the beginning or "punch-in" point of the musical selection. If the tape is to have time code added after all tracks are recorded, enough blank tape should be left at the beginning of the selection to accomplish this. As mentioned in Chapter 17, some automation systems may require up to 25 or 30 seconds of time code to calibrate the system.

For easy identification, the master takes on an analog reel may be separated from each other by a length of white leader tape. This type of editing is quite simple. The engineer listens to the beginning of the desired take, stops the tape, rewinds it to an appropriate point before the beginning of the music and physically cuts it. A length of white leader tape is wound around an empty reel, and the beginning of the master tape is spliced to the end of the leader. The take is then wound onto the master reel. At the end of the take, the tape is again cut and leader tape is spliced in. The master reel is put aside and the editing engineer locates the next desired take and repeats the process, splicing the take onto the master reel. The procedure is illustrated in Figure 19-1.

Musical Editing When the master take is to be a composite of segments of several takes, the editing process is somewhat more complex, although the principle remains as just described. However, great precision of editing may be required, as the beginning of one segment is spliced directly to the end of another. Since the segments come from differ-

Figure 19.1 The editing process. The takes that are to be used for mixdown work are removed and stored on a separate reel.

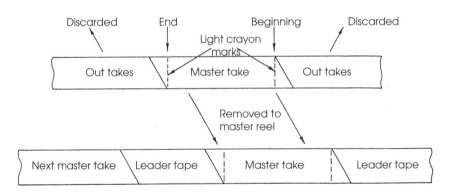

ent takes, the engineer must listen carefully to the proposed splice point before cutting the tape. Slight differences in phrasing, tempo, level, pitch, etc., that are not objectionable in themselves may become extremely noticeable if the takes are spliced together. Careful listening will usually reveal whether the splice can be made. However, if an attempted splice turns out to be unacceptable, the segments can be restored to their normal positions and a new splice point sought. It is here that digital editors excel.

In the event that a master tape containing a track of continuous SMPTE time code must be edited, the code must be replaced after the editing is completed as any discontinuities or discrepancies in the code will cause the automation system to provide false data to the mixdown monitor bus VCAs. Previously stored automation data may be lost in some systems, while in others the data can be reconstructed through merges or by starting the replacement code at the original SMPTE time code address.

Splicing Blocks

Figure 19-2 shows several commercially available splicing blocks. Note that each allows the tape to be cut on an angle. This is done to distribute the cut over a short segment of the tape. A 90 degree or "butt" splice will usually make an audible "thump" as it passes the playback head. This occurs because the change or shift in magnetic flux density strikes the playback head gap space simultaneously for the full width of the tape. An angle cut allows the splice to slide across the playback head, thus distributing the flux change gradually across the gap space. The time before and after the center of the splice is known as the "lead" and "lag" time respectively.

Although a 45 degree cut may be used and is desired on narrow width monaural tapes, it should be remembered that this angle distributes the cut over a length of tape that is equal to the width of the tape. Thus a 45 degree angle cut on a 2-inch tape takes up 2 inches of tape travel. This means that at a speed of 15 inches per second, the splice will take 2/15 of a second (133 milliseconds) to pass through all the tracks. This means that the edit on track 1 will occur 133 ms prior to the edit on track 24. This length of time could make the splice quite noticeable, and so a narrow angle cut is usually used. Narrow editing angles take advantage of the psychoacoustic fact that the ear is unable to distinguish signal transitions that occur at less than 10 ms. Figure 19-3 shows a 2-track tape with a 45 degree angle cut, illustrating the lead and lag between the two tracks.

Figure 19.2 Commercially available splicing blocks. The vertical cuts are for use with cutting digital tape and the angle cuts are used for splicing analog tapes.

Figure 19.3 A two-track tape with a 45° angle edit. Note the timing differential between the beginning of the cut on track 1 and the end of the cut on track 2.

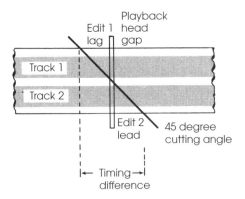

The preferred angles for splicing tape are:

¼-inch monaural	45 degrees
¼-inch two-track or quarter-track stereo	30 degrees
1-inch multi-track tape	22 degrees
2-inch multi-track tape	17 degrees

The above angles will result in splices with signal transitions below the 10 ms threshold.

Digital Editing   The same, and often much better, results can be obtained using a DAW. Instead of using a splicing block and a razor, the master tape is loaded into the DAW. Figure19-4 shows the edit screen of a 2-channel DAW with four virtual tracks. The master tape takes have been

Figure 19.4   The editing process using a DAW. The takes are rearranged onto another set of virtual tracks (Sek'd Samplitude).

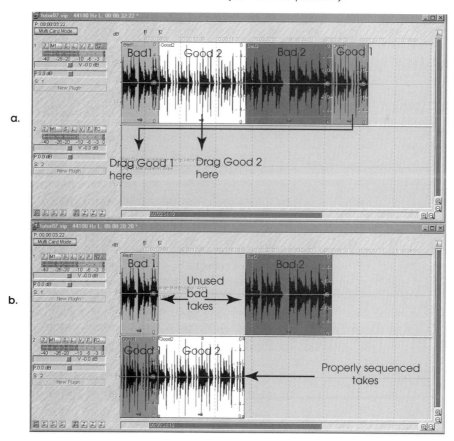

loaded into the system as shown in the top screen (a). Then using drag-and-drop techniques, the desired takes are taken from virtual tracks 1 and 2 and placed in the proper sequence on virtual tracks 3 and 4 as shown on the lower screen (b).

When doing musical editing on the DAW, the edit-in and edit-out points can be precisely located by zooming in to the waveform level as shown in Figure 19-5. In the top screen (a), the edit between Good1 and Good2 looks correct. However, when zoomed in to a higher resolution, it can be seen that the edit needs to be corrected.

One of the major advantages of the DAW editing system is that if the edit is not perfect, or the artist changes his mind regarding which take to use, it is a simple matter to "undo" or go back and redo the edit since digital editing is non-destructive. That is to say, since the DAW simply builds an EDL or instruction list instead of modifying the actual sound file, the original master recording is still intact. When editing with a razor blade and splicing block, it is often very difficult to undo an edit.

Track Editing

On a multi-track tape that is a product of many overdub sessions, it is very easy to wind up with unwanted material on small segments of various tracks. For example, there may be studio noises on some of the tracks during long pauses. Or a verse or instrumental solo may no longer be wanted once the complete recording has been reviewed.

It is obviously impossible to physically edit a track on a multi-track tape without affecting all the remaining tracks, so electronic editing measures are necessary. One method entails removing some of these segments by simply switching the appropriate tracks on and off on cue. In non-automated mixdown situations this must be done manually. If there are several such segments, it may be difficult to coordinate all the channel-on/channel-off cues while also concentrating on the mixdown itself. In this case, it may be wiser to review each track carefully and to erase the unwanted segments before beginning the mixdown work. However, before doing any erasing, the engineer should make certain that there is sufficient clearance before and after the segment to be erased so that wanted material is not accidentally lost. When this tape "housekeeping" is properly accomplished, only wanted usable material remains on the tape.

With an automation-assisted mixdown system, this track editing can be programmed with the Mute-Write function of the mixdown automation system. Any number of channels can be switched on or

THE MIXDOWN SESSION 529

Figure 19.5 The ability to zoom into the actual sample duration allows precision editing (Sek'd Samplitude).

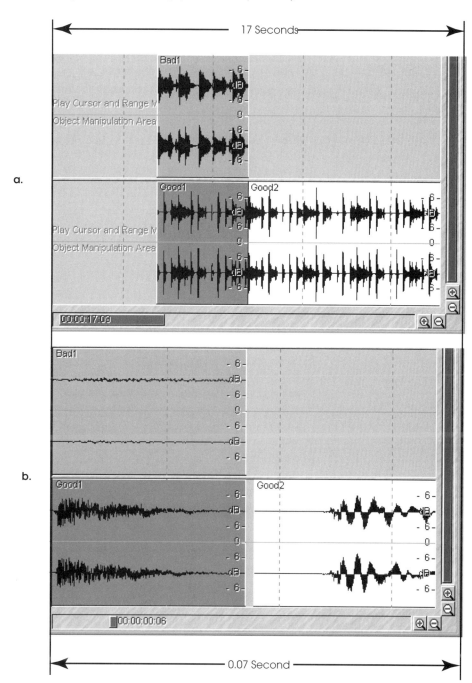

off throughout the duration of the musical selection. Once programmed, the channel-on/channel-off status will be repeated automatically on each successive mixdown pass unless changed by the engineer. One of the advantages of this method over the former "housekeeping" method is that in the former system, although the track is empty of unwanted material, it is still contributing noise to the monitor/mixdown bus. The later system maintains a quieter mixdown by keeping the unwanted channels muted when not required.

If the tracks were recorded on a multi-channel DAW, it is a simple matter to go in and delete the unused material from the EDL or playlist. This can often be done graphically or in a text-based fashion. Here again, though, the original sound files are not changed, and the material can be reinserted in the EDL if so desired.

■ TRACK ASSIGNMENT AND PANNING

In many cases, during recording, instruments are assigned to various console channel outputs without regard to their eventual left-to-right orientation. Each track may be heard on the left, center, or right simply by panning the signal sent to the console's monitor path. However, during the mixdown session, the monitor path is used to feed the two-track mix bus, and the monitor pans are used to send the various multi-track signals to these two console outputs. One represents the left track and the other the right track. Signals that are to be heard in the center are routed equally to both left and right tracks, or the pan-pot may be used for intermediate positions between left and right. The action of the pan-pot was described in Chapter 16.

Mixing for surround playback adds several more dimensions to the possible choices of panning. Now instead of a simple pan-pot, the engineer may have a joystick that allows the signal to be moved between and/or combined in up to eight loudspeakers. In a typical mixdown to a 5.1 format, the joystick will pan between the left, center, right, left-rear and right-rear loudspeakers. Additionally there is the LFE channel. Although the LFE channel is primarily used for sound effects, and as such is prevalent in audio for visual work, the LFE channel can be used in audio-only productions when required. If a 7.1 mix is done, two additional channels, either the left-side and right-side channels or the left-center and right-center channels, are added for a total of eight mixdown channels.

■ PREPARING FOR MIXDOWN

Before beginning the mixdown work, the tracks may be reviewed individually, while equalization, reverberation, and other types of signal processing are tried. Although there is nothing wrong with spending a little time verifying the contents of each track, there is little point in spending a lot of time evaluating each track out of context from the total program. By the time of the mixdown session, it really makes no difference what each isolated track sounds like. The signal processing that sounds right when the track is heard by itself will rarely remain effective once the track is mixed with others. What really matters is the contribution of each track to the total recording.

For example, two guitar tracks heard individually might be processed at great length, only to find that when they are combined together, they are noticeably out of synchronization with each other. It makes better sense to spend some time listening to groups of tracks to hear the effect of one instrument against another. The basic tracks may be monitored first, to work out position and balance. Then, rhythm instruments that were added during overdub sessions may be added and mixed in.

Later, strings, brass and synthesizer parts may be reviewed as an ensemble. Once a basic balance for this group is worked out, they are mixed in with the rhythm section. The solo vocal performance is often added last. It is a good idea, however, to carefully listen to the vocalist before spending too much time working out the ideal instrumental balance. Depending on the performance, the instrumental balance may have to be substantially modified once the vocalist is added.

During these mixdown preparations, decisions as to grouping, as mentioned in Chapters 16 and 17, can be made. For example, once the initial balance adjustments and panning for the drum kit are decided, the entire set of drum tracks can be grouped to, say, Group No. 1. A group master can be assigned, and then whenever a balance decision affecting the entire drum set is made, moving the group master will raise or lower the level of the entire drum mix without affecting the drum kit's internal balance or panning.

When an automation system is in use, these initial adjustments concerning balance, equalization and other processing are usually done in what is called the rehearsal or monitor mode. This allows the engineer to set up all the signals returning from the multi-track tape recorder and the external processing equipment before the initial Write pass is made. With tape-based automation systems, this prevents cumulative offsets between the data and the music, and with disk-based systems (hard-disk or floppy/zip disk) this prevents filling up the storage space with multiple revisions. Refer to Chapter 17 for a more detailed discussion on this subject.

■ ASSISTANCE DURING MIXDOWN

In a complex non-automated mixdown session, there may be more changes to be made than can be handled by one engineer. However, the role of the assistant must be carefully spelled out in advance, to keep confusion at a minimum. One person must remain in complete charge of the mixdown, while the other makes whatever changes have been assigned to him or her. If the two engineers attempt to function independently of each other, each will be confused by the action of the other, and little or nothing constructive will be accomplished. Mixing is pretty much like sailing: there can be only one captain at a time.

It is at this stage in the mixdown process that automated mixdown systems justify their existence. After initial console adjustments are made, the first Write pass is made. This stores the fader level and the mute status for each channel in the mixdown path. After a Read pass to verify that the initial data were stored properly, the system is then switched to the Update mode. This allows the engineer to make changes to individual track levels and mutes that are remembered by the automation system during subsequent passes of the master tape. More sophisticated digital systems may allow automation of nearly every parameter of each channel.

When the engineer is satisfied with the final mixed version of the selection, the system is once again placed in the Read or Playback mode, and upon playback, all the stored levels, level changes, equalization (if the system implements it) and mutes are executed by the automation system. This makes it possible for a single engineer to handle the mixdown of as many tracks as necessary. By routing the reverberation and effects returns to automated channels assigned to the mixdown path, the engineer can automate these effects as well.

■ RECORDING AND MONITOR LEVELS

As the mixdown session begins, the engineer establishes a comfortable listening level in the control room. As more and more tracks are added, the monitor level rises accordingly, and may have to be turned down to maintain a reasonable sound level. However, the recorded level also rises with each additional track and may become excessive once all the tracks have been added to the monitor path and sent to the two-track tape recorder.

The two-track master fader can of course be brought down somewhat. However, this does nothing to protect the amplifier stages just before it from high-level overload. In most cases, the individual input faders should each be brought down to maintain a safe recording level. This can be inconvenient, since it means that as each group of tracks is brought into the mix, all the monitor faders must be readjusted—an awkward procedure at best. Some automation systems allow the automation master fader (different from the two-track master fader) to be brought down prior to nulling (refer to Chapter 17) all the faders for another update pass. This automation master lowers all of the gains simultaneously. This is a stopgap measure at best since, at some point, the maximum travel point on the individual linear faders is going to be reached.

As a practical alternative, the mixdown session may begin with the monitor loudspeaker levels at a higher than normal setting. With the monitor level up, the engineer will not be tempted to raise the monitor path faders quite so high, thus keeping the recording level down. As more and more tracks are added, the control room monitor level may be brought down, and as the recording level gradually increases, it should remain within safe limits. A little practice will determine the optimum starting setting for the monitor control level.

As the mixdown continues, listening levels have a way of creeping gradually upward. (Some engineers swear that they can tell how long they have been mixing by looking at the position of the control room monitor level control.) But, as the listening level increases, the engineer should keep in mind the implication of the equal loudness curves discussed in Chapter 2. Music mixed at a high listening level will sound quite different when played back later on at a lower level. The engineer who disregards this important fact does so at his peril.

Monitoring levels in recording studios have never been standardized, and as a result levels from place to place have fluctuated greatly. Now, however, there is finally a standard monitor level that is becoming accepted in the sound industry. It is the Dolby and SMPTE standard used for motion picture theatre sound levels and for home DVD video playback with Dolby AC-3 (5.1) encoded surround audio. This monitor level is specified as a sound pressure level (spl) of 85 dB measured with an RMS meter at the listening position when mixing for the cinema, and an spl of 79 dB RMS when mixing audio for the home theatre system. This continuity of level allows the engineer to monitor the mixdown level at the same spl as the end listener.

■ MONITOR LOUDSPEAKERS FOR THE MIXDOWN SESSION

Presumably, the well-equipped recording studio will have the best monitor system possible within its budget limitations. At this point in the book, the advantages of a well-designed, wide-range monitor system should require no further discussion. However, a valid argument may be made for using a reasonably inexpensive monitor speaker at various times during the mixdown session. Despite the impressive technology discussed in the preceding chapters, it is unlikely that the majority of the recorded-music-buying public will be listening over state-of-the-art professional playback equipment. There are more portable radios in this world than can be counted, and their sound quality is—to put it politely—horrible. Yet this is where much of the final product will be heard, and it really doesn't matter what it sounds like on the studio's super sound monitor system. The question is: what does it sound like on the beach—or in the car? One doesn't have to be an expert to realize that many mixing subtleties will be lost on a two-inch speaker in a plastic case.

For this reason, it is good to have an inexpensive speaker system available in the control room, to get some idea of what the ultimate mixdown will sound like in the hands of the consumer. Some ambitious studios even rig up a feed through a small radio receiver so that the engineer and the producer may get some impression of what the CD or compact cassette buyer will hear when he brings the final product home, or hears it in his car.

It is an exercise in futility to attempt a mixdown that will sound ideal on all systems. A review of Chapter 5 should clear up any misconceptions on that point. However, the engineer should have at least some idea of what the master tape may sound like once it leaves the studio.

■ THE MASTER

Now that the two-track master has been mixed from the composite multi-track master, any additional musical editing that could not be done earlier can be accomplished. Since the whole take has been mixed to provide a continuity of style and rhythm, it is often better to edit out extra verses, choruses, etc., after the mixdown has been completed.

Next, white leader tape of the appropriate length should be placed between succeeding selections or, in the case of a digital tape, the appropriate spacing should be inserted between selections. The timing of this may vary, depending upon whether the product is going to be produced at the manufacturing facility into a compact disc, a compact cassette, a DVD-A or a DAT (Digital Audio Tape).

When the tape is simply a two-channel master, be it analog or digital, it is rare that the channels are reversed when the final product is manufactured, although the channels should be well defined as to left and right. However, in the case of a 5.1 or other surround master, it is imperative that the channel sequence on the master tape be defined in writing on the packaging. Most surround masters are currently delivered to the mastering facility on the DTRS or ADAT format with the DTRS type the preferred medium at this time. Although it is usually easy to notice that the left and right channels of a recording are reversed, it may not be so apparent in the rear channels. There have been a number of standards since the advent of 5.1 surround formats but the SMPTE/ITU (International Telecommunication Union) recommended standards are becoming more the norm than the exception. It is also the recommended layout for the Dolby AC-3 system (Dolby Digital). DTS is one of the competing film surround technologies, as is SDDS. Below is a chart that lists the various track assignments by discipline.

	1	2	3	4	5	6	7	8
SMPTE and ITU Standard, Dolby recommended	Left Front	Right Front	Center	LFE	Left Surround	Right Surround	(Left Stereo	Right Stereo)
Preferred by Film Companies	Left Front	Center	Right Front	Left Surround	Right Surround	LFE	(Left Stereo	Right Stereo)
DTS Standard	Left Front	Right Front	Left Surround	Right Surround	Center	LFE	(Left Stereo	Right Stereo)
SDDS Standard (Sony Dynamic Digital Sound)	Left Front	Left Center	Center	Right Center	Right Front	LFE	Left Surround	Right Surround

As time passes, the SMPTE/ITU recommended standard should become the only standard, as this formatting allows digital video systems with 4 channels to have the information necessary for simple stereo plus dialogue. If the master is a 5.1 mix, an optional stereo mix may be placed on channels 7 and 8 except in the SDDS (Sony Dynamic Digital Sound) recommendation, or if the channels are used in other 7.1 configurations.

Most mastering facilities will require reference tones at the head of the tape, whether it is a two-track digital master or an analog master. In the analog domain it is important that the reference tones be recorded on the same machine that produced the two-track master. This allows the mastering engineer to adjust the frequency balance, operating level, tape head azimuth, and overall response of his playback equipment to match that of the two-track mixdown tape recorder. This way, what was heard in the mixdown studio will be accurately transferred to the appropriate final playback medium. Reference tones should include a 1kHz signal for level purposes, a high-frequency tone of 10kHz or 15kHz for azimuth and frequency response adjustments, and a low-frequency tone of 50Hz or 100Hz so that the low-frequency response of the system can be adjusted. If noise reduction is used on the reel, these calibration tones should be included also. All of these tones should be separated from the first selection by a length of white leader tape. Digital masters should have a segment of digital black (all bits 0) of at least two minutes if using tape, and a 1kHz tone recorded at a reference level of -18 dBfs before the start of the program material. A mixdown master sheet is shown in Figure 19-6. Note that some of the specifications are different from those mentioned above, and it is the prudent engineer who checks the required specifications of the intended final destination.

Finally, if providing an analog tape, it should be wound at low speed onto a reel and stored "tails out" with the end of the final tail of white leader secured to the reel with adhesive tape to prevent the tape unwinding during transport. Digital reel-to-reel tapes should be stored "heads out," that is, rewound to the beginning. Additionally, the protective band that goes between the flanges to prevent the tape from being contaminated should be placed around the reel. It is also important to label the box containing the tape, as well as the reel or the cassette itself, with the appropriate recording information. This information should include the tape speed, operating level, track format, order of tones, and timings for the listed selections. Sampling rates, quantizing rates, time code type, and track layout should be included with digital tapes.

Figure 19.6 A track sheet for defining a surround master tape or disk (Courtesy of Surround Associates).

11333 Ventura Blvd.
Studio City, CA 91604
(818)487-9298 voice (818)980-7447 fax
ASSOCIATES

Date:

Project ______________ **Project #** ______________

Client ______________ **Producer** ______________

Studio ______________ **Engineer** ______________

Sample Rate	❑ 48kHz	❑ 44.1kHz	❑ Other	
Bit Resolution	❑ 16bit	❑ 20bit	❑ 24 bit	
Reference Level SPL	❑ 85dB	❑ 79dB	❑ Other	
Tape Format	❑ DA-88	❑ Optical	❑ Hard Disc	❑ Other
Time Code Format	❑ 29.97DF	❑ 30 NDF	❑ Other	
Surround Level SPL	❑ Equal to Front		❑ -3dB to Front	
Leader Contents	❑ Audio Slate ❑ 2 Pop ❑ Other		❑ 1kHz tone @ -20dBFS ❑ Pink Noise @ -20dBFS	
LFE Lowpass Filter	❑ On		❑ Off	
❑ Original Master	❑ Simultaneous Production Master			❑ Protection Dub

Track	1	2	3	4	5	6	7	8
❑	L	R	C	LFE	LS	RS	LT	RT
❑ Othe								

Time Code	Program Title	Mix	Extraction
: : :		❑	❑
: : :		❑	❑
: : :		❑	❑
: : :		❑	❑
: : :		❑	❑
: : :		❑	❑
: : :		❑	❑
: : :		❑	❑
: : :		❑	❑
: : :		❑	❑
: : :		❑	❑
: : :		❑	❑

Notes:

■ MASTERING

Although mastering of the final product is best done by a facility dedicated strictly to that purpose, many studios are now doing their own final tape preparation. One of the most important jobs of the mastering engineer is to do any final equalization, level adjustment, time-based processing and cleanup (fade-ins and outs from digital black or the previous selection) to make the finished product a cohesive continuous product instead of a collection of songs with different sounds and levels. Most CD pressing plants require either an Exabyte Master, a U-matic Master, or a Red Book compliant PMCD (Pre-mastered Compact Disc). All three of these masters will contain not only the program material sequenced in the proper order and spaced correctly, but will also contain the PQ codes required by the CD specification for identifying and locating the tracks and index points. These are also added at the mastering stage of production.

At the pressing plant, the output of the digital master is sent to the Laser Beam Recorder which cuts the glass master used for pressing the CDs. The Sony PCM1630 U-matic type master has previously been discussed in chapter 14. The Exabyte Tape (an 8mm data-type tape cartridge) uses a process called DDP (Disc Descriptor Protocol) to emulate the 1630 U-matic type of tape. It is more robust and smaller than a U-matic tape, and has a much higher level of error prevention and correction properties. As the number of U-matics in use continues to diminish, Exabyte DDP masters and the larger version DLT (Digital Linear Tape, principally used for DVD masters) become the preferred mastering media.

The PMCD is a compact disc containing the same information as an Exabyte or U-matic master but written to a CD-R or RW blank. Not all CD recorders write Red Book compliant CDs. Many write the disc to the Orange Book specification which includes linking blocks for track transitions. These are not acceptable as PMCDs and the pressing plant will convert these to the proper media, at an additional charge, before cutting the glass master. A Red Book PMCD is recorded in what is called the Disc-at-Once (DAO) mode as opposed to the Track-at-Once (TAO) mode. Figure 19-7 is an editing screen from a DAW showing the PQ marks and a printable TOC (Table of Contents) listing.

Figure 19.7 The PQ code writing screen on a DAW (Sek'd Red Roaster 24).

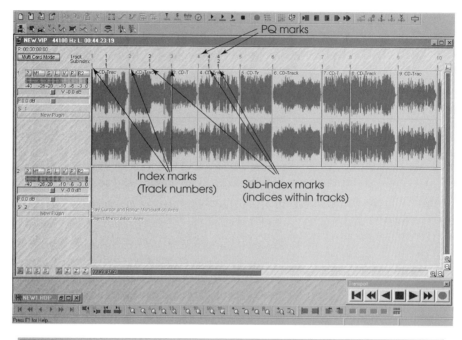

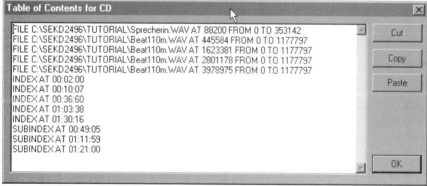

MIXDOWN SUMMARY

Following is a brief summary of the steps required to produce a finished two-track master tape from a completed multi-track master.

Non-Automated

1. Prepare the mixdown machine. The machine to be used should be cleaned, de-magnetized and loaded with the selected type of magnetic tape. The record and playback alignment should be checked and any discrepancies corrected. If recording to digital media for compact disc release, verify that sample rate is set to 44.1kHz with 16-bit quantization.

2. Record the calibration and alignment tones onto the mixdown machine.

3. Prepare the console for mixdown. Place the console in the "mix" or line-in mode so that the output of the multi-track recorder is routed to the monitor path on the console. Make sure that all console adjustments entered during the tracking sessions have been returned to their normal positions.

4. Assign monitor channel path inputs to groups (and select group masters) and set pan-pots for the desired instrumental placement in the stereo or surround image.

5. Play the multi-track tape as many times as necessary to achieve the desired basic mix.

6. Add equalization, reverberation and effects while monitoring the playback of the entire mix, making adjustments as necessary.

7. Changes that are to be made during the actual recording of the two-track mixdown master are noted in the score or on the lead sheet, so that the engineer will be reminded to make these adjustments as the mixdown progresses.

8. The two-track or surround mixdown machine is placed in the record mode and the mixdown of the multi-track tape and any effects are committed to tape.

9. The mixdown tape is rewound and played while the engineer and producer follow the score or lead sheet to verify that the mix is correct. The tape is edited, leadered, and labeled. Often, a safety copy of the mixdown master is made and stored or kept in a sep-

arate location. If digital editing or mastering is required, the master mix tape is copied into the editing DAW, edited, mastered (with PQ codes) and written to an Exabyte tape, DLT tape or PMCD. If desired, the mix could be recorded directly into the DAW during the mixdown process. However, the prudent engineer will also record the mix onto a back-up tape or disk. Remember the first rule of computer editing systems: all hard disks crash, and the hard disk you are using will crash at the most inappropriate time.

Automated The process for an automated mixdown is very similar to the steps listed above. The following steps would be added.

2a. Prepare the multi-track master to receive and send automation data or SMPTE time code, either by preparing two channels for record and playback with tape-based systems, or by striping a selected (or dedicated) channel on the multi-track recorder with SMPTE time code for use with computer disk-based systems.

4a. Assign the automation system to the rehearsal or monitor mode.

6a. Place the console in the Write mode and record the control settings of the basic mix with equalization and effects to the automation storage system. Save the mix.

7a. Place the console in the Update mode. Make and store any changes made during the mix as revisions of the basic Write pass. The multi-track may be rewound and played as many times as necessary until all fader levels, mutes and other automatable settings are updated. Reverberation and effects returns should be also automated and stored at this point.

7b. When all changes and updates have been made, the console is placed in the Read or Playback mode. When the multi-track tape is played, the various monitor path channel controls will be controlled by the automation system.

Glossary

A Weighting—*See* Weighting, "A".

A-B-ing—Comparing two programs, by frequently switching from one to the other.

Acetate—*See* Cellulose Acetate.

ACN—Active Combining Network.

Active Combining Network—*See* Combining Amplifier.

Acoustic Baffle—*See* Baffle, Acoustic.

Acoustic Center—In a loudspeaker, the point at which a soundwave appears to originate.

Acoustic Delay Line—*See* Delay Line, Acoustic.

Acoustic Intensity—*See* Intensity, Acoustic.

Acoustic Lens—A high frequency speaker attachment system, designed to provide a wide radiation angle.

Acoustic Phase Cancellation—*See* Phase Cancellation.

Acoustic Power—*See* Power, Acoustic.

Acoustic Suspension System—A sealed loudspeaker cabinet, in which the enclosed volume of air acts as an acoustic resistance to the speaker cone.

Active Device—A network or circuit capable of supplying a power or voltage gain; for example, an amplifier.

A/D Converter—*See* Analog-to-Digital Converter.

Address Bits—In SMPTE time code, a bit group assigned to various time, sync, or user-defined functions.

AES—Audio Engineering Society

Alias Frequency—An erroneous frequency created when sampling a signal whose frequency is higher than the Nyquist frequency.

Aliasing—The creation of erroneous, or "alias" frequencies.

Alignment—The adjustment of a device to bring it into conformance with published specifications. *See* Azimuth Alignment.

Alignment Tape—*See* Test Tape.

Ambient Noise—*See* Noise, Ambient.

Analog Recording—The traditional method of recording, in which the recorded waveform is an analog of the acoustical/electrical source signal.

Analog-to-Digital Converter—A device which converts (encodes) a quantized analog level into a digital word.

Anechoic Chamber—A room designed to provide a reflection-free environment, for testing microphones, loudspeakers, and other devices.

ANSI—American National Standards Institute.

Asperity Noise—*See* Noise, Asperity.

ATF—Automatic Track Finding. The method used in azimuth recording to synchronize the transport with the tape. Used in place of control track signals.

Attack Time—The time it takes for the gain of a signal processing device to change in response to the input signal level.

Attenuation Pad—A resistive network inserted in a microphone or other line to lower the level by a specified number of decibels.

Auto-Locator—A trade name of the MCI Company, to describe their tape transport control system.

Azimuth—The angular relationship between the head gap and the tape path.

B Weighting—*See* Weighting, "B".

Back Coating—A thin coating applied to the back of a magnetic recording tape, generally to reduce both slippage and the build-up of static charges.

Back Plate—The fixed rear element in the capacitor/diaphragm of a condenser (capacitor) microphone.

Baffle, Acoustic—Any partition, designed to be an acoustic obstruction to the passage of sound waves.

Baffle, Folded—A speaker cabinet that is completely open at the rear. So-called because it is formed by folding over the sides of an open baffle. *See* Horn, Folded.

Baffle, Infinite—Theoretically, a baffle so large that sound originating on either side of it never reaches the opposite side. A practical example would be a speaker mounted in an opening in a wall.

Baffle, Open—Simply a flat partition, of less than infinite dimension, in which a hole has been cut for mounting a speaker.

Balance—The relative level of two or more instruments, signal paths, or recorded tracks.

Balancing—Adjusting the relative levels of instruments or recorded tracks.

Balanced Line—A line consisting of two conductors plus a shield. With respect to ground, the conductors are at equal potential, but opposite polarity. *See* Unbalanced Line.

Ballistics—A property of a meter movement, referring to its ability to precisely respond to the envelope of the signal being measured.

Band-Pass Filter—*See* Filter, Band-Pass.

Band-Reject Filter—*See* Filter, Band-Reject.

Bandwidth—The arithmetic difference between the upper and lower cut-off frequencies of an audio system.

Base—In magnetic recording tape, the plastic or other film upon which the magnetic oxide is coated.

Basic Tracks—In multi-track recording, those tracks that are recorded first. In general, the rhythm tracks (guitar, bass, drums, etc.).

Bass Reflex Enclosure—A loudspeaker enclosure, with an open port cut into the front baffle. Also called a vented enclosure.

BCD—Binary-Coded Decimal.

Bi-Amplification—The process of separating the audio bandwidth in tow, with a separate amplifier for low and high frequencies.

Bi-Directional Microphone—A microphone that is sensitive to front- and rear-originating sounds, and relatively insensitive to side-originating sounds. Also called a Figure-8 microphone, after the shape of its polar pattern.

Bias—A very high frequency current applied to the record head to linearize the transfer characteristic of magnetic recording tape.

Bias Beats—An audio frequency signal that may be created if two slightly different bias frequencies are combined. The audio frequency is the arithmetic difference between the bias frequencies.

Bias Frequency—The frequency of the applied bias signal. Generally, about 150,000 to 180,000 Hz.

Bias Oscillator—A fixed frequency oscillator built into the tape recorder to supply the bias current.

Bias Trap—A filter designed to block the bias frequency, thus preventing it from overloading the record or playback amplifiers in a tape recorder.

Binary-Coded Decimal—A decimal number, in which each digit is encoded into its binary equivalent. Thus, 943 = 1001 0100 0011.

Binary Digit—*See* Digit, Binary.

Binary Groups—*See* User-assigned Bits.

Binary Numbering System—*See* Numbering System, Binary.

Binaural Recording—A recording technique using two omni-directional microphones, one on either side of an acoustic baffle designed to simulate the characteristics of a listener's head. For optimum results, the recording must be monitored over headphones.

Binder—The medium in which magnetic particles are suspended to form the oxide coating in magnetic recording tape.

Bi-Phase Modulation—The encoding scheme employed in the SMPTE time code, in which a modulated square wave continuously varies in frequency between 1200 Hz and 2400 Hz.

Bit—An abbreviation for BInary DigiT.

Bit Stream—A continuous sequence of binary digits, being transmitted along a data bus.

Bit-Packing Density—A reference to the number of bits per unit length, which may be stored on a digital tape.

Block, Data—A section of digital data, usually separated from adjacent data blocks by a gap.

Block-Coded Format—A digital signal in which the bit stream is written (i.e., recorded) in data blocks, separated by gaps.

Blumlein Pair—Any two microphones, arranged for a stereophonic pickup, according to the methods developed by Alan Dower Blumlein c. 1930.

Bouncing Tracks—The technique of transferring several previously recorded tracks to a single unused track on the same tape. The previously recorded tracks may then be erased and re-used.

Breathing—An audible rising and falling of background noise that may become objectionable when using a compressor. Also called pumping.

Broadcast Mode—A console mode in which the input signal is routed separately to both the channel path and the monitor path. Thus, adjustments to the signal feeding the multi-track recorder in the channel path will not affect the monitor path signal, which is presumably being used for a live stereo broadcast.

Buffer—An intermediate storage device, in which data may flow in, and out again, at different rates of flow. Usually provides a constant load to the source, but reduces the source impedance to the next stage.

Bulk Eraser—A strong electro-magnet, used to erase an entire roll of tape at once. *See* Degausser.

Bus—A common signal line, or junction, at which the outputs of several signal paths may be combined. (Frequently misspelled as Buss.)

Bus, Channel—Any output bus, usually so-called to distinguish it from a monitor/mixing bus. Carries signal to the multi-track.

Bus, Data—A wire, or group of wires, across which digital data is transmitted. An n-bit parallel output would require n-wire for mixing and monitoring purposes.

Bus, Mixing—On an in-line console, a bus (or buses) used for mixing and monitoring purposes.

Bus, Monitor—*See* Bus, Mixing.

Bus Selector Switch—A multi-position switch, which permits a signal path to be routed to one or more buses.

Byte—A sequence of bits, shorter than a digital word. Unless otherwise noted, a byte is usually comprised of eight bits.

C Weighting—*See* Weighting, "C".

C.C.I.R.—Comite Consultatif International Radio (International Radio Consultative Committee).

Cancellation—The severe attenuation that occurs when two identical signals of opposite polarity are combined.

Capacitance—An opposition to a change in voltage.

Capacitor—An electronic component that opposes a change in voltage. Parallel conductors in a signal line may take on the properties of a capacitor, thereby attenuating high frequencies.

Capacitor Loudspeaker—*See* Loudspeaker, Electrostatic.

Capacitor Microphone—*See* Microphone, Capacitor.

Capstan—On a tape recorder, the motor-driven spindle that drives the tape past the heads.

Capstan Idler—The rubber coated wheel that forces the tape against the capstan when the tape recorder is in the play mode. Also called a pinch roller or puck.

Capstan Motor—The motor that drives the capstan. The capstan is often the extended shaft of the capstan motor.

Cardioid Microphone—*See* Microphone, Cardioid.

Cathode Ray Oscilloscope—A test instrument, providing a visual display of the waveform being measured.

Cellulose Acetate—A plastic film, used as a base material in the production of magnetic recording tape.

Center, Acoustic—*See* Acoustic Center.

Center Frequency—In a peaking equalizer, the frequency at which maximum boost (or attenuation) occurs.

Center Tap—The electrical center of a transformer winding.

Center Tap, Artificial—When two identical precision resistors are wired in series across a transformer winding, the point at which the resistors are joined together becomes the electrical equivalent of the actual center tap.

Channel—A single audio signal. In a console, the signal-processing path for such a signal. In analog recording, Channel and Track are often used synonymously. In digital recording, one or more Tracks may be required for each audio Channel. *See* Tracks.

Channel Bus—*See* Bus, Channel.

Channel Path—The signal path through an in-line console in which the signal is routed more or less directly from the input to the output, usually bypassing the equalizer and the VCA fader.

Closed Loop Tape Path—A tape transport system in which the tape passes through two capstan/capstan idler systems; one on each side of the head assembly. Called an Isoloop system by the 3M Company.

Close Miking—The technique of placing microphones extremely close to the instruments they are picking up, thereby eliminating almost all but the direct sound of the instrument(s).

Cocktail Party Effect—The ability of the brain to pick out one conversation from many going on simultaneously.

Code—Any digitally-transmitted or recorded signal.

Code, CRC—*See* Cyclic Redundancy Check Code.

Code, Parity—*See* Parity Code.

Coding System—Any system in which a series of analog levels are encoded into a digital format. The SMPTE time code is a Bi-phase modulation coding system.

Coercivity—The field strength required to bring a saturated tape to complete erasure. Coercivity is abbreviated as Hc, and is measured in oersteds.

Coherence—The instantaneous polarity relationship between two complex sound waves.

Coherent Signals—Two complex waveforms that are—most of the time—of the same polarity.

Coincident Microphones—Two or more microphones on the same vertical axis. A stereo microphone.

Coloration—A distortion in frequency response, usually associated with off-axis signals picked up by a cardioid microphone.

Color-Frame Flag—In the SMPTE time code, the bit (#11) which is assigned a 1 to indicate the color-frame mode, for electronic editing purposes.

Combining Amplifier—An amplifier at which the outputs of two or more signal paths are mixed together, to feed a single track of a tape recorder.

Compander—The contraction of COMpressor/exPANDER, often used in describing the action of a noise reduction system.

Complementary Signal Processing—A signal processing technique, in which some processing is done before recording, with equal-and-opposite (complementary) processing during playback. Well known examples are noise reduction systems and tape recorder pre- and post-emphasis. *See* Non-Complementary Signal Processing.

Composite Equalization—*See* Equalization, Composite.

Compression Driver—A loudspeaker transducer with a relatively narrow throat, designed for maximum efficiency coupling with a horn assembly.

Compression Ratio—In a compressor, the ratio of dB change in input level to dB change in output level.

Compression Threshold—*See* Threshold, Compression.

Compressor—An amplifier whose gain decreases as its input level is increased.

Compressor, Program—A compressor that acts on an entire program, rather than on a single instrument or track.

Compressor, Voice-Over—A speech-activated compressor, in which an announcer's voice automatically drops the level of the regular program. Often used in broadcasting and paging systems.

Condensation—The instantaneous crowding together of air particles during the positive-going half cycle of a sound wave. The opposite of rarefaction.

Condenser Loudspeaker—*See* Loudspeaker, Electrostatic.

Condenser Microphone—*See* Microphone, Condenser.

Console, In-Line—*See* In-Line Console.

Console, Recording—The enclosure containing the various input, output, signal routing, and monitoring controls required for recording.

Contact Microphone—A microphone which is directly attached to an instrument. Generally, it responds to the mechanical vibrations of the instrument to which it is attached.

Continuous Jam Sync—*See* Jam Sync, Continuous.

Converter, Analog-to-Digital—*See* Analog-to-Digital Converter.

Cottage Loaf Microphone—A super- or hyper-cardioid microphone.

Coupling—The transfer of energy from one system to another. Often used to describe the interface between a loudspeaker and the surrounding air.

CRC—Cyclic Redundancy Check.

Critical Distance—The distance from a sound source at which the level of the direct and the reverberant field are equal.

Cross-Fade—A method of fading out one program segment while fading in the next segment.

Crossover Frequency—The single frequency at which both sides of a crossover network are down 3 db.

Crossover Network—A one input/two (or more) output network, in which the audio bandwidth is separated into two (or more) bands. Frequently used with multi-speaker systems.

Crossover Network Phase Shift—*See* Phase Shift.

Crosstalk—In a signal path, the unwanted detection of a signal from an adjacent signal path.

Cue System—That part of the console, plus associated circuitry, by which the engineer may route a headphone monitor feed to musicians in the studio. Also known as foldback.

Current—The rate of flow of electricity, measured in amperes.

Cross-Interleaving—*See* Interleaving, Cross-.

Cut-Off Filter—*See* Filter, Cut-off.

Cut-Off Frequency—In a high- or low-pass filter, the frequency at which the output level has fallen by 3 dB.

Cycles Per Second—The number of complete oscillations of a vibrating object, per second. The unit by which frequency is measured. Also called hertz, abbreviated Hz.

Cyclic Redundancy Check Code—A recorded code containing data derived from the audio data, and recorded at regular intervals along the tape. The code is used for error-detection purposes.

D/A Converter—*See* Digital-to-Analog Converter.

Damping—Acoustical, electrical, or mechanical opposition to a moving system, as in a speaker voice coil assembly.

DASH—Digital Audio Stationary Head. The predominant method of fixed head digital recording.

DAT—Digital Audio Tape Recorder. Also a tape medium for the digital audio tape recorder.

Data Block—*See* Block, Data.

Data Bus—*See* Bus, Data.

Data Stream—A series (or stream) of pulses, containing digitally-encoded data, such as time code, sync code, parity code, and CRC code. The term is often used to describe the entire digital signal; that is, the data and the encoded audio as well.

DAW—Digital Audio Workstation. A computer containing software and hardware for storing, editing, and processing digital audio information.

dB—*See* Decibel.

Dead—Description of a sound in which reverberant information is severely attenuated, or missing completely.

Decay—The fall-off in amplitude when a force applied to a vibrating device is removed.

Decay Time—The time it takes for echoes and reverberation to die away.

Decibel—A unit of level equal to ten times the logarithm of the ratio of two powers.

For a power ratio, $N_{db} = 10 \log (P_{out}/P_{in}) = 10 \log (P_a/P_b)$

For a voltage ratio, $N_{dB} = 20 \log (V_{out}/V_{in}) = 20 \log (V_a/V_b)$

In the second set of formulae, the subscript a represents a measured value, while the subscript b represents a standard reference level. Depending on the reference level used, the decibel symbol will be immediately followed by a letter as:

dBA—An "A" weighted decibel level. *See* Weighting, "A".

dBB—A "B" weighted decibel level. *See* Weighting "B".

dBC—A "C" weighted decibel level. *See* Weighting "C".

dBm—A decibel level in which P_b in the formula above represents one milliwatt of power dissipated in a 600-ohm line. P_a must also be measured in a 600-ohm line.

dBV—A decibel voltage level, in which the reference, Vb, is 1.0 volt. A reference of 0.775 volts is also frequently used.

Decimal Digit—*See* Digit, Decimal.

Decoder—A device used to convert the coded digital-word pulse stream into an analog signal. A D/A converter.

Decoding—The process of applying complementary signal processing to restore a signal to its normal state, as in the playback mode of a noise reduction system.

De-Esser—A compressor designed to minimize sibilants.

Degausser—A device for demagnetizing the heads and other surfaces on a magnetic tape recorder. *See* Bulk Eraser.

Delay—The time interval between a direct signal and its echo(es).

Delay Line, Acoustic—A delay line, in which the delay is accomplished acoustically, as in a long tube with a speaker at one end and a microphone at the other.

Delay Line, Digital—A delay line, in which the delay is accomplished electronically, via an analog/digital and digital/analog conversion.

Demagnetization—The erasure of a magnetic tape, or the degaussing of the tape recorder heads.

De-Multiplexer—A device which converts a sequential series of n bits into a single n-bit digital word. A serial-to-parallel converter.

Density, Bit-Packing—*See* Bit-packing Density.

Depth Perception—The ability of the listener to perceive the apparent relative distances of various instruments in a recording (or live concert).

Diaphragm—The moving membrane in a microphone or loudspeaker.

Diffraction—The bending of a sound wave, as it passes over an obstacle. The angle of diffraction is a function of wavelength.

Digit, Binary—Any one of the two digits (0, 1) used in the Binary Numbering System.

Digit, Decimal—Any one of the ten digits (0, 1, 2, 3, 4, 5, 6, 7, 8, 9) used in the Decimal Numbering System.

Digit, Hexadecimal—Any one of the sixteen digits (0 through 9, and A, B, C, D, E, F) used in the Hexadecimal Numbering System. Note that since the universally-recognized decimal system does not contain enough digits, the letters A through F are also used, to express the hexadecimal equivalents of decimal 10 through 15.

Digit, Octal—Any one of the eight digits (0, 1, 2, 3, 4, 5, 6, 7) used in the Octal Numbering System).

Digital Delay Line—*See* Delay Line, Digital.

Digital Recording—A technique in which an analog signal is converted into a digital-format signal prior to recording. On playback, the digital data is converted back to the analog format.

Digital-to-Analog Converter—A device which converts (decodes) the digital word into its equivalent quantized analog level.

Direct Mode—A method of routing a single channel directly to its associated output channel, bypassing the channel pan-pot and sometimes the channel bus itself, in which case other input channels cannot be routed to the same output channel.

Direct Output—A console output, taken directly from an input module and bypassing the pan pots and bus selector switches.

Direct Pickup—A transformer pickup of a musical instrument, in which the instrument's amplifier output is fed directly to the console, via a matching transformer.

Director Radiator—*See* Radiator, Direct.

Direct Sound—The sound that reaches the listener via a straight line path from the sound source. A sound with no echoes or reverberation. The sound heard from a transformer pickup.

Directional Characteristic—The polar response of a transducer.

Dispersion—The splitting of a complex sound wave into its various frequency components, as the sound wave passes from one medium to another.

Displacement—The distance between the position of a moving object such as a speaker diaphragm, and its original position.

Distant Miking—The placement of a microphone, or microphones, relatively far from the sound source, thus picking up a larger proportion of reflected sound.

Distortion—An unwanted change in waveform as it passes through an electronic component, or from one medium to another.

Distortion, Harmonic—The appearance of harmonics of the applied input signal, as measured at the output of an electronic component.

Distortion, Intermodulation—Distortion in the form of unwanted frequencies corresponding to the sums and differences between various components of a complex waveform.

Distortion, Percent—The amount of distortion, measured as a percentage of the total waveform amplitude.

Distortion, Third Harmonic—The presence of the third harmonic (3 X f) of an applied input signal, as measured at the output of an electronic component.

Distortion, Transient—Distortion produced when an audio system is unable to accurately reproduce a transient.

Dither—A low level of noise added to digital recording or transmission media that is used to linearize digital audio and reduce noise from quantization. Usually an amount of ½ to ⅓ of the LSB or least significant bit.

Dolby Tone—A reference tone, recorded at the head of a Dolby encoded tape, for alignment purposes.

Domain—10^{18} molecules of ferric oxide; the smallest physical unit that may be considered a magnet.

Doubling—Mixing a slightly delayed signal with a direct signal, to simulate the effect of twice as many recorded instruments. Also, a deficiency of some speaker systems, in which low frequencies may be reproduced up one octave.

Drift—In a tape recorder, a long term deviation from the specified tape speed.

Driver, Compression—*See* Compression Driver.

Drone Cone—A passive radiator.

Drop-Frame—In SMPTE time code, a system in which 108 frames-per-hour are discarded, or dropped, to compensate for the discrepancy between color and monochrome frame rates (29.97 and 30 frames-per-second, respectively).

Drop-Out—On a magnetic recording tape, a momentary drop in output level, usually caused by an imperfection in the oxide coating.

Dry Recording—The practice of recording a signal without applying artificial reverberation.

Dry Sound—A description of a sound which lacks reverberant information. The direct sound of a musical instrument.

Dual Diaphragm Microphone—*See* Microphone, Dual.

DVD—The Digital Versatile Disc. A compact disc type that holds a great amount of data by decreasing the pit size and spacing. Capable of double-layer, double-sided data storage. Currently a play-only medium, and used principally for CD-ROM and video data storage.

Dynamic Filter—*See* Filter, Dynamic.

Dynamic Loudspeaker—*See* Loudspeaker, Dynamic.

Dynamic Microphone—A moving coil or ribbon microphone. *See* Microphone, Moving Coil.

Dynamic Range—In a musical instrument, a measure of the span between the quietest and loudest sounds it is capable of producing. In a tape recorder, the dB interval between the noise level and the level at which 3% distortion occurs.

Dynamic Ribbon Microphone—*See* Microphone, Ribbon.

Dynamic Signal Processing Device—A signal processing device whose operating parameters change as a reaction to the program content. For example, a compressor or expander. *See* Static Signal Processing Device.

Dyne—A unit of force. At the threshold of hearing, the acoustic force per unit area is 0.0002 dynes/cm^2.

EBU—European Broadcasting Union.

Echo—A repetition of a sound. One, or a few at most, repetitions of an audio signal.

Echo, Post—A signal routed to an echo send line from a point after the input fader. The position on the echo send switch which accomplishes this. An after-the-signal tape echo, caused by print-through.

Echo, Pre—A signal routed to an echo send line from a point before the input fader. The position of the echo send switch which accomplishes this. A before-the-signal tape echo, caused by print-through.

Echo Return—The signal path and the associated controls that affect the signal sent to an artificial echo and/or reverberation system.

Echo Send—The signal path and the associated controls that affect the signal sent to an artificial echo and/or reverberation system.

Echo Tape—The tape used to create artificial echoes in a tape delay system.

Edit Switch—On a tape recorder, a switch that puts the machine in the play mode, while the tape motor remains disabled. Thus, the segment of tape being played spills off the machine, and may be easily discarded.

Editing, Electronic—The process of assembling a final master tape by transferring segments of the required takes, in sequence, onto a fresh roll of tape.

Editing, Razor Blade—The process of cutting and splicing a magnetic tape to remove or rearrange certain segments. Producing a master tape by splicing together segments of several different takes.

Efficiency—In an audio system, the ratio of power output to power input.

Electret Microphone—*See* Microphone, Electret.

Electrical Phase Cancellation—*See* Phase Cancellation.

Electrostatic Loudspeaker—*See* Loudspeaker.

Electrostatic Microphone—*See* Microphone.

Elevated Level Test Tape— *See* Test Tape, Elevated Level.

Enclosure, Bass Reflex—*See* Bass Reflex Enclosure.

Enclosure, Sealed—*See* Sealed Enclosure.

Enclosure, Vented—*See* Vented Enclosure.

Encoder—A device which converts a quantized analog level into an encoded digital word. An A/D Converter.

Encoding—The application of some form of signal processing before recording, that will be removed via complementary processing (decoding during playback).

Energy Conversion—The process of changing a signal from one form of energy to another, as in a loudspeaker, which converts electrical energy into acoustical energy.

Energy Distribution Curve—A graph of energy vs. frequency for a typical voice, musical instrument, or program.

Energy Transfer—The delivery of power from a generator instrument.

Envelope—The overall shape of the waveform of a musical instrument.

Equal Loudness Contours—A series of graphs of sensitivity (of the ear) vs. frequency at various loudness levels. Also known as the Fletcher-Munson curves.

Equalization—An intentional modification of an audio system's frequency response.

Equalization, Composite—The net frequency response of an audio system in which two or more sections of an equalizer are in use.

Equalization, Playback—Equalization applied in the playback circuit of a tape recorder to produce a flat frequency response.

Equalization, Record—Equalization applied in the record circuit of a tape recorder.

Equalization, Room—The practice of tailoring the frequency response of a signal delivered to a speaker to correct for certain frequency response anomalies created by the room.

Equalizer—Any signal processing device that is used to change the frequency response anomalies created by the room.

Equalizer, Active—An equalizer containing active components, such as vacuum tubes or transistors.

Equalizer, Graphic—An equalizer with a series of slide controls, arranged so as to give a graphic representation of the resulting frequency response.

Equalizer, Parametric—An equalizer in which the frequency selector control is continuously variable over a wide range.

Equalizer, Passive—An equalizer containing only passive components, such as resistors, capacitors and inductors.

Equalizer Phase Shift—*See* Phase Shift, Equalizer.

Equalizer, Shelving—An equalizer that supplies a constant amount of boost or attenuation at all frequencies beyond the equalizer's turnover frequency.

Equivalent Circuit—A network designed to duplicate the operating parameters of some other network. For example, a resistive/capacitive network may be used to simulate a microphone line, for testing purposes.

Erase Head— *See* Head, Erase.

Erase Oscillator—A fixed frequency oscillator built into the tape recorder to supply erase current.

Eraser, Bulk— *See* Bulk Eraser.

Error Concealment—An error "repairing" scheme, in which data is interpolated from data in adjacent blocks to be used in place of the defective data.

Error Correction—A coding system which provides a method of correcting errors which occur in the digital data stream.

Error Detection—A coding system which provides a method for detecting errors which occur in the digital data stream.

Error Signal—A voltage that is proportional to the difference between an actual and a desired condition (as in a servo motor system). The error signal brings the system back to the desired operating condition.

Expander—An amplifier whose gain decreases as its input level is decreased.

Expansion, Peak—The use of an expander in a playback system to restore peaks that may have been compressed during recording. The peak expander may also be used to widen the dynamic range of programs that were not compressed earlier.

Expansion Ratio—In an expander, the ratio of dB change in input level to dB change in output level.

Expansion Threshold— *See* Threshold, Expansion.

Exponent—A superscript placed to the right of a number, indicating the power to which the number is to be raised. In the expression 9^3—729, the superscript 3 is the exponent.

Fade-out—Ending a recording by lowering the level, generally as the musicians play the last few measures over and over.

Fader—A variable level control in a signal path. Sometimes called a mixer.

Fader, Master—A single fader which regulates the level of all tracks being recorded.

Fader, VCA— *See* VCA Fader.

Feedback—The return of some portion of an output signal to the system's input.

Feedback, Acoustic—An audible howl or squeal, produced when a portion of a speaker's output is picked up by a nearby microphone.

Ferric Oxide—*See* Gamma Ferric Oxide.

Fifth—A musical interval of five diatonic degrees. The interval between a fundamental frequency and its third harmonic is equal to an octave plus a fifth.

Figure-8 Microphone— *See* Microphone, Figure-8.

Filter—An equalizer designed to attenuate certain frequencies or bands of frequencies.

Filter, Anti-Aliasing—A low-pass filter, designed to remove high-frequency components above the Nyquist frequency, which would create aliasing errors.

Filter, Band-Pass—A filter that attenuates above and below a desired bandwidth.

Filter, Band-Reject—A filter that attenuates a desired bandwidth, while passing frequencies above and below the bandwidth.

Filter, Cut-Off—A filter that sharply attenuates frequencies beyond a specified frequency.

Filter, Dynamic—A filter whose bandwidth changes in response to the program level.

Filter, Flutter— *See* Flutter Filter.

Filter, High-Frequency—A filter that attenuates high frequencies.

Filter, High-Pass—A filter that passes high frequencies, while attenuating those below a specified frequency.

Filter, Low-Frequency—A filter that attenuates low frequencies.

Filter, Low-Pass—A filter that passes low frequencies, while attenuating those above a specified frequency.

Filter, Notch—A filter designed to attenuate a relatively narrow band of frequencies.

Filter, Proximity Effect—A filter built into a cardioid microphone to attenuate low frequencies when the microphone is used close-up.

Filter, Telephone—A narrow band-pass filter, used to simulate the sound of a telephone transmission.

Flanging—A variable comb filter effect, created by mixing a direct signal with the same signal slightly delayed. To create the effect, the delay time is continuously varied.

Fletcher-Munson Curves—The equal loudness contours.

Flutter—A high-frequency speed variation of an audio signal, generally caused by irregularities in the tape path.

Flutter Filter—A low-friction surface, placed in the tape path to minimize scrape flutter. Also called a scrape flutter filter.

Flutter, Scrape—Flutter caused by mechanical vibrations of the tape as it passes over various surfaces in the tape path.

Flux—Magnetic lines of force.

Fluxivity—The measure of the flux density of a magnetic recording tape, per unit of track width.

Fluxivity, Reference—A specified fluxivity, as recorded on a test tape.

Foldback System—A cue system.

Folded Baffle—*See* Baffle, Folded.

Folded Horn—*See* Horn, Folded.

Frames—On film, the series of still pictures photographed along the length of the film. On videotape, the series of single television pictures, each comprised of two interlaced fields, recorded on the tape.

Frame Rate—The speed at which frames are recorded or photographed.

Free Space—A reflection-free environment, as in an anechoic chamber. Also called full space.

Frequency—The number of vibrations per unit time, measured in hertz (cycles per second).

Frequency Response—A graph of amplitude vs. frequency.

Fringing—A rise in low-frequency response when a tape is reproduced by a playback head that is narrower than the record head that was used to produce the tape.

Full Space—*See* Free Space.

Full Track—A tape with a single track, recorded across its entire width.

Fundamental—The primary frequency of vibration of a sound source.

Gain-Before-Threshold—The dB gain of a compressor, when the input signal level is below threshold.

Gain Reduction—In a compressor, the decrease in gain when the input signal level is above threshold.

Gain Riding—Manually adjusting the gain in a signal path in an effort to decrease dynamic range.

Gamma Ferric Oxide—The type of ferric oxide compound that is used in the manufacture of magnetic recording tape.

Gap, Head—*See* Head Gap.

Gap Space—The gap dimension, measured in the direction of tape travel.

Gauss—A unit of measurement of a tape's remanent magnetization.

Generation—A copy of a tape. The original recording is a first generation tape. A copy is a second generation; a copy made from the second generation tape is a third generation, and so on.

Gobo—A sound absorbing panel, used in the studio to acoustically separate one instrument from another.

Golden Section—A ratio of room height to width to length, first recommended by the ancient Greeks for optimizing acoustics in spaces. The golden section is a ratio of 1 to 1.62 to 2.62, as in a room with a 10-foot ceiling, 16.2-foot width and 26.2-foot length.

Graphic Equalizer—*See* Equalizer, Graphic.

Group Master—A single fader assigned to a group of VCAs.

Grouping—The practice of controlling selected VCAs with a single VCA fader. The fader becomes a sub-master group fader, controlling the signal level through all the VCAs that are under its control.

Guard Band—The spacing between tracks on a multi-track tape or tape head.

Harmonic—A whole number multiple of a fundamental frequency.

HDM—High-Density Modulation.

Head—On a tape recorder, the transducer used to apply and/or detect magnetic energy on the tape.

Head, Erase—The head that is used to apply a gradually diminishing magnetic force to the tape, thus erasing it just prior to recording.

Head Gap—The space between pole pieces in a head.

Head Losses—Losses in frequency response that are a function of head design limitations.

Head, Playback—The head that is used to detect the tape's magnetic field.

Head, Record—The head that is used to apply a magnetic force to the tape.

Head Shield—A metal shield around the playback head, designed to protect it from stray magnetic fields.

Head, Sync—The record head, when used for playback during Sel-Sync sessions.

Headphone Leakage—*See* Leakage, Headphone.

Headroom—In magnetic recording tape, the dB difference between standard operating level (+4dBm) and the 3% distortion point.

Hertz—Cycles per second. The unit of measurement of frequency.

Hexadecimal Digit—*See* Digit, Hexadecimal.

High-Density Modulation—Any PCM system in which the coding scheme permits a high bit-packing density.

High-Frequency Filter—*See* Filter, High-Frequency.

High-Output Tape—A high-sensitivity tape.

High-Pass Filter—*See* Filter, High-Pass.

Hiss—*See* Tape Hiss.

Horn—A speaker system, so-called because of its characteristic shape. The horn design provides an efficient coupling of the diaphragm to the surrounding air mass.

Horn, Folded—A speaker system in which the horn is folded on itself to conserve space. *See* Baffle, Folded.

Horn Loaded System—Any speaker system in which a horn is used.

Horn, Multi-Cellular—A horn cluster, designed to provide a wide radiation angle.

Hub—The center of a tape reel, around which the tape is wound.

Hyper-Cardioid Microphone—*See* Microphone, Hyper-Cardioid.

Hysteresis Loop—A graph of magnetizing force vs. remanent magnetization.

Idler, Capstan—*See* Noise, Impact.

Image Shift—An undesired change in the apparent location of a recorded sound source.

Impact Noise—*See* Noise, Impact.

Impedance, High—Generally, a circuit with an impedance of several thousand ohms or more.

Impedance, Low—Generally, a circuit with an impedance of 600 ohms or less.

Impedance, Matching Transformer—*See* Transformer, Impedance, Matching.

Incoherent Signals—Two complex waveforms that are—most of the time—of opposite polarity.

Indirect Radiator—*See* Radiator, Indirect.

Inductance—An opposition to a change in current.

Inductor—An electronic component that opposes a change in current.

Inertance—The acoustical equivalent of inductance.

Infinite Baffle—*See* Baffle, Infinite.

In-Line Console—A console in which all controls (input, output and monitor) for a channel are placed in-line at the location traditionally associated with the input fader.

Input/Output Module—The section containing the input, output, and monitor functions for a specific channel.

Insertion Gain/Loss—A change in signal level, as a result of inserting an electronic component (amplifier, signal processing device, pad, etc.) in a line.

Intensity, Acoustic—A measure of acoustic power per unit area. At the threshold of hearing, the acoustic intensity is 0.000000000001 watts/m^2 (meter squared) or 1×10^{-12} watts/m^2.

Interface—The proper inter-connection of two networks, components, or systems.

Interleaving—An error concealment scheme, in which sequential data words are first separated and then recorded in odd and even groups (for example, 1, 3, 5, 7, 9, 11, 2, 4, 6, 8, 10, 12).

Interleaving, Cross- —An error concealment scheme, in which the interleaved data are again separated and re-sequenced (for example, 1, 5, 9, 3, 7, 11, 2, 6, 10, 4, 8, 12).

I/O—Input/Output.

I/O Module—Input/Output Module. Contains all necessary circuits and controls between microphone and recorder for a single audio channel. Also contains the circuits and controls between the multi-track returns and the stereo mixdown buses.

Isoloop—The registered trade name of the 3M Company, describing their closed loop tape path. *See* Closed Loop Tape Path.

Isolation—The acoustic (or electrical) separation of one sound source from another.

Jack Bay—In a recording console or equipment rack, a strip of female input and output sockets in conjunction with patch cords for signal routing purposes.

Jack Field—A jack bay.

Jam Sync—The practice of synchronizing, or "jamming," a time-code generator's output to an external time-code signal.

Jam Sync, Continuous—The time code is continuously synchronized to the external reference time code signal.

Jam Sync, One-Time—The time code is synchronized to the external reference time code signal just once, and then continues to be generated independently of the external reference.

Kepex—A trade name of Allison Research, Inc., to describe its KEyable Program EXpander.

Keying Input—On a signal processing device, an input for a control input.

Keying Signal—A control signal, routed to the keying input.

Kilo—A prefix, abbreviated k, for thousand; 10k ohms = ten thousand ohms.

Leader Tape—Non-magnetic tape, spliced between segments of magnetic tape, to visually indicate the beginning and end of the recording.

Leakage—Extraneous sounds, picked up by a microphone. Generally used to describe the unwanted sound of one musical instrument as heard by a microphone in front of another instrument.

Leakage, Headphone—The transmission of sound from a headphone to a nearby microphone.

Lens, Acoustic— *See* Acoustic Lens.

Level—The magnitude of a signal, expressed in decibels.

Level Sensing Circuit—Any circuit that converts an audio signal into a control voltage, which may be used to regulate the operating parameters of a signal processing device.

Limiter—A compressor whose output level remains constant, regardless of its input level. Generally, a compressor with a compression ratio of 10:1 or greater.

Limiter, Program—*See* Compressor, Program.

Limiting Threshold—*See* Threshold, Limiting.

Line—A transmission line, or any signal path.

Line, Balanced—*See* Balanced Line.

Line Level—A signal whose level is at or about +4 dBm.

Line Matching Transformer—An impedence matching transformer, used to match the impedance of one line to another.

Line Microphone— *See* Microphone, Line.

Line Pad—An attenuation network, designed for insertion in a line.

Line, 600-ohm—A transmission line with a characteristic impedance of 600 ohms.

Line, Unbalanced—*See* Unbalanced Line.

Live Recording—A recording made at a concert. Sometimes used to describe any recording that is done all at once, as opposed to a Sel-Sync session.

Load Resistor—A resistor placed across a line to meet impedance matching requirements. A resistor placed across the output terminals of an amplifier for testing purposes. The resistor takes the place of a normal load.

Loading—Placing a load across a line. Often used to describe the effect on a circuit of a load that is equal to, or less than, the characteristic impedance of the line to which it is connected.

Lobes—The side and rear protrusions on some uni-directional polar patterns, denoting slight sensitivity increases at various off-axis angles. The front and rear segments of a bi-directional polar pattern.

Log—a logarithm.

Logarithm—The logarithm of a number is that power to which 10 must be raised to equal the number. The logarithm of 1,000 is 3.

Loudness—The subjective impression of the intensity of a sound.

Loudspeaker—A transducer that converts electrical energy into acoustical energy.

Loudspeaker, Capacitor—An electrostatic loudspeaker.

Loudspeaker, Condenser—An electrostatic loudspeaker.

Loudspeaker Cone—The diaphragm of a moving coil loudspeaker.

Loudspeaker, Damping—*See* Damping.

Loudspeaker, Electrostatic—A loudspeaker in which the diaphragm is one plate of a capacitor.

Loudspeaker, Moving Coil—A loudspeaker in which the diaphragm is attached to a voice coil suspended in a magnetic field.

Loudspeaker Polar Pattern—*See* Polar Pattern, Loudspeaker.

Loudspeaker Radiation Pattern—*See* Polar Pattern, Loudspeaker.

Loudspeaker, Ribbon—A loudspeaker in which the diaphragm is a ribbon.

Loudspeaker Transient Response—*See* Transient Response.

Loudspeaker Voice Coil—The moving coil to which the loudspeaker diaphragm is attached.

Low-Frequency Filter—*See* Filter, Low-Frequency.

Low-Pass Filter—*See* Filter, Low-Pass.

M-S Recording—A coincident microphone technique, in which the M (middle) microphone is cardioid, pointing toward the middle of the orchestra, and the S (side) microphone is a Figure-8, with its dead sides on the same axis as the front of the cardioid.

Magnetic Field—The magnetic flux surounding a magnet or a section of magnet recording tape.

Magnetic Flux—*See* Flux.

Magnetic Recording Tape—A recording medium, consisting of magnetic particles, suspended in a binder, and coated on a plastic or other film base.

Masking—The process by which a sound source becomes inaudible, due to the presence of some other sound source in the immediate area.

Master Fader—*See* Fader, Master.

Master Module—On an in-line console, a single module providing various master controls for mixing bus outputs, cue lines, send and return lines, etc.

Master Tape—A completed tape, used in tape-to-disc transfer, or from which other tape copies are produced.

Master Transport—The tape or film transport to which all other transport systems are locked in synchronization.

Matching Transformer—*See* Transformer, Impedance Matching.

Matrix—A transformer network, in which the outputs of an M-S microphone pair are combined additively and subtractively, to produce left and right output signals for stereo reproduction.

Memory—A buffer, into which program segments may be transferred in order to be digitally manipulated.

Meter, Peak Reading—A meter whose ballistics allow it to closely follow the peaks in a program. Also called a peak program meter.

Meter Sound Level—A decibel-calibrated meter, used to measure sound pressure levels.

Meter, VU—A meter calibrated to read volume units.

Microbar—A unit of pressure, equal to 1 dyne/cm2.

Microphone—A transducer that converts acoustical energy into electrical energy.

Microphone, Bi-Directional—A microphone with a bi-directional polar pattern. *See* Polar Pattern, Bi-directional.

Microphone, Blumlein Pair—*See* Blumlein Pair.

Microphone, Capacitor—A microphone in which the diaphragm is one plate of a capacitor.

Microphone, Cardioid—A microphone with a cardioid polar pattern. *See* Polar Pattern, Cardioid.

Microphone, Coincident—Two or more microphones on the same vertical axis. A stereo microphone.

Microphone, Condenser—The popular name for a capacitor microphone.

Microphone, Contact—*See* Contact Microphone.

Microphone, Dual-Diaphragm—A microphone with two diaphragms. The second diaphragm may be electronically combined with the first to produce more than one polar pattern. In another type of dual-diaphragm microphone, the two diaphragms are for high and low frequencies.

Microphone, Dual-Pattern—A microphone with two switchable polar patterns.

Microphone, Dynamic—A moving coil or ribbon microphone.

Microphone, Eletret—A microphone with a permanently charged capacitor/diaphragm.

Microphone, Electrostatic—A capacitor microphone.

Microphone, Figure-8—A microphone with a bi-directional polar pattern. *See* Polar pattern, Bi-directional.

Microphone, Hyper-Cardioid—A microphone with a hyper-cardioid polar pattern. *See* Polar Pattern, Hyper-Cardioid.

Microphone, Lavalier—A microphone designed to be worn on a cord around the neck. Primarily used for announcers, talk shows, etc.

Microphone Line—Any line between a microphone and the first stage of amplification.

Microphone, Moving Coil—A microphone in which the diaphragm is attached to a voice coil, suspended in a magnetic field.

Microphone, Multi-Pattern—A microphone in which the diaphragms are electrically or acoustically andjustable to derive several patterns from one microphone.

Microphone, Omni-Directional—A microphone with an omni-directional polar pattern. *See* Polar Pattern, Omni-Directional.

Microphone, Phase Shift—A microphone whose directional characteristics are the result of acoustic phase shifts within the microphone. A uni-directional or cardioid microphone.

Microphone Preamplifier—In a recording console, the first stage of amplification, which raises microphone levels to line level. Also, the amplifier built into a condenser microphone.

Microphone, Pressure—A microphone that responds to instantaneous variations in air pressure, caused by the sound wave in the vicinity of the microphone. An omni-directional microphone.

Microphone, Pressure Gradient—A microphone that responds to the difference in acoustic pressure between the front and rear of the diaphragm. A bi-directional microphone.

Microphone, Shotgun—A highly directional microphone, so-called because of its characteristic appearance.

Microphone, Stereo—A microphone with two separate transducing systems, built into one housing. The two outputs are kept separate, and are fed to two separate tracks on the tape recorder.

Microphone, Super-Cardioid—A microphone with a super-cardioid polar pattern. *See* Polar Pattern, Super-Cardioid.

Microphone, Ultra-Directional—A microphone with a uni-directional polar pattern. *See* Polar Pattern, Uni-directional.

Microphone Voltage Rating—*See* Open Circuit Voltage Rating.

Middle-Sides Recording—*See* M-S Recording.

MIDI—Muaic Instrument Digital Interface

Mil—One thousandth of an inch.

Milli—One thousandth.

Mixdown Session—A recording session, during which the many separate tracks of information on a multi-track tape are processed and combined (that is, mixed down) to form a two- or four-track program, which is then recorded on a second machine.

Mixer—*See* Fader.

Mixing Bus—*See* Bus, Mixing.

Modulation Noise—*See* Noise, Modulation.

Monitor—A loudspeaker in a control room or other listening area, so-called because its primary purpose is to monitor the recorded performance, signal transmission, etc.

Monitor Bus—*See* Bus, Mixing.

Monitor Module—On an in-line console, a single module providing monitor switching and master level controls for studio and control room monitors.

Monitor Path—The signal path through the console in which the signal is routed through the equalizer and VCA fader to the monitor/mixing buses.

Monophonic—Pertaining to an audio system in which the entire program is heard from a single sound source.

Motion Sensing—On a tape recorder, a system which prevents tape damage when the play button is depressed while the machine is in either rewind or fast forward. The motion sensing system brings the tape to a complete stop before going into the play mode.

Moving Coil Loudspeaker—*See* Loudspeaker, Moving Coil.

Moving Coil Microphone—*See* Microphone, Moving Coil.

Multi-Cellular Horn—*See* Horn, Multi-Cellular.

Multi-Microphone Technique—The practice of using many close-up microphones, as opposed to a coincident pair, or similar stereo pickup.

Multi-Pattern Microphone—*See* Microphone, Multi-Pattern.

Multiplexer—The device which converts an n-bit digital word into a sequential series of n bits, for transmission or recording on a single channel. A parallel-to-serial converter.

Multi-track—Referring to a tape recorder, recording console, etc., in which there are more than two tracks of recorded information. Generally, Multi-Track implies eight or more tracks.

Mute—An on/off switch in a signal path which interrupts the signal flow. Commonly found in each input module on a recording console.

N.A.B.—National Association of Broadcasters.

Nanoweber—A unit of magnetic flux. The flux density, or fluxivity, of a test tape is measured in nanowebers per meter.

Narrow Band Noise—*See* Noise, Narrow Band.

Neck—The narrow end of a horn, where it connects to the driver.

Network, Crossover—*See* Crossover Network.

Network, Weighting—*See* Weighting Network.

Newton—A unit of force. The sound pressure level at the threshold of hearing is 0.00002 newtons/m^2.

Noise, Ambient—The long-term noise within any environment, in the absence of extraneous sound sources.

Noise, Asperity—literally, roughness noise. An increase in noise level or a narrow bandwidth on either side of a recorded frequency.

Noise, Background—The noise level of the surrounding environment.

Noise Filter—A filter, such as a notch or cut-off filter, designed to filter out narrow band noise, or noise at either frequency extreme.

Noise Gate—An expander whose threshold is set to attenuate low level signals, such as leakage, rumble, etc.

Noise, Impact—Noise that is a function of a mechanical contact with, or by, a moving object.

Noise Level—The amplitude of a noise. Usually refers to the decibel level of a steady state noise.

Noise, Modulation—Noise components across the entire audio bandwidth that are produced by any audio signal.

Noise, Narrow Band—Noise that is confined to, or measured across, a relatively narrow bandwidth.

Noise, Pink—Wideband noise that maintains constant energy per octave.

Noise, Quantization—The distortion that manifests itself as a result of quantizing an analog waveform into a series of discrete voltages.

Noise, Quiescent—The noise of an audio system in a static condition; that is, with no applied signal.

Noise Reduction System—A signal processing system designed to attenuate noise components within an audio system.

Noise, Residual—The noise level of a tape after it has been erased.

Noise Voltage—A noise measured on a voltage scale.

Noise, White—A wideband noise that contains equal energy of the audio bandwidth. *See* Noise, Pink.

Non-Complementary Signal Processing—Signal processing that is done either before or after recording. *See* Complementary Signal Processing.

Notch Filter—*See* Filter, Notch.

NTSC—National Television Standards Committee.

Numbering System (with base n)—Any counting system, in which a series of n digits is used to express all quantities.

Numbering System, Binary—n = two. *See* Digit, Binary.

Numbering System, Decimal—n = ten. *See* Digit, Decimal.

Numbering System, Hexadecimal—n = sixteen. *See* Digit, Hexadecimal.

Numbering System, Octal—n = eight. *See* Digit, Octal.

Nyquist Frequency—The highest frequency that may be accurately sampled. The Nyquist frequency is one-half the sampling frequency.

Nyquist Rate—The sampling frequency that is required in order to accurately sample all frequencies within a specified bandwidth. The Nyquist rate is twice the highest frequency which must be sampled.

Octal Digit—*See* Digit, Octal.

Octave—The interval between any two frequencies, f_1 and f_2, when $f_2 = 2f_1$.

Oersted—A unit of magnetic force, symbolized by the letter H.

Off-Axis—Not directly in front (of a microphone or loudspeaker, for example).

Off-Axis Coloration—In a microphone, a deterioration in frequency response of sound arriving from off-axis locations. In a loudspeaker, a deterioration in perceived frequency response when the listener is standing off-axis.

Offset—A user-defined fixed time interval by which a slave transport may be programmed to track the master transport.

Ohm—The unit of resistance to current flow.

Omni-Directional Microphone—A microphone that is equally sensitive to all sounds, regardless of the direction from which they arrive.

On-Axis—Directly in front (of a microphone or loudspeaker, for example).

One-on-One Recording—The practice of recording one instrument on one track with one microphone, or minor variations thereof.

One-Time Jam Sync—*See* Jam Sync, One-Time.

Open Baffle—*See* Baffle, Open.

Open Circuit Voltage Rating—The output voltage of a microphone when it is not connected to a load or when the load is about twenty times the impedance of the microphone itself.

Open Tracks—On a multi-track tape recorder, tracks that have not yet been used.

Operational Level—The nominal level at which an audio system operates. *See* Standard Operating Level.

Oscillator—A signal generator, whose output is a pure sine wave. Generally, the output frequency may be varied continuously, or in discrete steps, over the audio frequency bandwidth.

Oscillator, Bias—*See* Bias Oscillator.

Oscillator, Erase—*See* Erase Oscillator.

Oscillator, Vari-Speed—*See* Vari-Speed Oscillator.

Oscilloscope—*See* Cathode Ray Oscilloscope.

Out Take—A take, or section of a take, that is to be removed or not used.

Overdubbing—Producing a recording by mixing previously recorded material with a new material. The musicians listen to the previously recorded tape over headphones, while the old and new program material are recorded onto a second tape recorder. (The term is now often used synonymously with Sel-Sync, since the original practice described as overdubbing has fallen into disuse.)

Overload—The distortion that occurs when an applied signal exceeds the level at which the system will produce its maximum output level.

Overtone—A whole number multiple of a fundamental frequency.

Oxide—In magnetic recording tape, a solution of magnetic particles suspended in a binder.

Oxide, Gamma Ferric—*See* Gamma Ferric Oxide.

Packing Density—*See* Bit-packing Density.

Pad—*See* Attenuation Pad.

Pan-Pot—A potentiometer used to vary the proportion of an audio signal routed to two or more locations.

Parallel Output—A group of n output terminals, across which an n-bit digital word appears simultaneously.

Parity Code—An error-correction code created by adding several audio data codes. The parity code will be added to an incorrect data code in order to correct it.

Parametric Equalizer—*See* Equalizer, Parametric.

Pass Band—The band of frequencies that are not attenuated by a filter.

Passive Device—A network or circuit containing only passive components, such as resistors, capacitors and inductors.

Passive Radiator—*See* Radiator, Passive.

Patch Bay—*See* Jack Bay.

Patch Cord—A short length of cable, with a coaxial plug on each end, used for signal routing in a jack bay.

Patch Point—any socket in a jack bay.

Path Length—The total point-to-point distance between sound source and the listener.

PCM—Pulse-Code Modulation.

Peaks—The instantaneous high-level transients of an audio signal.

Peak Program Meter—*See* Meter, Peak Reading.

Peak Reading Meter—*See* Meter, Peak Reading.

Peak Expansion—*See* Expansion, Peak.

Phantom Power Supply—*See* Power Supply, Phantom.

Phase—The instantaneous relationship between two measured signals, when both are derived from a single, pure sine wave input. When the two outputs are always of the same polarity, the signals are said to be "in phase." When the outputs are always of opposite polarity, the signals are 180 degrees out of phase, or "out of phase." Other polarity relationships are expressed as a number of degrees of phase shift. The word "phase" is often used to describe the relationship between complex waveforms.

Phase Cancellation—The attenuation that occurs when two waveforms of equal frequency and opposite polarity are combined. The attenuation may be total when the waveforms are also of equal amplitude.

Phase Shift—The angular displacement, measured in degrees, between two sine waves of the same frequency.

Phase Shift, Crossover Network—The phase shift introduced by a crossover network.

Phase Shift, Equalizer—The phase shift introduced in a signal path by the insertion of equalization.

Phase Shift Microphone—*See* Microphone, Phase Shift.

Phasing—A variable comb filter effect, created by mixing a direct signal with the same signal passed through the phase shift network.

Phon—A unit of loudness level, related to the ear's subjective impression of signal strength. At 1,000Hz, the phon rating corresponds to the measured sound level. At all other frequencies, a sound level which to the listener seems to be of the same loudness as the 1,000Hz tone is given the same phon rating, regardless of the actual measured sound level.

Pinch Roller—A capstan idler.

Pink Noise—*See* Noise, Pink.

Playback Equalization—*See* Equalization, Playback.

Playback Head—*See* Head, Playback.

Plus-One Frame—The addition of one frame count to the SMPTE time-code readout, to compensate for the time code being one frame count behind the actual tape (or film) position.

Point Source—Theoretically, a source of sound, of infinitely small dimension, located in free space.

Polar Pattern—The graph of a transducer's directional sensitivity, measured over a 360° circumference drawn around the transducer. *See* Polar Pattern, Loudspeaker, and Polar Pattern, Microphone.

Polar Pattern, Bi-Directional—A polar pattern with axes of maximum sensitivity at 0° and 180°, and minimum sensitivity at 90° and 270°.

Polar Pattern, Cardioid—A uni-directional polar pattern, with the axis of minimum sensitivity at 180°. The pattern is so-called because of its characteristic shape.

Polar Pattern, Cottage Loaf—A hyper- or super-cardioid polar pattern.

Polar Pattern, Figure-8—A bi-directional polar pattern, so-called because of its characteristic shape.

Polar Pattern, Loudspeaker—A graph of a loudspeaker's measured output level at various points on a 360° circumference drawn around the speaker.

Polar Pattern, Hyper-Cardioid—A uni-directional polar pattern, slightly narrower than a regular cardioid pattern, and with a lobe in the rear. Axes of minimum sensitivity are at about 110° and 250°.

Polar Pattern, Microphone—A graph of a microphone's relative output level for sound sources originating at various points on a 360° circumference drawn around the transducer.

Polar Pattern, Omni-directional—A circular polar pattern, indicating equal sensitivity (or measured output level) at all angles on a 360° circumference drawn around the transducer.

Polar Pattern, Super-Cardioid—A uni-directional polar pattern, slightly narrower than a regular cardioid pattern, with a lobe in the rear that is somewhat wider than the one on a hyper-cardioid pattern. The axes of minimum sensitivity are at about 125° and 235°.

Polar Pattern, Uni-directional—A polar pattern of a microphone that is most sensitive to sounds originating directly in front of it.

Polarity—Referring to the positive or negative direction of an electrical or magnetic force.

Polarizing Voltage—The charging voltage applied to the capacitor/diaphragm of a condenser (capacitor) microphone.

Polyester—A plastic film, used as a base material in the production of magnetic recording tape.

Pop Filter—A wind screen.

Port, Speaker—An opening in the front baffle of a loudspeaker cabinet.

Ports, Rear- and Side-entry—Openings to the rear and side of a uni-directional microphone, allowing sound waves to reach the rear of the diaphragm.

Post-Echo—*See* Echo, Post-.

Post-Emphasis—Playback equalization.

Potentiometer—A network consisting of a resistor and a wiper arm. The resistance, measured from either end of the resistor to the wiper arm, is continuously variable, according to the position of the wiper arm.

Power—Rate of flow of energy, developed by an acoustical or electrical system.

Power, Acoustic—The sound energy produced by a sound source.

Power, Electrical—The electrical energy produced or dissipated in a circuit.

Power Supply—A circuit supplying d.c. power to an amplifier or other electronic system.

Power Supply, Microphone—A circuit supplying d.c. power to a condenser (capacitor) microphone.

Power Supply, Phantom—A circuit that supplies d.c. powering to condenser microphones, using the same conductors as the audio signal.

Preamble—A system-defined digital word, which appears at regular intervals to indicate the beginning of each new block of digital data.

Preamplifier—In an audio system, the first stage of amplification, usually designed to boost very low level signal to about the line level.

Preamplifier, Microphone—*See* Microphone, Preamplifier

Pre-Echo—*See* Echo, Pre-.

Pre-Emphasis—Record equalization.

Pre-Roll—The technique of beginning a playback or recording several moments early, in order to allow for multi-machine synchronization, cross-fade edits, count-offs, etc.

Presence—An equalization boost in the middle or upper middle frequency range, often used to give a voice a more close-up effect.

Pressure Gradient Microphone—*See* Microphone, Pressure Gradient.

Pressure Level, Sound—The acoustic pressure, expressed in decibels. The sound pressure level at the threshold of hearing is 0dB SPL.

Pressure Microphone—*See* Microphone, Pressure.

Pressure, Sound—The acoustic pressure of a sound wave. The sound pressure at the threshold of hearing is 2×10^5 newtons/m^2.

Print-Through—The transfer of a signal from one layer of magnetic tape to an adjacent layer.

Program Compressor—*See* Compressor, Program.

Program Limiter—*See* Compressor, Program.

Protection Copy—A copy of a master tape, generally filed as a protection against the damage or loss of the master tape.

Proximity Effect—In a cardioid microphone, a rise in low-frequency response when the microphone is used at very close working distances.

Psychoacoustics—The study of the brain's perception of, and reaction to, all aspects of sound. (i.e., intensity, time of arrival differences, reverberation, etc.).

Puck—*See* Capstan Idler.

Pulse-Code Modulation—A modulation process in which an analog signal is digitally encoded as a series of pulses.

Pumping—*See* Breathing.

Punching In—The practice of recording a track, or tracks, in small segments, say, one phrase at a time. So-called because the engineer spends so much time "punching" the record button.

Pure Tone—A single frequency sine wave, with no harmonics present.

Q—In a band-pass equalizer, the ratio of center frequency to bandwidth.

$$Q = \frac{fc}{bandwidth}$$

Quantization—The process of converting an infinitely variable waveform into a finite series of discrete levels. The resultant signal displays a "stepped" waveform which is said to be quantized.

Quantization, N-Bit—The process in which each quantization level is expressed as an n-bit digital word. In most digital tape recorders, n=16.

Quantization Noise—*See* Noise, Quantization.

Quiescent Noise—*See* Noise, Quiescent.

Radian—The angle subtended by an arc that is equal in length to the radius of the circle of which it is a part (approximately 57.29°).

Radiation Pattern—The polar pattern of a loudspeaker.

Radiator, Direct—A loudspeaker diaphragm that is coupled directly to the surrounding air mass of the listening room.

Radiator, Indirect—A loudspeaker diaphragm that is coupled to the surrounding air mass by an acoustic impedance matching transformer; that is, a horn.

Radiator, Passive—An unpowered loudspeaker cone, placed in the port of a vented enclosure system. Also called a drone cone, or a slave cone.

Rarefaction—The instantaneous spreading apart of air particles during the negative going half cycle of a sound wave. The opposite of condensation.

Ratio, Compression—*See* Compression Ratio.

Ratio, Expansion—*See* Expansion Ratio.

Ratio, Signal-to-Noise—*See* Signal-to-Noise Ratio.

Read—Playback (read head = playback head).

Rear Entry Port—*See* Ports, Rear- and Side-Entry.

Record Equalization—*See* Equalization, Record.

Record Head—*See* Head, Record.

Record Head, Advance—A record head positioned in advance of the playback head, as in the traditional analog recording format.

Record Head, Sync—In digital recording, an additional record head, positioned after the playback head, and used in cross-fade punch-in recording.

Recording Console—*See* Console, Recording.

Recovery Time—*See* Release Time.

Reference Level—A standard level, such as 0 VU, + 4dBm, 0 dBV, to which other levels may be compared.

Reference Tone—A single frequency tone, recorded at the head of a tape and used for alignment purposes when the tape is replayed at a later date.

Reflected Sound—Sound waves that reach the listener after being reflected from one or more surfaces.

Refraction—The change of direction, or bending, of a sound wave as it passes from one medium to another.

Regenerated Time Code—*See* Time Code.

Rehearse Mode—A technique in which the act of punching-in (or -out), or of electronic editing is simulated prior to actually being executed.

Reinforcement—An increase in amplitude when two sound waves combine additively.

Release Time—The time it takes for a signal processing device, such as a compressor or expander, to return to its normal gain-before-threshold, once the applied signal is removed or attenuated.

Reluctance—Opposition to a magnetic force. Reluctance is analogous to resistance in a purely electrical circuit.

Remanence—The magnetization left on a tape when a magnetic force is removed. Remanence is measured in lines of flux per quarter inch of tape width. Also called remanent flux.

Remanent Flux—*See* Remanence.

Reproducing Characteristic—The frequency response of the post-emphasis circuit in a tape recorder. The C.C.I.R. and the N.A.B. have both published recommended reproducing characteristics.

Residual Magnetization—The magnetization remaining in a magnetic material once an applied magnetic force is removed.

Residual Noise—*See* Noise, Residual.

Resistance—The opposition of a circuit to a flow of direct current. Resistance is measured in ohms, abbreviated Ω, and may be calculated from the formula, R=E/I.

Resistor—An electronic component that opposes current flow.

Resonance—The condition of a system when the applied frequency is equal to the natural frequency of vibration of the system.

Resonant Peak—The increase in amplitude that occurs at the resonant frequency.

Restored Time Code—*See* Time Code, Restored.

Retentivity—A measure of a magnetic tape's flux density after a saturation-producing magnetic field has been withdrawn.

Reverberation—Many repetitions of an audio signal, becoming more closely spaced (denser) with time.

Reverberation System—Any electronic or acoustical device used to simulate the natural reverberation of a large concert hall, or to produce a reverberant effect.

Reverse, Channel/Line—A mode in which a selected module switches to the opposite mode from that selected for the rest of the console.

Ribbon Loudspeaker—*See* Loudspeaker, Ribbon.

Ribbon Microphone—*See* Microphone, Ribbon.

Room Acoustics—The properties of a room that affect the quality of a sound source in the room (i.e., reverberation, resonance modes, etc.).

Room Equalization—*See* Equalization, Room.

Room Modes—Increases in amplitude at resonant frequencies that are a function of the dimensions of a room.

Room Resonance—*See* Resonance, Room.

Room Resonance Modes—*See* Room Modes.

Rotation Point—The point at which the transfer characteristic of a compressor or expander intersects the unity gain curve.

Rumble—Unwanted low-frequency signals that are a function of mechanical vibrations.

Sampling—The process of examining an analog signal at regular intervals, which are defined by the sampling rate.

Sampling Frequency, Sampling Rate—The frequency, or rate, at which the analog signal is sampled, usually expressed in hertz. Thus, a sampling rate of 50 kHz indicates 50,000 samples per second.

Saturation—The condition of exceeding a tape's magnetic capacity.

Saturation Distortion—The distortion created by driving a magnetic tape beyond its saturation point.

Saturation Point—The level beyond which any further increase in applied signal strength will cause no further increase in fluxivity.

Scrape Flutter—*See* Flutter, Scrape

Scrape Flutter Filter—*See* Flutter, Filter.

Sealed Enclosure—A loudspeaker cabinet with no vents or ports, as in an acoustic suspension system.

Self-Erasure—The condition whereby a record head tends to partially erase a high-level, high-frequency signal as it is being recorded.

Sel-Sync—A trade name of The Ampex Corporation used to describe the process of using the record head for playback of previously recorded tracks, while simultaneously recording new material on open tracks. Sel-Sync is an abbreviation of selective synchronization.

Selective Synchronization—*See* Sel-Sync.

Sensitivity—The acuity of the ear, or the response of a transducer, to various properties of a sound wave, such as frequency. level, angle of arrival, etc. On magnetic recording tape, an indication of the tape's relative output level, as compared to some specified reference tape.

Separation—A measure of the degree of segregation of one signal from another.

Serial Output—A single output terminal at which each bit in an n-bit word appears sequentially.

Shelving Equalizer—*See* Equalizer, Shelving.

Shield—Any device used to reduce the effects of spurious electrical or magnetic fields on a signal path or system.

Shielded Cable—Any cable in which the conductors are protected by a surrounding braided or foil shielding.

Shock Mount—A suspension system which mechanically isolates a microphone from its stand or boom, thus protecting the microphone against mechanical vibrations.

Sibilance—A hissy type of distortion often produced by the presence of high energy level in words containing "s" sounds.

Sibilant—A description of those consonants that are uttered with an "s" sound (s, z, sh, zh, etc.).

Side Chain—A secondary signal path through which a signal may be processed before recombination with the primary signal path. As an example, the companding action in a Dolby Noise Reduction System takes place in the system's side chain.

Side-Entry Port—*See* Ports, Rear- and Side-Entry.

Signal Flow Chart—A block diagram of a recording console or other audio system, showing the various possible signal paths through the system, but not detailing the actual electronic components making up each part of the system.

Signal Generator—A test instrument whose output may be one or another of the following; sine wave, square wave, sawtooth, ramp voltage, etc.

Signal Processing Device—An audio system (equalizer, compressor, expander, et al.) used to modify some characteristic of the signal passing through it.

Signal Routing—The process of devising a signal path through a console or other audio system, using bus assignment switches, or patch cords.

Signal-to-Noise Ratio—The ratio of the signal voltage to the noise voltage, usually expressed as the decibel difference between the signal level and the noise level.

Sine Wave—The waveform of a single frequency.

Skew—A deflection of a tape, as it passes over an improperly aligned head or tape guide.

Slap-back—An audibly distracting echo from a reflective surface in a room.

Slate—The recording of an announcement of the take number at the beginning of a recording. Also, a low-frequency audio signal recorded on tape used as a take marker.

Slave Cone—*See* Radiator, Passive.

Slave Transport—One or more tape or film transports, whose movements are locked in synchronization with a single master transport.

SMPTE—Society of Motion Picture and Television Engineers.

SMPTE Time Code—*See* Time Code, SMPTE.

Solo-In-Place—A method of muting all input signals, except the one being soloed, so that this signal (and its artificial reverberation, if any) are heard "in place." Must not be used during actual recording.

Solo Switch—On a recording console's input module, a switch that turns off the normal monitor system, and instead routes the appropriate input signal directly to the monitor.

Sound Level—*See* Pressure Level, Sound.

Sound Level Meter—*See* Meter, Sound Level.

Sound Pressure—*See* Pressure, Sound.

Sound Pressure Level—*See* Pressure Level, Sound

Sound Source, Spherical—The shape of a sound wave radiating away from an ideal sound source. Practical sound sources usually produce sound waves resembling a segment of a spherical sound wave.

Spectrum, Frequency—A distribution of frequencies. For example, the frequencies within the audio bandwidth may be called the audio frequency spectrum.

Spherical Radiator—An ideal sound source. *See* Sound Source, Ideal.

Splice—The point at which two pieces of magnetic tape are joined together, as in editing.

Splicing Block—A device used for positioning and holding down a section of magnetic tape while making splices.

Splicing Tape—An adhesive tape used for joining spliced tapes.

Split Feed—Any network that enables a signal to be routed to two or more separate outputs.

Spring Reverberation System—An artificial reverberation system using springs to simulate the sound of natural reverberation.

Square Wave—A waveform comprised of a fundamental sine wave frequency and its odd-numbered harmonics. So-called because of its characteristic square shape.

Standard Operating Level—A specified reference level. In recording applications, standard operating level is defined as 0VU = +4dBm. In broadcasting, 0 VU = +8dBm.

Standing Wave—An apparently stationary waveform, created by a reflection back towards the sound source. At certain points along the standing wave path, the reflected and direct waves will always cancel, while at other fixed points the waves will reinforce each other.

Static Signal Processing Device—A signal processing device whose operating parameters are not affected by the signal passing through the device. *See* Dynamic Signal Processing Device.

Stationary Wave—A standing wave.

Steel Plate Reverberation System—An artificial reverberation system using a steel plate to simulate the sound of natural reverberation.

Steradian—The solid angle which, on a sphere, encloses a surface equal to the square of the radius of the sphere.

Stereo Microphone—*See* Microphone, Stereo.

Stereophonic—An audio system which reproduces spatial information, giving the listener the illusion of width and depth.

Stereosonic Recording—A stereo recording made with two bi-directional microphones whose axes are at 90° to each other. The microphones are usually aimed at the extreme right and left edges of the sound source to be recorded.

Stretched—Encoded, as in a noise reduction system.

Sub-Master—Any tape used in the production of a master tape. For example, if a master tape is an equalized (or otherwise processed) copy of an earlier tape, the earlier tape is called the sub-master.

Super-Cardioid Microphone—*See* Microphone, Super-Cardioid.

Supply Reel—On a tape recorder, the reel from which tape winds as it passes the head assembly.

Sweetening Session—A recording session during which strings, brass, chorus, etc. may be added to a previously recorded tape, usually containing basic rhythm tracks.

Sync Head—The record head, when used as a playback head during a Sel-Sync session.

Sync Level—Pertaining to the output level of the sync head.

Sync Output—The output of the sync head, in the Sel-Sync mode.

Sync-Word Bits—In SMPTE time code, a permanently-assigned 16-bit word (0011 1111 1111 1101) which is used to identify the end of each frame.

Take—An uninterrupted segment of a recording.

Take-Up Reel—On a tape recorder, the reel on which tape is wound as it leaves the head assembly.

Take-Up Reel Motor—On a tape recorder, the motor that is used to supply take-up tension to the tape on the take-up reel and for fast forward operation.

Talkback System—The communication system by which control room personnel may communicate with musicians in the studio, usually over the regular studio monitor system.

Tangency—The tangential relationship of the tape with the convex surface of the head, as measured at the location of the head gap.

Tape Delay System—A delay system using an auxiliary tape recorder. The delay is a function of the tape transit time as it travels from the record head to the playback head.

Tape Headroom—*See* Headroom.

Tape, High Output—*See* High Output Tape.

Tape Hiss—A low-level, wide-spectrum noise heard when a recorded tape is played back.

Tape Overload—*See* Overload and Saturation.

Tape Saturation—*See* Saturation.

Tape Sensitivity—*See* Sensitivity.

Telephone Filter—*See* Filter, Telephone.

Tension, Hold-Back—The torque applied by the supply reel motor to keep the tape from freely spilling off the supply reel, and to maintain good tape-to-head contact.

Tension, Supply Reel—*See* Tension, Hold-Back.

Tension Switch—On a tape recorder, a two-position switch which changes the torque applied by the reel motor(s).

Tension, Take-Up—The torque applied by the take-up reel motor to maintain a smooth wind as the tape leaves the capstan/capstan idler assembly.

Test Tape—A tape containing a series of test tones at a standard reference fluxivity. The test tape is used to verify the performance of the tape recorder's playback system.

Test Tape, Elevated Level—A test tape with a higher-than-normal reference fluxivity. Such test tapes usually read about 3 dB higher than standard test tapes.

Third Harmonic Distortion—*See* Distortion, Third Harmonic.

Threshold, Compression—The level above which a compressor begins functioning.

Threshold, Expansion—The level below which an expander begins functioning.

Threshold, Limiting—The level above which a limiter begins functioning.

Threshold of Hearing—The lowest level sound that an average listener with good hearing can detect.

Threshold of Pain—The sound level at which the listener begins to experience physical pain.

Throat—The opening at the narrow end of a horn, where it is attached to a compression driver.

Tight Sound—Subjective expression, describing the sound picked up by a microphone placed very close to an instrument.

Tilt—A misalignment of a tape recorder head, around its vertical axis.

Time Code—Any data signal containing coded time-domain information, stored on a tape as the recording is being made.

Time Code, Regenerated—A newly-generated time code, which began with a one-time Jam-Sync.

Time Code, Restored—A newly-generated time code, which remains in continuous synchronization with the reference time-code signal.

Time Code, SMPTE—The time-code standard, as adopted by the SMPTE, in which each recorded tape segment is assigned a unique 80-bit digital word code, defining the hour, minute, second and frame number at which that segment of tape was recorded.

Time Delay System—Any signal processing device in which there is a time delay between input and output.

Tracks—The recorded paths on a magnetic recording tape. In digital recording, a single audio channel may require more than one recorded track on a digital tape recorder. *See* Basic Tracks.

Tracking—The ability of a meter movement, or other dynamic device, to precisely follow the envelope of the applied waveform.

The process of completing a recording session, track by track, as in a Sel-Sync session.

Tracking Error—An unwanted error introduced in an audio system, when the output level or frequency response of the system deviates from the input signal.

Transducer—Any device which converts energy from one system to another. A loudspeaker is an electro-acoustical transducer.

Transfer Characteristic—Any curve on a graph which is a plot of input vs. output.

Transfer Characteristic, Linear—A transfer characteristic which may be drawn as a straight line.

Transferring Tracks—*See* Bouncing Tracks.

Transformer—An electrical network consisting of two or more coils, used to couple one circuit to another.

Transformer, Impedance Matching—A transformer used to match the impedance of one line or network with another.

Transient—A relatively high-amplitude, suddenly decaying, peak signal level.

Transient Distortion—*See* Distortion, Transient.

Transient Response—A measure of an audio system's ability to accurately reproduce transients.

Transport System—In a tape recorder, the system of motors, tape guides, etc., used to move tape past the head assembly.

Turnover Frequency—In a shelving equalizer, the frequency at which the equalizer begins to flatten out, or shelve. Defined as the frequency at which the level is 3 dB above (or below) the shelving level.

Tweeter—A high-frequency loudspeaker.

Ultra-directional Microphone—*See* Microphone, Ultra-directional.

Unassigned Address Bits—In SMPTE time code, four bits (27, 43, 58, 59) which have been defined as permanent zeroes, until otherwise assigned by the SMPTE.

Unbalanced Line—A line consisting of two conductors, one of which is at ground potential. The unbalanced line is often in the form of a single conductor plus shield, with the shield serving as the second conductor. *See* Balanced Line.

Unstretched—Decoded, as in a noise reduction system.

Uni-directional Microphone—A microphone that is most sensitive to front-originating sounds.

Unity Gain—A gain of × 1. That is, output level = input level.

User-Assigned Bits—In SMPTE time code, eight groups of four bits each, which have been reserved to record user-assignable information (take number, session data, etc.). Sometimes called Binary Groups.

Vari-Speed Oscillator—Any oscillator used to drive a tape recorder's capstan motor at various speeds, to effect pitch and tempo changes.

Variable Frequency Oscillator—Any oscillator whose frequency may be varied, but usually used to describe a Vari-Speed Oscillator.

VCA—*See* Voltage-Controlled Amplifier, or Voltage-Controlled Attenuator.

VCA Fader—A fader used to regulate the control voltage applied to a VCA. Instead of directly regulating the audio signal itself, the VCA fader is placed in the control path to the VCA, and not in the audio signal path.

Velocity of Sound—The speed at which sound travels away from a sound source. The velocity of sound is 1,087 feet per second at a temperature of 32° F.

Vented Enclosure—A loudspeaker enclosure, with an open port cut into the front baffle. Also called a bass reflex enclosure.

Voice Coil—The coil winding attached to the diaphragm of a dynamic microphone or loudspeaker.

Voice-Over Compressor—*See* Compressor, Voice-Over.

Voltage—The difference in potential between two points in an electrical circuit.

Voltage-Controlled Amplifier—An amplifier whose gain is a function of an externally-supplied (DC) control voltage.

Voltage-Controlled Attenuator—A resistive network, or an amplifier whose gain is less than unity, whose attenuation is a function of an externally-supplied (DC) control voltage.

Voltage Rating, Microphone—*See* Open Circuit Voltage Rating.

Volume Indicator—*See* Meter, VU.

Volume Unit—A unit of measurement related to the ear's subjective impression of program level or loudness.

Watt—A unit of power.

Wave, Standing—*See* Standing Wave.

Wave, Stationary—*See* Standing Wave.

Waveform—A graph of a signal's amplitude vs. time. The waveform of a pure tone is a sine wave.

Wavelength—The length of one complete cycle of a sine wave.

Wavelength Response—A graph of amplitude vs. wavelength.

Weber—A unit of magnetic flux.

Weighting—Filtering a frequency response, prior to measurement.

Weighting, "A"—A filtering network, corresponding to the ear's sensitivity at 40 phons.

Weighting, "B"—A filtering network, corresponding to the ear's sensitivity at 70 phons.

Weighting, "C"—A filtering network, corresponding to the ear's sensitivity at 100 phons.

Weighting Network—A filter used for weighting a frequency response, prior to measurement.

Wet Recording—The practice of recording artificial reverberation along with the direct signal.

Wet Sound—Subjective description of a sound with a high proportion of reverberation present.

White Noise— *See* Noise, White.

Wind Screen—An acoustically transparent filter, placed over a microphone of reverberation present.

Woofer—A low-frequency loudspeaker.

Wow—A low-frequency fluctuation in tape speed that results in an audible "wow," especially noticeable on sustained tones.

Write—Record (write head=record head).

X-Axis—The horizontal axis on a graph or on a cathode ray oscilloscope.

X-Y Recording—A stereo recording made with two cardioid microphones located in the same vertical plane, with their axes about 90-135° to each other.

Y-Axis—The vertical axis on a graph or on a cathode ray oscilloscope.

Y Connector—Any 2:1 adapter placed in a line to permit a split feed.

Z-Axis—The other horizontal axis on a graph or on a cathode ray oscilloscope that is used to define the third dimension, as in height, width, and depth.

Zero Reference Level—*See* Reference Level.

2-Mix—The stereo program output, often the monitor mix.

Suggested Reading

Baert, Luc, Luc Theunissen, and Guido Vergult, Sony Service Centre (Europe), ed. *Digital Audio and Compact Disc Technology.* 2nd ed. Oxford: Newnes, 1992.

Borwick, John, ed. *Sound Recording Practice.* 3rd ed. Oxford: Oxford University Press, 1987.

Craven, Gerzon, "Lossless Coding for Audio Discs." Journal of the Audio Engineering Society 44, no. 9 (September 1996): 706–20.

Guenette, David, and Parker, Dana. *CD, CD-ROM, CD-R, CD-RW, DVD, DVD-R, DVD-RAM: The Family Album.* Online. *http://www.onlineinc. com/emedia/AprEM/parker4.html.* Internet, June 4, 1997.

Hitachi Corporation. *Information Systems and Electronics.* Online. *http://www.hitachi.com/Pfinder/5014.html.* Internet, June 4, 1997.

Kefauver, Alan. *The Audio Recording Handbook.* Baltimore: MII Publishing, 1997.

Kefauver, Alan, and John Woram. *The New Recording Studio Handbook.* New York: ELAR Publishing, 1989.

Minasi, Mark. *PC Upgrade and Maintenance Guide.* 4th ed. San Francisco: Sybex, 1994.

Parker, Dana, and Robert Starrett. *CD-ROM Professional's CD-Recordable Handbook.* Wilton: Pemberton Press, 1996.

Philips Corporation. *DVD Standards.* Online. *http://www.sel.sony. com/SEL/consumer/DVD/specs.html.* Internet, June 4, 1997.

Pioneer Corporation. *An Introduction to DVD Recordable (DVD-R).* Online. *http://www.km-philips.com/dvd/dst_01.html.* Internet, June 4, 1997.

Pohlmann, Ken. *Principles of Digital Audio.* 2nd ed. Indianapolis: Sams, 1989.

Pohlmann, Ken, ed. *Advanced Digital Audio.* Indianapolis: Sams, 1991.

Pohlmann, Ken. *The Compact Disc: A Handbook of Theory and Practice.* Madison, Wis: A-R Editions, 1989; 2nd ed. *The Compact Disc Handbook,* 1992.

Robinson, D., and Dadson, R. *British Journal of Applied Physics* 7 (1956).

Sony Corporation. *About DVD.* Online. *http://www.sel.sony.com.SEL/consumer/dvd/specs.html.* Internet, July 4, 1997.

Steiglitz, Ken. *A Digital Signal Processing Primer.* Menlo Park, Calif.: Addison-Wesley, 1996.

Strasser. *Thin Film Technology for Data Storage Disks.* Tape/Disk Business, February 1997.

Watkinson, John. *The Art of Digital Audio.* 2nd ed. Oxford: Focal Press, 1994.

Woudenberg, Eric. *The MiniDisc™ Community Page.* Online. *http://www.hip.atr.co.jp/~eaw/minidisc.html.* Internet, May 17, 1996.

Index

A

"A" Weighting, 35, 40
A-B'ing, 441
AC-3, *see* Surround Sound
ACN, *see* Active Combining Network
A/D Converter, 344, 446
Acetate, 276
Accent microphones, 118
Acoustic
    attenuation, 90
    baffle, 101
        in binaural recording, 101
    bass, microphone for, 136, 144
    center, 170
    delay line, 193
    direct box, 137
    guitar, microphone placement for, 139
    phase cancellation, 29
    phase shift, in multispeaker systems, 170
    power, measurement of, 7, 11
    reverberation chamber, 197
    separation, 128
    specifications, of microphone, 56
    suspension system, 156
    transformer, 162
Active Combining Network, 454, 456
Active equalizer, 213
ADAT, 374
Address bits, 411
Advance record head, digital, 360
AES, reproducing characteristics, 302
    tapes, alignment for, 314
AES/EBU digital interface, 472
Alesis Digital, 374
Alias frequency, 334
Aliasing, 334
Alignment
    elevated level test tape, 315
    elevated level, 315
    for IEC (CCIR) tapes, 319
    playback, 316
    procedures, tape recorder, 314
    tape, 314

Ambient noise level, 4
Amplifier
    bandwidth, 20
    combining, 464
    playback, 272
    record, 272
Amplitude domain, 221
Anechoic chamber, 79, 80
Analog-to-digital converter, 344, 446
    recording, 344
    compared to digital, 352
    compared to sample rate converter, 446
Anti-aliasing filter, 344
Asperity noise, 151, 245
Assemble editing, 384
Assigned bits, SMPTE, 411
ATF, *see* Automatic Track Finding
Attack time, of compressor, 230
Attenuation pad, microphone, 90
Autolocator, 312
Automatic track finding, 371
Automation, console
    cues, 495
    data storage, 483
        hard disk, 496
        tape-based, 486
    functions, 487
        mute write, 489
        off-line, 496
        read, 490
        update, 490
        write, 489
    grouping, 491
    hard disk, 496
    master, 494
    merges, 493, 495
    off-line, 496
    rehearsal, 496
    solo-in-place, 493
    snapshot, 494
Automation master, 493
Azimuth
    alignment, 317, 324
    head, tape recorder, 317, 324

B

"B" Weighting, 35, 40
Back coating, 271
Back plate, in condenser microphone, 51
Baffle
    acoustic, 101
    folded, 156
    infinite, 152
    loudspeaker, 152–57
    open, 153
    resonance, frequency of, 153
Balanced line, microphone, 88
Ballistics, meter, 33, 34, 143, 519
Band pass filter, 212
Bandwidth, 209
    amplifier, 20
    equalizer, 209
Base tape, 271
Basic tracks, 506
Bass, microphone placement for, 136, 144
    direct pickups for, 144
Bass
    reflex enclosure, 157
    rolloff switch, microphone, 86
Beta format recorder, 368
Bi-amplification, 172
Bias
    A.C., 285
    calibrate control, 320
    current, 282
    D.C., 282
    frequency, 284
    and harmonic distortion, 286
    introduction to, 282
    level adjustment, 320
    level, effect of, 287
    level, optimum, 287
    oscillator, 286
    summary of, 287
Bias signal, no required in digital, 336
Bi-directional microphone, 59
    polar pattern of, 62

use for greater separation, 60
Binary
    coded digital system, 336
    groups, 336
    numbering system, 337
BInary digiTS, 336
Binaural
    hearing, 57
    recording, 101
Bi-phase modulation, 414
BITS, 336
Blauert and Laws criteria, 170
Block coded data, 346
Blumlein microphone, *see* Stereo microphones
Bouncing tracks, 508
Boundary microphones, 134
Braking system, in tape recorder, 308
Brass, microphone placement for, 144
Braunmuhl-Weber condenser microphones, 104, 173
Breathing, in compressor, 227
Broadcast mode, 464
Broadcast reference level, 38
Burst error, 347
Bus
    channel, 437
    mixing, 437, 454, 456
Bus/tape monitoring, 449

C

"C" Weighting, 35, 40
CCIR equalization, *see* IEC equalization
Cable, microphone, 86
Calibration controls, 316
Cancellation, phase, 29–32
Capacitance, of microphone cable, 84
Capacitor microphone, 51
    see also Condensor microphone listings
Capstan, 273, 305
Capstan idler, 273
Cardioid microphone, 61
    compared with supercardioid microphone, 64
    off-axis coloration, 54
    polar pattern, 64
    proximity effect in, 82
    rear and side entry ports, 59
Cathode ray oscilloscope, *see* Oscilloscope

Cellulose acetate, 276
Center
    acoustic, 170
    frequency of equalizer, 210
    of noise signal, 170
    tap, transformer, 53
Channel
    assignment switches, 449
    bus, 451, 469
    cut switch, 460
    path, 451
    trim, 451
Channel/line
    monitor switch, 449
    switch, 449
Channel/monitor switch, 449
Character generator, video, 417
Chorusing, 204
CIRC, *see* Cross Interleave Reed-Solomon code
Cleaning, of heads and tape guides, 313
Close miking, 517
    introduced, 56
    techniques, 128
Clutch system, in tape recorder, 308
Coaxial loudspeakers, 170
Code
    cyclic redundancy check, 346
    parity block, 346
Coding systems, 337
Coercivity, 276
Coherence and phase, 28–32
Coincident pair microphones
    see Stereo microphones
Color frame
    code bit, 413
    flag, 413
Coloration, off-axis, 81
    in drum setup, 141
Combining amplifiers, 464
Combining filtering, 126
Combining network, active, 442, 451
Compact disc, 393, 395, 398
Compact disc recordable, 398
Compander, 252
Complementary signal processor, 249
Compliance, 147
    of sealed enclosure, 156
Composite equalization, 210
Compression
    driver, 163
    program, 230

ratio, 227
ratio, variable, 227
threshold, 222
Compressor
    attack time, 230
    definition, 222
    gain reduction, 251
    as a noise reduction device, 251
    release time, 230
    voiceover, 234
Compressors and limiters, 222, 232
Concealment, error, 350
Condenser microphone, 53–56
    directional patterns
        see Microphone polar pattern
    overload, 90
    phantom powering of, 53–55
    power supplies, 53
    sensitivity, 56, 81, 87
Connectors
    microphone, 90
    recording console, 88
Console, *see* Recording Console
Console, in-line, *see* Recording Console
Contact
    microphone, 133
    tape-to-head, 323
Continuous jam-sync, 423
Count-offs, 519
Coupling, diaphragm to air, 160
CRC code, *see* Cyclic Redundancy Check code
Crossfade, electronic, 360
Cross Interleave Reed-Solomon code, 347
Cross-interleaving, 347
Crossover
    frequency, 168
    in bi-amplification system, 172
    network, 168
Crosstalk
    digital, 353
    and time code, 417
Cue systems, *see* Recording Consoles
Cycles per second (hertz), 15
Cyclic Redundancy Check code, 346

D

D/A converter, 344
DASH, 360, 361
DAT, 368
Damping, loudspeaker, 146

Data, block coded, 346
DAW, see Digital Audio Workstation
dB, 3
dbx Noise Reduction System, 253
    block diagram, 254
    pre- and post-emphasis, 253
    specifications, 256
    tracking errors, 256
DCC, see Digital Compact Cassette
Decay
    definition, 188
    and reverberations, 188
    time, variable, 188
Decca Tree, 114
De-multiplexer, 351
Decibel
    across non-standard resistance, 13
    introduced, 3
    power formula, 3, 7
    reference levels, 3
        dBm, 12–13
        dBu, 14
        dBv, 14
        dBV, 14
    specifications, in equipment, 34
    voltage formula, 7
Decimal number system, 336
Decoding, noise reduction, 265
De-esser, 233
Delay
    definition, 188
    effects, 193
Delay systems
    acoustic, 193
    analog, 193
    digital, 193
    tape, 193
Demagnetization, of head and tape guides, 313
Demodulation, 348
Demultiplexer, 351
Depth perception, listener's, 97
Diaphragm
    loudspeaker, 145
    microphone, 47
    to air coupling, 160
Diffraction, 24
    and goboes, 130
    and loudspeakers, 148
    sound waves, 24
Digital
    audio, introduction to, 331
    condenser microphone, 56

    clock, 474
    crosstalk, 353
    design, basic of, 333
    dropouts, 412
    dynamic range, 353
    editing, 527
    encoding, 332, 336
    error
        concealment, 350
        correction, 346
        detection, 346
    harmonic distortion, 353
    playback, 348
    print-through, 353
    punch-ins, punch-outs, 380
    recording, 342
    record modulation, 339
    recorders
        DASH, 360
        F1 (Beta VHS), 368
        PD, 363
        PRO-DIGI, 363
        reel-to-reel, 376
        U-matic, 367
    reverberation systems, 193
    signal processing, 242
    tape duplication, 354
    tapes, physical properties, 380
    -to-analog converter, 351
    word, 339
    wow and flutter, 310, 354
Digital Audio Tape Recorder, 368
Digital Audio Workstation, 388, 527
Digital Compact Cassette, 535
Digital console, 472
Digital console clock source, 474
Digital delay line, 193
Digital Versatile Disk, 396
Digital Video Disk, see Digital Versatile Disk
Digits, binary, 336
Direct
    assignment switch, 449
    boxes, 137
    output, recording console, 462
    pickup, 134
    radiators, summary of, 157
    sound
        from loudspeaker, 174
        in room, 187
Direction, perception of, 28
Directional, characteristics of
    loudspeaker, 148

    microphone, 56
Disc allocation, 515
Distortion
    and bias level, 286
    crossover, 168
    harmonic, 277
    microphone, 90, 101
    off-axis, 128
Dither, in digital systems, 342
Dolby Noise Reduction System
    AC-3, see Surround Sound
    type 'A', 258
        block diagram, 260
        side chain in, 260
        tracking errors in, 262
    type 'B', 262
    type 'C', 262
    type 'SR', 263
        block diagram, 265
        tracking errors in, 264
Doubling, 200
Driver, compression, 163
Drone cone, 160
Dropframe, time code, 411, 419
Drop outs, 345
Drums, microphone placement for, 140, 142
Dry sound, from contact microphone, 188, 197
DTRS, 372, 535
Dual-diaphragm microphone, 69, 78
    off-axis response, 81
    polarizing voltage, 79
    proximity effect, 82
    schematic, 73
    single pattern, 81
Dummy head, 133
Dump switch, 409, 469
DVD, see Digital Versatile Disk
DVD-A, see Digital Versatile Disk
DVD-V, see Digital Versatile Disk
Dynamic
    microphone, 47
    overload, 90
    range
        of audio equipment, 18
        of digital systems, 353
        and equalization, 218
        of human voice, 20
        of magnetic recording tape, 222
        of musical instruments, 16
        restrictions on, 222

signal processors
    analog, 193, 222
    digital, 193
Dynamic range, digital, 353
Dyne, 88

E

Early sound field, 193
Echo, 188, 193
    controls and paths in recording console, 193
    definition, 188
    and delay, 193
    print through, 289
    and reverberation, 188
    reverberation system, complete, 194
    tape, 193
Echo, send, *see* Recording Consoles Reverberation Sends
Eddy currents, 298
Edit
    command, 384
    memory, 384
    rehearse mode, 384
Edit Decision List, 338, 530
Edit switch, 309
Editing, 380
    digital, 527
    electronic, 382, 424
    mixdown session, 523
    multi-channel digital, 380
    musical, 524
    razorblade digital, 381
    and SMPTE time code, 383–84
    track, 528
EDL, *see* Edit Decision List
Efficiency
    loudspeaker, 147
    of direct radiator system, 147
Electret microphone, 52
Electric guitar, microphone placement for, 134
Electrical
    phase cancellation, 29–32
    power measurements, 10
    separation, 128
    specifications, microphone, 87
Electro-mechanical systems, 192
Electronic editing, 328, 424
    and SMPTE time code, 383
Electronic keyboard instruments
    microphone placement for, 135

Electrostatic microphone
    see Condenser microphone
Elevated level test tape, 315
Enclosure
    bass reflex, 157
    sealed, 156
    vented, 160
Encoded tapes, monitoring, 265
Encoding, noise reduction, 253, 263
Energy
    conversion in signal path, 145
    distribution curve, of human voice, 20
    of music, 294
Envelope, 240
Equal loudness contours, 25
    graph of, 26
    and monitor levels, 35
    significance of, 25
Equalization, 207
    composite, 215
    and dynamic range, 218
    high-frequency, 215
    high-frequency, playback, 207
        adjustment of, 318
    low-frequency, 214
    low-frequency, playback, 207
        adjustment of, 318
    mid-frequency, 215
    mid-frequency, playback, 299
        record, adjustment of, 318
    room, 181
    shelving, 207
    tape recorder, 321
Equalizers, 210
    active, 213
    graphic, 210
    parametric, 210
    passive, 213
    phase shift, 218
    in recording console, 447
Equivalent circuit, microphone, 87
Erase
    current, 305
    head, 304
Error
    concealment, 350
    correction, 345, 349
    detection, 349
Errors, frame rate, 408–9
Expander
    definition, 234
    as a noise gate, 236, 291

    range, 234
    release time, 238
    for special effects, 240
Expansion
    ratio, 234
    threshold, definition of, 234
Exponent, 4–6

F

Fade-out, 240, 521
Fader
    automation, 533
    exchange switch, 448
    master, 454, 533
    monitor, 441
    monitor system, 441
    two-mix, 453
    VCA, 452
Faders, input, 434
Fast forward switch, 309
Feedback
    in bouncing tracks, 508
    from headphones, 514
Ferric oxide, 271
Ferrous metal, 271
Field strength, magnetic, 271
Fifth, musical interval of, 18
Figure 8 microphone
    polar pattern of, 59
    used for greater separation, 69
Filter
    anti-aliasing, 344
    bandpass, 212
    cutoff, 215
    flutter, 310
    high-frequency, 215
    high-pass, 82, 214
    low-frequency, 215
    low-pass, 448
    noise, 250
    notch, 211
    output, 352
    passive, 261
    pop, 92
    proximity effect, 82
    scrape flutter, 310
Filter, digital, low-pass, 333
Flags, color and drop frame, 413
Flanging and phasing, 202
Fletcher-Munson curves, 25
Flutter, 310, 354
    filter, 310, 354

Flux lines, 274
    and playback head, 274
Fluxivity, 313
    on high output tapes, 313
    reference, of test tape, 313
Foldback system, 459
    monitoring in recording console, 440
Folded baffle, 156
Folded horn, 163
Frame rates, 408, 410
Frames, noise reduction, 267
Frequency, 15
    alias, 334
    vs. listening level, 25
    and logarithms, 18
    of noise, 245
    range of musical instruments, 16
    response
        of audio equipment, 18
        of ear, 18
    sampling, 333
    of voice, 20
Frequency domain, 207–15
Fringing, 318
Full space, 147

G

Gain
    before threshold, compressor, 222
    reduction, compressor, 222
    riding, 222
Gain riding, VCA, 482
Gamma ferric oxide, 271
Gap, 274
    losses, 297
    space, 297
Generation tape, in overdubbing, 503
Generator, timecode, 417
Goboes, 130
Golden section, 177
Graphic equalizer, 210
Group
    level control, 467
    master control, 467
    selection, 467
Grouping, VCA, 466, 491
Guitar
    acoustic, microphone placement for, 137
    direct pickup for, 134

    electric, microphone placement for, 139

H

Hamming codes, 350
Hard disk systems, 377, 387
Harmonic distortion, 277, 353
    and bias level, 277
    digital, 353
Harmonizer, 204
HDM, 339, 362
Head, tape recorder
    alignment, 316
    cleaning, 313
    demagnetization, 313
    erase, 304
    gap, 274
    losses, 297
    playback, 274
    record, 274
    sel-sync, 506
Headphone
    leakage, 514
    monitoring, 514
    selection, 514
Headroom, 277
    on high output tapes, 314
    improvement with noise reduction system, 253
Height, tape recorder head, 323
Helical scan, 367
Hertz, 15
Hexadecimal numbering system, 336
HI Logic state, 353
High density modulation, 362
High frequency
    equalization, 215
    playback equalization, 299
High impedance microphones, 84
High output tape, 288
    flux level on, 288
High-pass filter, 214
Hiss, 245, 332
Horn
    folded, 163
    loaded systems, 162–67
    multicellular, 163
    straight, 163
    throat, 162
Hypercardioid microphone, 64
Hysteresis loop, 279

I

I/O module
    input section of, 437
    output section of, 437
Idler
    capstan, 273, 308
    scrape flutter, 310
    take-up, 308
IEC (CCIR), 43, 301
    metering standards, 43
    reproducing characteristic, 319
    tapes, alignment for, 319
Image shift, 98
Impact noise, microphone, 91
Impedance
    matching transformer, 137
    microphone, 84
Incoherent waveforms, 40
Indirect radiator, 162
Infinite baffle, 152
In-line console, *see* Recording console
Input level control, in recording console, 448
Inputs, recording console
    line, 458
    microphone, 458
Input section, of I/O module, 446, 475
Input/Output module, *see* I/O module
Insert editing, 380, 383
Insert recording, 510
Intensity, acoustic, 3
Interleaving, 345, 347
Interpolation, 350
Interval, musical perception of, 18
Inverse square law, 10
Isolate mode, 489
Isolation booth, 130
    for acoustical separation, 130

J

Jack bay, 446
Jam sync
    continuous, 423
    one-time, 423

K

Kepex (KEyable Program EXpander), 241
Key input, 240
Keying signal, 240

L

Lavalier microphone, 135
Leader tape, use of, 524
Leakage
    headphone, 514
    microphone, 126
    microphone, minimizing, 128
Level
    operating, adjustment of, 314
    playback, adjustment of, 316
    quantization, 335
    record, adjustment of, 319
    record, for best signal-to-noise ratio, 519
Limiter, definition, 222
Limiters and compressors, 222
Limiting, 222
    program, 222
    threshold, 222
Line
    amplifier in recording console, 464
    inputs, in recording console, 446, 458
    matching transformer, 137
Lines, microphone, 88
Lissajous pattern, see Oscilloscope
Listening
    level, effects of changes in, 533
    room, 147
Lobe, in polar pattern, 64
Log paper, 18
Logarithm, definition, 4
Logarithms, 4
    and frequency response, 18
Logic states, HI and LOW, 353
Longitudinal time code, 406
Loudness, measured in phons, 29
Loudspeaker
    acoustic center, 170
    acoustic phase shift, 170
    acoustic suspension system, 156
    bass reflex, 157
    bi-amplification, 172
    coaxial, 170
    compliance, 147
    compression driver, 163
    controls, on recording console, 453
    coupling, 160
    crossover frequency, 168
    crossover network, 168
    damping, 146
    diaphragm, 146
    diffraction, 148
    direct radiator, 152, 160
    drone cone, 160
    efficiency, 147
        of direct radiator system, 160
    folded baffle, 156
    folded horn, 163
    horn-loaded systems, 162–67
    horn, straight, 163
    indirect radiators, 162
    infinite baffle, 152
    monitoring, in recording console, 515
    multi-speaker system, 168
    open baffle, 153
    passive radiator, 160
    polar pattern, 148
    resonance, 145
    room interface, 174
    sealed enclosure, 156
    slave cone, see Loudspeakers, drone cone
    transient response, 146
    tuned port, 157
    tweeter, 168
    vented enclosure, 157
    voice coil, 145
        of compression driver, 168
    woofer, 165
Low frequency
    filter, 214
    playback equalization, 215
        adjustment of, 316
Low impedance microphone, 84
LOW logic state, 353
Low-pass filter, 448
Low-pass filter, digital, 448

M

MADI, 472
MAF, 25
M-S Recording, 106
Magnet, artificial, 271
    and domains, 275, 278
    permanent, 271
    temporary, 271
Magnetic domains, see particles
    field, 275
    field strength, 276
    flux, 274
    vs. frequency, 291
    particles, 276
    recording tape, 272, 275
        flux level, 276
        headroom, 277
        high output, 288
        hiss, 245
        magnetic properties, 271
        noise, 245
        operating parameters, 276
        physical properties, 275
        properties, 275
    saturation, 280
Magnetism
    basic principles, 271
    residual, 271
Magnetization, remanent, 280
Magneto-optical media, 398
Masking
    of noise, 541
    of reverberation, 192
Master
    fader, 454
        automation, 541
        two-mix, 535
    module, in in-line console, 442
Mastering, 538
Master reel, preparation of, 442
Master/slave machines, 419
Matching transformer, 137
Matrix system, MS, 106
Mechanical system, tape recorder, 323
Memory, edit, 384
Merges, 495
Metering
    ballistics, 33
    dB, 3
    monitoring
        in recording console, 38
    peak, 36, 43
    standards, 43–44
    sound level, 35
    VU, 33
        and transient peaks, 34
Microbar, 7
Microphone
    accent, 118
    accessories
        extension tubes and swivel, 141
        shock mounts, 93
        wind screens, 92

acoustical specifications, 56
attenuation pads, 90
balanced lines, 88
bass rolloff switch, 86
bi-directional, 59
binaural head, 133
boundary, 134
Braunmuhl-Weber capsule, 104
cables, 68
capacitor, see Microphone, condenser
condenser, 50, 52
   capacitor, 51
   digital, 56
   overload, 90
   power supply, 53
contact, 133
diaphragm, 47
digital condenser, 56
directional characteristics, 56
distortion, causes and prevention, 90
dual-diaphragm, 69, 78
   off-axis response, 81
   proximity effect, 82
   schematic, 73
   single pattern, 81
dynamic, 47–49
   moving coil, 47–49
   overload, 90
   ribbon, 50
electret, 52
electrical specifications, 87
equivalent circuit, 87
impedance, 84
inputs, on recording console, 445, 458
lavalier, 135
lines, 88
moving coil, 48
multi-placement techniques, 100, 123
off-axis coloration, 81
off-axis sounds, 59
omni-directional, 59
on-axis sounds, 56
overload, 90
open circuit voltage rating, 88
phase cancellations, avoiding, 124
phase reversal, at rear of figure-8, 59
phase shift, 61

placement
   acoustic bass, 144
   brass, 144
   drums, 140, 142
   guitars, 137, 139
   multi-microphone, 123
   percussion, 140
   piano, 142
   stereo, 101–18
   strings, 144
   woodwinds, 144
plugs, 445
polar patterns, 62
   bi-directional, 62, 64, 66
   cardioid, 64, 67
      compared with supercardioid, 67–69
   combining, 62
   figure-8, 64
   hypercardioid, 66
   measuring, 79
   multiple, 69
   omni-directional, 65
   shotgun, 67–68, 72
   supercardioid, compared with cardioid, 67–69
   ultra-directional, 68
polarity, 88
power rating, minimum, 87
pre-amplifier, 445
   built-in, 81
   in recording console, 445
pressure, 59
pressure gradient, 59
pressure zone, 134
proximity effect, 82
   in drum microphones, 140–41
rear- and side-entry ports, 59
ribbon, 50–51
RF, 52
sensitivity, 59, 87
shock mounts, 93
specifications
   acoustical, 56
   electrical, 84
stereo, 101
supercardioid, compared with cardioid, 67–69
systems, 104
   coincident, 104
      Blumlein, 105

      M-S, 106
      X-Y, 105
   near-coincident, 110
      Decca Tree, 114
      NOS, 112
      ORTF, 112
techniques, close miking, 97
techniques, surround sound, 119
unbalanced lines, 88
uni-directional, 59
usage, rules of, 123
vibration noises, 91
voltage rating, open circuit, 88
wind noises, 91
wind screens, 92
Microphone/Line selector switch
   in recording console, 446, 458
Mid-frequency equalization, 212–15
Middle-Sides record, 106
MIDI, 425
   interference from, 137
   and SMPTE time code, 426, 429
MIDI equipment, 427
MIDI interconnection, 427
Milliwatt, and the dBm, 12
Mini-Disk, 399
Mixdown session, 497, 523
   assistance during, 532
   monitor levels, 533
   monitor speakers, 534
   preparation, 531
   recording levels, 533
   summary, automated, 541
   summary, non-automated, 540
Mixing bus, 454, 456
Modulation
   bi-phase, 414
   high density, 362
Modulation noise, 245
   and bias level, 320
Module
   I/O, 437
   master, 440
Monitor
   headphone, 514
   levels, during mixdown, 533
   mix to cue, 469
   module, in-line console, 437
   pan to channel buses, 469
   pan to sent buses, 469
   pan pot, 437, 453
   path, 440

reverberation on, 440
section, in recording console, 440
speakers, for mixdown, 534
VCA fader, 482
Motors, supply and takeup, 308
Moving coil
loudspeaker, 145
microphone, 48
M-S recording, 106
Multi-cellular horn, 163
Multi-microphone placement techniques, 123
Multi-pattern microphone, 69
Multiplexer, 345
Multi-speaker system, 168
Multi-track rotary head system, 371
Musical Instrument Digital Interface, 425
Mute switch, 460
Mute write mode, 489

N
NAB, 317
hubs, 317
reproducing characteristics, 317–19
tapes, alignment for, 317
Nanoweber, 314
Near-coincident microphone systems, 110
NOS, 112
ORTF, 112
Network
crossover, 168
weighting, 35
Noise, 245
ambient, 4
analog, 245
asperity, 245
and bias level, 320
digital tape, 352
filter, 250
frequency of, 245
gate, 236
impact, 91
level, fluctuating, 247
masking of, 245
microphone line, 91
modulation, 245
and noise reduction principles, 245
pink, 247
principles of, 247

quantization, 335
ratio, signal to, 248
reduction systems, 245, 249
dbx Type I, 253
dbx Type II, 256
Dolby 'A', 258
Dolby 'B', 262
Dolby 'C', 262
Dolby 'SR', 263
dynamic system, 251
static system, 250
Telcom C-4, 263
tracking error in, reduction, 256
residual, 247
signal-to-noise ratio, 248
summary of, 265
tape, 352
wide-band, 506
white, 246
Non-complementary signal processors, 249
NOS microphone system, 112
Notch filter, 211
Numbering systems, 336, 518
Nyquist
frequency, 333
rate, 333

O
Octal numbering system, 336
Octave definition, 18
Oersted, 280
Offset, timecode, 418
One-on-one technique, 423, 437
One-time jam sync, 423
Operating level, tape recorders, 315
Optical disk, 392
ORTF microphone system, 112
Oscilloscope, 41, 324
for azimuth alignment, 324
Outputs, parallel and serial, 345
Overbias, 320
Overdubbing, 365, 503, 523
Oxide tape, 271

P
Pad, microphone attenuation, 90
Pan pot, use of, 453
Panning, 530
Parallel output, 345
Parametric equalizer, 210
Parity block code, 346

Pass band, 212
Passive equalizer, 213
Passive radiator, 160
Patch bay, 446
Patch points, in recording console, 446, 458, 476
Path length, 124
PCM, see Pulse code modulation
PD, see Pro-Digi
Peak program meter, 36
Peak reading meter, 36, 43
Percussion instrument, microphone placement for, 140
Perspective, listening, 97, 114
Phantom power supply, 53
Phase
cancellations, 29–32
cancellations, avoiding, 29, 124
and coherence, 28
distortion, crossover, 168
in and out of, 30
reversal, at rear of figure-8 microphone, 59
shift
at crossover frequency, 168
and equalizers, 218
and microphones, 59
in multispeaker systems, 170
Phasing and flanging, 202
Phon, 25
Piano, microphone placement for, 142
Pinch roller, 305
Pink noise, 247
Playback
alignment, 316
amplifier, 272
circuit, 348
equalization, 299, 301
high-frequency adjustment, 316
low-frequency adjustment, 316, 536
level adjustment, 316
play mode, 490, 511
process, 296, 346
Plugs, microphone, see Connectors, microphone
Plus-one frame, 413
Point source, 147
Polar pattern
loudspeaker, 148
microphone, see Microphone polar pattern

Polarity, microphone, 88
Polarizing voltage, in condenser microphones, 51
Polyester, base film, 271
Pop filter, 92
Ports
 rear and side entry, 57
 loudspeaker, 157
 tuned, 157
Post-emphasis, in dbx Noise Reduction System
 and headroom improvement, 253
Power, measurement of, 7–15
Power rating, of microphone, 87
Power requirements, of monitor systems, 179
Power supply, microphone, 53
PPM, see Peak program meter
Preamble, 345
Preamplifier
 in microphone, 51, 53
 in recording console, 253, 445
Pre-emphasis
 in dbx Noise Reduction System, 253
 and headroom improvement, 253
Pre-fader switch, on recording console, 461
Pre-roll, 386
Print-through, in magnetic tape, 289, 353
Pro-Digi, 363
Program compression, 363
Proximity effect, 82
Pumping, 227
Punch-in, 421, 510
 digital, 380
Pulse Code Modulation, 337–38, 365

Q

Q, 209
Quantization, 335
 levels, 341
 noise, 335

R

Rack, tape recorder head, 323
Radiation loudspeaker
 angle, 148, 163
 into full space, 162
 pattern, 148
Radiator
 direct, 152, 160
 indirect, 162
 passive, 160
Ratio
 compression, 227
 expansion, 234
 room, 177
 signal-to-error, 342
RDAT, see DAT
Reader, time code, 416
Read mode, 490
Real time spectrum analyzer, 246
Rear entry ports, microphone, 59
Record
 alignment, 319
 amplifier, 272
 calibrate control, 316
 circuit, 347
 equalization, 295, 321
  adjustment of, 318
 head, 274
  used as playback head, 333
 level, adjustment of, 316
 process, 294
  digital, 344
Recording console, 433
 ACN, 442
 analog, 291
 assign switch, 449
 automation systems, 481
  data systems, 483
  hard disk, 496
  tape based, 486
  and time code, 486
 auxiliary return controls, 457
 auxiliary send controls, 455
 basic recording console, 433
  in-line, 437
  split configuration, 437
 basic console layout, 435
 broadcast mode, 464
 bus selector switch, 449
 channel
  ACN, 454
  assign switches, 449
  bus, 451, 460
  cut switch, 460
  fader, 452
  path, 451
  trim, 451
 channel/monitor equalization switch, 447
 combining amplifiers, 464
 component parts, 445
 cue system, 469, 515
  assign switches, 455, 469
  reverberation, to cue, 455
  send controls, 440, 459
 delay section, 447
 digital console
  auxiliary send bases, 476
  clock, 474
  input pool, 475
  output pool, 478
  patch points, 476
  signal flow paths, 472
  system configuration, 474
 direct assignment switch, 449
 direct outputs, 462
 dump switch, 271
 echo controls and signal paths, 193
 echo send controls, 193
 equalization switch, 440
 equalizers, 447
 fader
  channel, 449
  exchange switch, 448
  monitor, 452
  mute switch, 460
  VCA, 482
 foldback, 459
 group selects, 467
 grouping, 466
 headphone monitoring, 514
 I/O module, 437
 in-line, 442
 input section, 434, 437
 line amplifiers, 464
 line-in/line-out switch, 452
 line inputs, 458
 loudspeaker monitoring, 515
 master module, details, 434, 440
 meters, see metering
 microphone
  inputs, 445
  pre-amplifiers, 445
  trims, 451
 MIDI interface, 429
 mix-to-cue, 469
 monitor module, 440
 monitor pan to channel buses, 453, 469
 monitor section, 434, 440

monitoring
    headphones, 514
    loudspeakers, 515
mute switch, 460
output section, 434, 437
output bus, 437, 449, 462
pan pot, 437, 449
patch points, 446, 458
PFL, *see* pre-fade listen
pre-fade listen, 461
preparation
    for mixdown, 531
    for overdubbing, 503
    for recording, 517
preview, *see* solo and PFL
returns, 457
reverberation, 520
    in channel path, 457
    in monitor path, 457
send control, 454
send level, 457
sends, 457
    to cue, 440, 445
signal flow diagram, 444
    combining circuits, 464
    echo and cue send, 440
    input section, 434
    patch points, 446
    simplified, 484
    solo function, 460
signal flow path, summary of, 458
signal processing, 520
simplified I/O module
    solo function, 493
        channel solo, 493
        monitor solo, 493
        solo-in-place, 493
split configuration, 484
subgroups, 481
submasters, 481
VCA, 452
Recording session, 517
    preparing for, 517
Record modulation, for digital recording, 347
Recovery time
    compressor, 230
    expander, 238
Reed-Solomon code, 347
Reel size switches, 306
Reference fluxivity, 313
Reference levels, 3
    broadcast, 38

decibel, 3
studio, 38
voltage, 8
zero, 3, 4
Reflected sound, 187
    from loudspeaker, 174
Reflections, 24
    early, 193
    late, 194
    simulation of early, 193
Reflective surfaces, 187
Refraction, 24
Regenerated time code, 423
Rehearse mode, 531
Reinforcement, by standing waves, 176
Release time
    compressor, 230
    expander, 238
Reluctance, 271
Remanence, 276
    and fluxivity, 195
Remanent flux, 195
Remanent magnetization, 280
Remote control
    of record/playback mode, 511
    of tape transport system, 311
Reproducing characteristic
    AES, 302
    IEC (CCIR), 301
    NAB, 317
Residual noise, 247
Resonance
    loudspeaker, 145
    peak, minimizing, 181
    room, 145
Resonance frequency, loudspeaker, 145
Restored time code, 423
Retentivity, 276
Reverberation
    artificial, 520
    chamber acoustic, 188, 197
    computers, 199
    controls and signal paths
        in recording console, 440
    and decay, 188
    definition, 188
    digital systems, 188
    electro-mechanical systems, 192
    plate, 195
    spring system, 197
    stereo, 197

using, during recording, 520
Reverse switch
    fader, 453
    line-in/line-out, 452
    phase, 448
Rewind switch, 309
RF microphone, 52
Ribbon microphone, 50
Robinson-Dadson equal loudness contours, 25
Roll-off switch, bass, 86
Room
    acoustics, 187
    dimension ratios, 177
    equalization, 181
    modes, 177
    ratios, 177
    sound, effect surfaces on, 150, 174, 177
Rotary head recorders, 365, 371
Rotation point, 226
Rumble, building, 245

S

Safe mode, 317
Sample
    and hold, 333, 344, 351
    a waveform, 339
Sampling, 205
    frequency, 333
    rate, 358
    rate converters, 358, 446
Saturation, tape, 278, 280
    vs. frequency, 280
Scrape flutter, 310
    filter, 310
    idler, 310
SDAT, *see* DAT
Sealed enclosure, 156
Seating plan, for recording, 517
Send control, 454
Self-erasure, at high frequencies, 297
Sel-sync process, 506, 512
Sensitivity
    and bias, 282
    of ear, 24
    microphone, 87
    tape, 277
    tape vs. bias level, 277, 282
Separation
    microphone placement for, 128
    with figure-8 microphone, 129

using isolation booths and goboes, 130
Serial
    output, 345
    storage device, 345
Servo motor, 310
Shelving equalization, 207
Shielded cable, 88
Shock mount, microphone, 93
Shotgun microphone, 67
Sibilants, minimizing, 91–92, 233
Side chain
    in compressors, 234
    in Dolby Noise Reduction System, 258, 262
    in expanders, 234
    in noise gates, 236
Side-entry ports, microphone, 59
Signal flow diagrams
    see Recording console signal flow diagram
Signal flow path, summary of, 458
Signal processing devices, 520
    acoustic delay line, 193
    compressors, 222–29
    chorus effects, 204
    digital delay line, 193
    de-essers, 233
    digital, 193
    dynamic complementary, 251
    dynamic non-complementary, 251
    echo reverberation system, complete, 194
    equalizers, 210
    expanders, 234
    filters
        band-pass, 212
        high-pass, 214
        low-pass, 448
        notch, 211
    flangers, 202
    harmonizers, 204
    limiters, 222
    noise gates, 236
    phasers, 202
    in recording consoles, 520
    reverberation systems, 520
        acoustic reverberation chamber, 197
        computer systems, 192
        digital systems, 193
        plate systems, 195
        spring systems, 195

static, complementary, 251
static, non-complementary, 251
tape delay system, 188
used during recording, 520
Signal-to-noise ratio, 248
    and recording level, 520
    digital, 248
Skewing tape, 323
Slating, 518
Slave cone, *see* Loudspeaker, drone cone
Slave/master machines, 419
SMPTE/EBU Time code, 407, 411
    and MIDI, 429
    structure, 407
SMPTE/ITU, 535
Solo-in-place, 460
Solo switch, 460
    to check headphone leakage, 514
Sound
    direct, 188
    and hearing, 24
    level, hearing, 24
    level meter, 35, 38
    pressure levels, table of, 9
    production of, 15
    reflected, 24
    refracted, 24
    source, ideal, 147
    velocity, 15
    and vibration, 15
    wave, spherical, 147
    wavelength of, 15
Sound level meter, 35
Spaced apart microphone systems, 114–17
Speaker level control, 454
Specifications, microphone
    acoustical, 56–82
    electrical, 87
Specifications, digital tape recorder, 357
Spectrum analyzer, 40
Spherical sound wave, 147
Splicing blocks, 381, 525
Split feed, 137
Spring reverberation system, 195
Square wave defined, 333
Standard operating level, 315
Standing waves, 176
Steel plate reverberation system, 195
Steradiam, 147

Stereo
    microphones, 101
        advantages of, 102
        disadvantages of, 102
        perspective, 87
        program compression, 363
        reverberation, 197
Strings, microphone placement for, 144
Super Audio CD, 396
Super-cardioid microphone, 64
    compared with cardioid, 64
    polar pattern, 64
Supply reel motor, 308
Supply side idler, 308
Surround sound, AC-3
    Dolby AC-3, 534
    Dolby Surround, 534
    microphone technique, 119
Sweetening session, 506
Sync record head, digital, 363
Sync-word bits, 413, 416
Synchronizers, timecode, 405, 419
System configuration files, 474

T

Takes, numbering of, 518
Takeup idler, 308
Takeup motor, 308
Tangency, tape recorder head, 324
Tape
    delay system, 193
    editing, 380
    magnetic, *see* Magnetic recording tape
    physical properties
        analog, 288
        digital, 288
    recorder
        alignment procedures, 315–22
        equalization, 440
        mechanical system, 275
        theory of operation, 275
        saturation vs. frequency, 280
        tension switches, 306
        -to-head contact, 323
        transport system, 305
            remote control for, 311
Telcom C-4 noise reduction, 263
Tension switches, tape, 306
Test tape, 314
    elevated level, 315

Third harmonic distortion
and bias level, 277
and headroom, 277
Three-to-one, microphone placement, 124
Transport system, tape, 305
Threshold
compression, 222
compression, definition, 222
expansion, 234
vs. frequency, 24
of hearing, definition, 3, 5
limiting, 232
of pain, 226
variable, 226
Throat
compression driver, 163
horn, 162
Tilt, tape recorder head, 323
Time code, 405
address bits, 411
applications, 421
copying, 423
dropframe, 409, 411
generator, 416
hardware, 416
implementation, 416
longitudinal, 408
offset, 418
reader, 416
recording, 417
regenerated, 423
restored, 423
SMPTE/EBU standards, 407
synchronizers, 419
using, 414
vertical interval, 409
Time domain, 111
Track assignment, 515, 530
Track editing, 528
Track width format
analog, 292
digital
DASH, 360–61
Pro-Digi, 363
magnetic tape, 291
Tracking errors, in noise reduction systems, 256
Tracks, 506
Transducers, 45

see also Loudspeakers, Microphones
Transfer characteristic
linear, 279
of magnetic recording tape, 278
Transferring tracks, 508–9
Transformer
center tap, 53
impedance matching, 137
Transition, waveform, 337
Transport system, tape, 304
Tuned port, 157
Tuning, 519
Turnover frequency, 214
Tweeter, 168

U

U-format recorders, 367
Ultradirectional microphone
see Shotgun microphone
Unassigned address bits, 414
Unidirectional, *see* Cardioid microphone
Update mode, 490
User assigned bits, 414

V

Variable speed operation, 310
and tuning, 311
VCA, 452, 482
and automation, 485
fader, monitor, 452, 482
grouping, 466, 491
Velocity of sound, 22
Vented closure, 157
with passive radiator, 160
Vertical interval time code, 415
VHS format recorder, 368
Vibration
of air column and instruments, 16
noises, microphone, 91
Virtual console, *see* Console, digital
VITC, *see* Vertical interval time code
Voice
characteristics of, 20
energy distribution curve, 23
frequency range of, 20

Voice coil
compression driver, 163
loudspeaker, 145
Voiceover compressor, 234
Voltage controlled amplifier, *see* VCA
Voltage rating, open circuit, of microphone, 88
Volume indicator, 33
reading transients, 140
Volume unit, 33–35
compared with dBm, 33
see also Metering

W

Wave, standing, 176
Waveform transition, 337
Waveforms, complex and coherent, 32
Wavelength
in air, 22
of bit stream, 354
formula, 22
vs. frequency, 15, 22
response of playback head, 296
of sound, 22
on tape, 22
Weighting, 35
networks, 35
Wide band noise, 40
White noise, 246
Wind noises, 91
Wind screens, 92
Woodwinds, microphone placement for, 144
Woofer, 168
Word, digital, 339
Wow, 310, 353
Wow and flutter, digital, 310, 354
Wrap, tape recorder head, 324
Write mode, 489

X

X axis, oscilloscope, 41
X-Y recording, 105

Y

Y axis, oscilloscope, 41

Z

Zenith, tape recorder head, 323